Applications Manual

Essentials
of
Anatomy and Physiology

Applications Manual

for

Essentials
of
Anatomy & Physiology

by

Frederic H. Martini, Ph.D.

Edwin F. Bartholomew, M.S.

Kathleen Welch, M.D.

with

William C. Ober, M.D.
Art coordinator and illustrator

Claire W. Garrison, R.N.
Illustrator

Ralph T. Hutchings
Biomedical photographer

Prentice Hall, Upper Saddle River, New Jersey 07458

Executive Editor: *David Kendric Brake*
Development Editor: *Laura Edwards*
Production Editor: *James Buckley*
Director, Production & Manufacturing: *David W. Riccardi*
Interior Designer: *Lee Goldstein, Lorraine Mullaney*
Cover Designer: *Heather Scott*
Buyer: *Ben Smith*
Page Layout: *Mike Bertrand*
Illustrators: *William C. Ober, M.D.; Claire W. Garrison, R.N.;*
         Ron Ervin; Tina Sanders; and Craig Luce

© 1997 by Prentice-Hall, Inc.
A Simon & Schuster Company
Upper Saddle River, New Jersey 07458

Printed in the United States of America

10 9 8 7 6 5 4 3 2 1

ISBN 0-13-532755-5

Prentice-Hall International (UK) Limited, *London*
Prentice-Hall of Australia Pty. Limited, *Sydney*
Prentice-Hall Canada Inc., *Toronto*
Prentice-Hall Hispanoamericana, S.A., *Mexico*
Prentice-Hall of India Private Limited, *New Delhi*
Prentice-Hall of Japan, Inc., *Tokyo*
Simon & Schuster Asia Pte. Ltd., *Singapore*
Editora Prentice-Hall do Brasil, Ltda., *Rio de Janeiro*

Contents

Preface

This *Applications Manual* is a unique companion volume to *Essentials of Anatomy and Physiology*. The creation of the manual was stimulated by the belief that an introductory text in this field should do more than explain the mechanics of physiological systems. It should also serve as the focal point for practical information that will be of value to students in their academic studies, their personal lives, and their professional careers. To meet these objectives, the applied material must be clear and concise, yet comprehensive enough to avoid errors caused by oversimplification. At the same time, the availability of this supplementary information should not interfere with the flow of essential text material concerning normal anatomy and physiology.

Few instructors are likely to cover all of the material in the *Applications Manual*. Indeed, some instructors may choose not to cover all of the material in the text. Because courses differ in their emphases and students differ in their interests and backgrounds, the goal in designing the *AM* has been to provide maximum flexibility of use. The diversity of applied topics in the text discussions and boxes, the *Applications Manual*, and the *New York Times* "Themes of the Times" articles provides instructors with a wide variety of ways in which to integrate the treatment of normal function, pathology, and other clinical or health-related topics. Boxed material and topics in the *Applications Manual* that are not covered in class can be assigned, suggested, used for reference, or left to the individual student. Each student will find and read those selections that deal with disorders affecting friends or family, address topics of current interest and concern, or include information relevant to a chosen career path.

TO THE STUDENT

This *Applications Manual* is organized in units dealing with a wide variety of applied topics.

- **A Foundation for Anatomy and Physiology** provides information on various strategies for learning the vocabulary of the sciences of anatomy and physiology and an introduction to the scientific method.

- **An Introduction to Diagnostics** discusses the basic principles involved in the clinical diagnosis of disease states.

- **Applied Research Topics** considers principles of chemistry and cellular biology that are important to understanding, diagnosing, or treating homeostatic disorders.

- The **Body Systems** section, organized to parallel the text chapter by chapter and system by system, includes more detailed discussions of many clinical topics introduced in the text, along with the discussions of additional diseases and diagnostic techniques not covered in the text. Each discussion is cross-referenced to the text; relevant chapter numbers in *Essentials of Anatomy and Physiology* are indicated by thumb tabs that appear in the page margins.

- The **Critical Thinking Questions** that occur at the end of each system help you to sharpen your ability to think analytically.

A full-color **Surface Anatomy and Cadaver Atlas** of dissection photographs allows you to visualize the internal structure of all major body regions and organs. A selection of surface anatomy photographs of live models is included for comparison with the dissection views.

There are many different ways in which the material in the *Applications Manual* can be used. For example:

- You can read the *Applications Manual* simultaneously with *Essentials of Anatomy and Physiology*, referring to topics as each is referenced in the text.

- You can read the *Applications Manual* separately, referring to the text for relevant background information as needed.

- You can use the *Applications Manual* as a reference, reading only those discussions that are of personal interest to you, that are relevant to your proposed career, or that you need to research for some special purpose.

As you examine the *Applications Manual* you will see that the specific topics are highlighted by icons similar to those used in *Essentials of Anatomy and Physiology*. These icons are designed to give you a preview of the primary focus for each passage. Here is a list of the icons found in the *Applications Manual*, with appropriate examples.

KEY TO ICONS USED IN THE APPLICATIONS MANUAL

= **Reference Material**
(The Nature of Pathogens, p. 23)

= **Topics Relating To the Diagnosis of Disease**
(Blood Tests and RBC's, p. 95)

= **Disorders and Clinical Syndromes**
(Eating Disorders, p. 141)

(R) = **Drugs and Treatment Methods**
(Stimulation of Bone Growth and Repair, p. 49)

= **Exercise and Sports-related Topics**
(Sports Injuries, p. 61)

ACKNOWLEDGMENTS

This was a complex project, and we would like to thank those who helped bring it to completion. Our students provided invaluable suggestions and support. Further, we would also like to express our thanks to David Brake, Executive Editor for Biology, for supporting this rather unconventional project; to Laura Edwards, Senior Development Editor; to to James Buckley, Mike Bertrand and the rest of the production staff who worked on the design, layout, and assembly of this manual.

Of course, any errors or omissions are attributable to the authors, rather than to the reviewers. Readers with comments, suggestions, relevant reprints, or corrections should contact the authors at the address below.

Frederic Martini, PhD
Edwin F. Bartholomew, MS
Kathleen Welch, MD
c/o Prentice Hall
1208 East Broadway, Suite 200
Tempe, AZ 85282

Photo Credits

Unless noted otherwise below, all photographs are courtesy of Ralph Hutchings.

Appleton & Lange, "Smith's General Urology", 13/E, Tanagho:
Figure A-50

Centers for Disease Control:
Figure A-43

Custom Medical Stock Photo:
Figure A-25, Figure A-28, Figure A-33a; 4.3a, 5.4b, and 6.2b in the Cadaver section

Medcom:
Figure A-18a, Figure A-18c

Medtronic, Inc.:
Figure A-33b

Monkmeyer Press:
Figure A-6c

National Health Institute:
Figure A-33d

PH Archives:
Figure A-3, Figure A-6a

Photo Researchers, Inc.:
Figure A-16, Figure A-33c

Martin M. Rotker:
Figure A-6b

Stuart Kenter Associates:
Figure A-18b

Michael J. Timmons:
2.5a in the Cadaver section

Patrick M. Timmons:
3.6, 5.4, 5.5b, 7.3c in the Cadaver section

United Nations:
Figure A-47

University of Toronto:
3.1a through 3.2b; 4.1a through 4.1c; 4.2b, 4.3c, 4.5a; 5.1a through 5.2b; 6.2e, 6.4b; 67.4d in the Cadaver section

Visuals Unlimited:
Figure A-24

A Foundation for Anatomy and Physiology

Learning the Language of Anatomy and Physiology

with Martha Newsome
Tomball College

Medical books bristle with difficult-looking terminology such as *rhinorrhagia* and *osteomyelitis.* Why are the words so difficult to understand? Is a doctor who writes "rhinorrhagia" on the chart just trying to be more impressive than one who records "a bloody nose"?

Not at all. Medical personnel learn and use a special language, not to make communication harder but to make it easier. The terms of this language offer two great advantages over words in common usage:

- *Precision:* Unlike many everyday words, the technical terms used by doctors, nurses, and researchers have very exact and specific meanings.

- *Universality:* A technical term will mean the same thing to a doctor in Lima as it does to one in Beijing.

Medical technicians, doctors, and nurses record and discuss patient information utilizing such terminology. The statement, "The patient broke his arm," is much too general for medical records. There are three large bones. And numerous smaller ones) in the upper and lower limb; the identity of the precise bone involved is obviously important both to the patient and to all those entrusted with his or her care.

Although a technical vocabulary is essential to facilitate diagnosis and treatment, but learning all these terms isn't easy. The length and unfamiliarity of the words are likely to be barriers for the beginning student. Your anatomy and physiology text introduces dozens of new terms in the first chapter alone. Regional terms such as *cephalic, brachial, gluteal, popliteal,* and others may seem impossible to learn. What is the best approach to this daunting task?

1. **Relax! Realize that not all information will be retained the first time you are exposed to it.**

Learning comes with time, experience, and repetition. If you are on a career pathway in the allied health professions, subsequent courses will reinforce the terminology introduced in this text. Every student who has gone on to become a successful nurse, doctor, technician, or therapist has overcome this challenge. Commitment is the key!

2. **Divide and conquer!**

Morphemes are the smallest parts of words that still have meaning. A study of morphemes in medical terminology is very useful to the anatomy and physiology student. For instance, an analysis of the word *automobile* will uncover two roots, *auto* and *mobile.* Determining the meaning of each of the roots is a great help in understanding the complete term. *Auto* is from a Greek word for "self" and *mobile* from a Latin word for "moving." The combined word is defined as a vehicle that moves by some sort of self-propulsion. When a term with one or more familiar roots is encountered in later reading, you can recall the meaning of the roots to deduce the definition of the term. This technique of relating one word to another is called *word association.* As an example, try determining the meaning of the word *autobiography* in this way. (*Bio* means "life" and *graphy* is defined as "writing or record.")

This type of exercise is of considerable benefit to the student learning the "foreign language" of anatomy and physiology. Try a medical term like *osteomyelitis* with the same approach. First, if possible, divide the term in question into separate parts to discern the meaning of each component. For example, osteomyelitis could be dissected into

osteo-
myel-
itis

When noting these component parts of the word, try word association to help deduce the definition:

- You may be familiar with such terms as *osteoporosis* or *osteoarthritis*, common disorders affecting the bones. Indeed, the word root *osteo* means "bone."

- The term *myel* is derived from *myelon*, meaning bone marrow. This is a less familiar term. You will probably need to focus your studies on word parts such as these that are not as easy to recognize.

- Most people recognize *appendicitis* as an inflammation or infection of the appendix. Many names of diseases contain the suffix *itis* connected to the term for the organ or tissue that is inflamed or infected. The term *itis*, therefore, means "inflammation or infection of."

The definition of *osteomyelitis* is thus "an inflammation of the bone and bone marrow."

Most of the word roots used in medicine are derived from Greek or Latin. Many of them occur again and again in various forms and combinations. When you master the most common ones, difficult words become easier to understand. A medical dictionary is a very helpful reference book for finding the meanings of word roots. Some of the more common ones, difficult words become easier to understand. A medical dictionary is a very helpful reference book for finding the meanings of word roots. Some of the more common prefixes, suffixes, and combining forms derived from foreign word roots are given in Table A-1.

A more complete list of word roots, prefixes, suffixes, and combining forms can be found on the endpapers of the text.

3. Repetition and association are the keys to learning word roots, and therefore medical terminology.

- Make a flash card for each unfamiliar term and its definition. Study these cards during a wait at the bank or the dentist's office.

- Create a visual association by drawing a picture next to the written word root. For example, a picture of a simple cell could be drawn next to the term *cyto-*.

- Associate a word with another memorable word or phrase. Often a silly one is best: For example, you may tell yourself that you'll break your arm if you forget the term *brachial*. You'll be surprised how well this can work. And once the term *brachial* is learned, you'll also know the location of such anatomical entities as the brachial artery and vein, the brachialis muscle, and the brachial nerve plexus.

- Categorize terms to facilitate learning. Terms describing color, for example, such as *erythros* (red), *leuko* or *alba* (white), *cyan* (blue), and *melan* (black), can be associated together.

- Practice spelling terms to reinforce your visualization of the roots and combining forms they contain.

Table A-1	An Introduction to Word Construction		
Word Root	*Definition*	*Example*	*Example Definition*
a-	without	anucleate	without a nucleus
bi-	two	bilateral	involving both sides of the body
-cyte	cell	osteocyte	bone cell
epi-	on	epicardium	layer of connective tissue on the heart
ex-	out; away from	exocytosis	movement out of the cell
hem-	blood	hemorrhage	bleeding
hyper-	above; excessive	hypertrophy	increase in the size of cell, such as muscle cell
hypo-	below; under	hypothyroidism	underproduction of thyroid hormone
gen-/genesis	producing; forming	osteogenesis	bone formation
intra-	within	intracellular	inside of the cell
inter-	between	intermolecular	between molecules
-lysis	breakdown	hemolysis	destruction or rupture of the red blood cell
macro-	large	macrocyte	cell that is larger than normal
micro-	small	microscopic	pertaining to something very small; to be viewed with microscope
peri-	around	periodontal disease	disorder of the gums surrounding the teeth
sub-	under; below	sublingual gland	salivary gland that opens under the tongue
trans-	across; through	transport	movement across or through

- Use the text or a medical dictionary to determine the proper pronunciation of words and practice vocalizing them. Words such as *pharynx, carotid,* and many others are frequently mispronounced. Some students will not ask questions in class for fear of mispronouncing a term. Overcome this anxiety and use the terminology as often as possible in the classroom. It is better to make the mistake in the classroom than in the clinic.

- Read the clinical discussions in this manual to develop a greater awareness of the role of word roots in the naming of various diseases.

Unfortunately, all anatomical terms are not so easily translated. Over the centures, anatomists have also named anatomical structures after mythical heroes (*Achilles tendon*) or their favorite professors (*Eustachian tube, Sylvian aqueduct*). For instance, the Fallopian tube was named for the Italian anatomist Gabriele Fallopio (1523-1562). Clinical conditions were named after either the most famous victim (*Lou Gehrig's disease,* a progressive motor paralysis) or the physician who first described it (*Parkinson's disease, Alzheimer's disease*).

Commemorative names are called eponyms, and after 300 or 400 years the eponym situation was out of control. National as well as professional egos were involved, and a common understanding of Latin could not help you to recognize the "disease of Philip" or the "ligament of Haller." In addition, when structures are named after people, it is difficult to remember them because there is no connection between the appearance and the name, as there often is when Latin or Greek roots exist.

A series of international meetings have attempted to correct this situation by replacing eponyms with more precise terms. This movement has been reasonably successful, and major anatomical structures such as bones, cartilages, muscles, nerves, ligaments, and tendons no longer bear the names of famous anatomists. The clinical literature still retains many eponyms, but many of the older eponyms are no longer in use. Although reports of new conditions often use the names of physicians or patients, descriptive terms are usually assigned after additional research has been done.

Unfortunately, many eponyms remain as labels for clinical tests, minor anatomical features, and histological details. A common clue to this type of naming is the capitalization of the term. Thus, the sheath covering nerve fibers of the peripheral nervous system is composed of Schwann cells, named for the nineteenth-century German anatomist, Theodore Schwann. These have been ignored or minimized in the text, and wherever possible they have been replaced by suitable alternative terms. For those interested in eponyms and historical details, Table A-2 provides information concerning the origins of the most common anatomical eponyms.

⧉ The Scientific Method

Your course in anatomy and physiology should do more than simply teach you the names and functions of different body parts. It should provide you with a frame of reference that will enable you to understand new information, draw logical conclusions, and make more intelligent decisions. A great deal of confusion and misinformation exists about just how medical science "works," and people make unwise and even dangerous decisions as a result. Nowhere is this more apparent than when a discussion drifts around to health, nutrition, and cancer. If you are going to be working in a health-related profession, or are even just trying to make sound decisions about your own life, you must learn how to organize information, evaluate evidence, and draw logical conclusions.

FORMING AN HYPOTHESIS

There is a lot more to science than the collection of information. You could spend the rest of your life carefully observing the world around you, but this won't reveal very much unless you can see some kind of pattern and come up with an idea, or hypothesis, that explains your observations.

Hypotheses are ideas that may be correct or incorrect; to evaluate one, you must have relevant data and a reliable method of data analysis. For example, you could propose the hypothesis that radiation emitted by planet X produces immortality. Could anyone prove you wrong? Not very likely, particularly if you didn't specify the location of the planet or the type of radiation. Would anyone believe you? If you were a "leading authority" on something (anything) a few probably would.

That's not as ridiculous as it might seem. For almost 1,500 years "everyone knew" that inhaled air went from the lungs through blood vessels to the heart. They knew this because Galen, the famous Roman physician, had said so. Since he was right in several other respects, all his statements were accepted as true, and contrary opinions were held in low esteem. To avoid making this kind of error, you must always remember to evaluate the hypothesis, not the individual who proposed it!

The evaluation process examines the hypothesis to see if it makes correct predictions about the real world. The steps in this process are diagrammed in Figure A-1. A valid hypothesis will have three characteristics: it will be testable, unbiased, and repeatable.

A testable hypothesis is one that can be studied by experimentation or data collection. Your assertion concerning planet X qualifies as an hypothesis, but it cannot be tested unless we find the planet and detect the radiation. An example of a testable hypothesis would be "left-handed airplane pilots have fewer crashes than right-handed pilots." That is testable because it makes a prediction about the world that can be checked, in this case by collecting and analyzing data.

AVOIDING BIAS

Suppose, then, that you went out and collected information about all the crashes in the world and discovered that 80 percent of all crashed airplanes were flown by right-handed pilots. "Aha!" you might shout, "The hypothesis is correct!" The implications are obvious: Ban all right-handed airline pilots, eliminate four-fifths of all crashes, and sit back and wait for your prize from the Air Traffic Safety Association.

Unfortunately, you would be acting prematurely, for your data collection was biased. To test your hypothesis adequately, you need to know not only how many crashes involved right-handed or left-handed pilots, but how many right-handed and left-handed pilots were flying. If 90 percent of the pilots were right-handed, but they accounted for only 80 percent of the crashes, then left-handed pilots are the ones to watch out for! Eliminating bias in this case is relatively easy, but in health studies there may be all kinds of complicating factors. Because 25 percent of us will probably develop cancer at some point in our lives, cancer studies will be used to exemplify the problems encountered.

The first example of bias in action concerns cancer statistics, which indicate that there are definite regional variations in cancer rates in the United States and abroad. For example, although the estimated U.S. yearly cancer death rate was 173 per 100,000 population in 1996, the rate in Utah was only 126 per 100,000, while the rate in the District of Columbia was 225 per 100,000. It would be very easy to assume that this difference is the direct result of country versus city living. But these data alone should not convince you that moving from the District of Columbia to Utah will lower your risk of developing cancer. To draw that conclusion, you would have to be sure that the

Table A-2	Eponyms in Common Use	
Eponym	*Equivalent Terms*	*Individual Referenced*
The cellular level of organization (Chapter 3)		
Golgi apparatus		Camillo Golgi (1844-1926), Italian histologist; shared Nobel Prize in 1906
Krebs cycle	Tricarboxylic or citric acid cycle	Hans Adolph Krebs (1900-1981), British biochemist; shared Nobel Prize in 1953
The skeletal system (Chapter 6)		
Colles' fracture		Abraham Colles (1773-1843), Irish surgeon
Haversian canals	Central canals	Clopton Havers (1650-1702), English anatomist and microscopist
Haversian systems	Osteons	
Pott's fracture		Percivall Pott (1713-1788), English surgeon
Volkmann's canals	Perforating canals	Alfred Wilhelm Volkmann (1800-1877), German surgeon
Wormian bones	Sutural bones	Olas Worm (1588-1654), Danish anatomist
The muscular system (Chapter 7)		
Achilles' tendon	Calcaneal tendon	Achilleus, hero of Greek mythology
Cori cycle		Carl Ferdinand Cori (1896-) and Gerty Theresa Cori (1896-1957), American biochemists; shared Nobel Prize in 1947
The nervous system (Chapters 8-9)		
Broca's center	Speech center	Pierre Paul Broca (1824-1880), French surgeon
Foramen of Luschka	Lateral foramina	Hubert von Luschka (1820-1875), German anatomist
Foramen of Magendie	Median foramen	François Magendie (1783-1855), French physiologist
Foramen of Munro	Interventricular foramen	John Cummings Munro (1858-1910), American surgeon
Nissl bodies		Franz Nissl (1860-1919), German neurologist
Purkinje cells	Basket cells	Johannes E. Purkinje (1781-1869), Czechoslovakian physiologist
Nodes of Ranvier		Louis Antoine Ranvier (1835-1922), French physiologist
Island of Reil	Insula	Johann Christian Reil (1759-1813), German anatomist
Fissure of Rolando	Central sulcus	Luigi Rolando (1773-1831), Italian anatomist
Schwann cells		Theodor Schwann (1810-1882), German anatomist
Aqueduct of Sylvius	Mesencephalic aqueduct	Jacobus Sylvius (Jacques Dubois, 1478-1555), French anatomist
Sylvian fissure	Lateral sulcus	Franciscus Sylvius (Franz de le Boë, 1614-1672), Dutch anatomist
Pons varolii	Pons	Costanzo Varolio (1543-1575), Italian anatomist
Sensory function (Chapter 10)		
Organ of Corti		Alfonso Corti (1822-1888), Italian anatomist
Eustachian tube	Auditory tube	Bartolomeo Eustachio (1520-1574), Italian anatomist
Golgi tendon organs	Tendon organs	*See* Golgi apparatus *under* the cellular level

Table A-2 Continued

Eponym	Equivalent Terms	Individual Referenced
Hertz (Hz)		Heinrich Hertz (1857-1894), German physicist
Meibomian glands		Heinrich Meibom (1638-1700), German anatomist
Corpuscles of Meissner		Georg Meissner (1829-1905), German physiologist
Merkel's discs		Friedrich Siegismund Merkel (1845-1919), German anatomist
Pacinian corpuscles		Fillippo Pacini (1812-1883), Italian anatomist
Ruffini's corpuscles		Angelo Ruffini (1864-1929), Italian anatomist
Canal of Schlemm		Friedrich S. Schlemm (1795-1858), German anatomist
Glands of Zeis		Edward Zeis (1807-1868), German ophthamologist

The endocrine system (Chapter 11)

Eponym	Equivalent Terms	Individual Referenced
Islets of Langerhans	Pancreatic islets	Paul Langerhans (1847-1888), German pathologist
Intersitial cells of Leydig	Interstitial cells	Franz von Leydig (1821-1908), German anatomist

The cardiovascular system (Chapters 12-14)

Eponym	Equivalent Terms	Individual Referenced
Bundle of His		Wilhelm His (1863-1934), German physician
Purkinje cells		*See under* The nervous system
Starling's law		Ernest Henry Starling (1866-1927), English physiologist
Circle of Willis	Cerebral arterial circle	Thomas Willis (1621-1675), English physician

The lymphatic system (Chapter 15)

Eponym	Equivalent Terms	Individual Referenced
Hassall's corpuscles		Arthur Hill Hassall (1817-1894), English physician
Kupffer cells		Karl Wilhelm Kupffer (1829-1902), German anatomist
Langerhans cells		*See* Islets of Langerhans *under* The endocrine system
Peyer's patches	Aggregate lymphoid nodules	Johann Conrad Peyer (1653-1712), Swiss anatomist

The respiratory system (Chapter 16)

Eponym	Equivalent Terms	Individual Referenced
Adam's apple	Thyroid cartilage (laryngeal prominence)	Biblical reference
Bohr effect		Niels Bohr (1885-1962), Danish physicist; won 1922 Nobel Prize
Boyle's law		Robert Boyle (1621-1691), English Physicist
Charles' law		Jacques Alexandre César Charles (1746-1823), French physicist
Dalton's law		John Dalton (1766-1844), English physicist
Henry's law		William Henry (1775-1837), English Chemist

The digestive system (Chapter 17)

Eponym	Equivalent Terms	Individual Referenced
Plexus of Auerbach	Myenteric plexus	Leopold Auerbach (1827-1897), German anatomist
Brunner's glands	Duodenal glands	Johann Conrad Brunner (1653-1727), Swiss anatomist
Kupffer cells	Stellate cells	*See under* The Lymphatic System
Crypts of Lieberkuhn	Intestinal crypts	Johann Nathaniel Lieberkuhn (1711-1756), German anatomist
Plexus of Meissner	Submucosal plexus	*See* Corpuscles of Meisser *under* Sensory Function
Sphincter of Oddi	Hepatopancreatic sphincter	Ruggero Oddi (1864-1913), Italian physician
Peyer's patches		*See under* The Lymphatic System
Duct of Santorini	Accessory pancreatic duct	Giovanni Domenico Santorini (1681-1737), Italian anatomist
Stensen's duct	Parotid duct	Niels Stensen (1638-1686), Danish physician/priest
Ampulla of Vater	Duodenal ampulla	Abraham Vater (1684-1751), German anatomist
Wharton's duct	Submandibular duct	Thomas Wharton (1614-1673), English physician
Foramen of Winslow	Epiploic foramen	Jacob Benignus Winslow (1669-1760), French anatomist
Duct of Wirsung	Pancreatic duct	Johann Georg Wirsung (1600-1643), German physician

The urinary system (Chapter 19)

Eponym	Equivalent Terms	Individual Referenced
Bowman's capsule	Glomerular capsule	Sir William Bowman (1816-1892), English Physician
Loop of Henle		Friedrich Gustav Jakob Henle (1809-1885), German histologist

The reproductive system (Chapters 20-21)

Eponym	Equivalent Terms	Individual Referenced
Bartholin's glands	Greater vetibular glands	Casper Bartholin, Jr (1655-1738), Danish anatomist
Cowper's glands	Bulbourethral glands	William Cowper (1666-1709), English surgeon
Fallopian tube	Uterine tube/oviduct	Gabriele Fallopio (1523-1562), Italian anatomist
Graafian follicle	Tertiary follicle	Reijnier de Graaf (1641-1673), Dutch physician
Interstitial cells of Leydig	Interstitial cells	*See under* The Endocrine System
Glands of Littre	Lesser vestibular glands	Alexis Littre (1658-1726), French surgeon
Sertoli cells	Sustentacular cells	Enrico Sertoli (1842-1910), Italian histologist

observed rates were the direct result of just a single factor, the difference in physical location. As you will find in later chapters, many different factors can promote cancer development. To exclude all possibilities other than geography, you would have to be certain that the populations were alike in all other respects. Here are a few possible sources of variation that could affect that conclusion:

- *Different population profiles:* Cancer rates vary between males and females, among racial groups, and among age groups. Therefore, we need to know how the populations of Utah and the District of Columbia differ in each respect.

- *Different occupations:* Because chemicals used in the workplace are implicated in many cancers, we need to know how the populations of each region are employed and what occupational hazards they face.

- *Different mobilities:* Because the region in which a person dies may not be the same as the one in which he or she lived and developed cancer, we need to know whether people with cancer in Utah stay in the state or go elsewhere for critical care, and whether people with cancer travel to the District of Columbia to seek treatment at special clinics.

- *Different health care:* Since cancer death rates reflect differences in patterns of health care, we need to know whether residents of Utah pay more attention to preventive health care and have more regular checkups, whether their medical facilities are better, and whether they devote a larger proportion of their annual income to health services than do residents in the District of Columbia.

You can probably think of additional factors, but the point is that avoiding experimental bias can be quite difficult!

A second example of the problem of bias comes from the collection of "miracle cures" that continue to appear and disappear at regular intervals. Pyramid power, pendulum power, crystals, magnetic energy fields, and psychic healers come and go in the news. Miraculous drugs are equally common, whether they are "secret formulas" or South American plant extracts discovered by Mayan colonists from other planets. The proponents of each new procedure or drug report glowing successes with patients who would otherwise have surely succumbed to the disease. And, of course, all these remedies are said to have been suppressed or willfully ignored by traditional therapists.

Even accepting that the claims aren't exaggerated, does it prove anything that 1 or 100 or even 1,000 patients have been cured? No, it doesn't, for a list of successes doesn't mean very much. To understand why this is so, consider the questions you might pose to an instructor who announced on the first day of class that he or she had given 20 A's last semester. You would want to know how many students were in the class: only 20, or several hundred? You would also want to find out how the rest of the class performed—20 A's and 200 D's might be rather discouraging. You could also check on how the students were selected. If only students with A averages in other courses were allowed to enroll, your opinion should be adjusted accordingly. Finally, you might check with the students and compare their grades with those given by other instructors teaching the same course.

With just a couple of modifications, the same questions could be asked about a cancer "cure":

- How many patients were treated, how many were cured, and how many died?

- How were the patients selected? If selection depended on wealth, degree of illness, or previous exposure to other therapeutic techniques, then the experimental procedure was biased from the start.

- How many might have recovered regardless of the treatment? Even "terminal" cancers sometimes simply disappear for no apparent reason. Such occurrences are rare, to be sure, but they do happen.

- How do the foregoing statistics compare with those of more traditional therapies when subjected to the same unbiased tests?

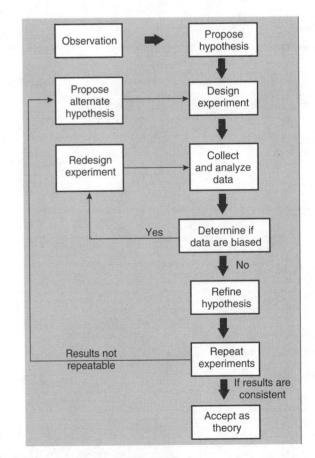

Figure A-1   The Scientific Method.
The basic sequence of steps involved in the development and acceptance of a scientific theory.

THE NEED FOR REPEATABILITY

Finally, let's examine the criterion of repeatability. It's not enough to develop a reasonable, testable hypothesis and collect unbiased data. Consider the hypothesis that every time a coin is tossed, it will come up heads. You could build a coin-tossing machine, turn it on, and find that in the first experiment of 10 tosses, the coin came up heads every time. Does this prove the hypothesis?

No, it doesn't, despite the fact that it was an honest experiment and the data supported the hypothesis. The problem here is one of statistics, sample size, and luck. The odds that a coin will come up heads on any given toss are 50 percent, or 1 in 2—the same as the odds that it will come up tails. The odds that it will come up heads 10 times in a row are about 1 in 2000—pretty small, but certainly not inconceivable. If that coin is tossed 50 times, however, the chance of getting 50 heads becomes just 1 in 4,500,000,000,000,000, a figure that most people would accept as "vanishingly small." To prove that the hypothesis "a tossed coin always lands heads up" is false, the coin need only come up tails once. So the truth could be revealed by running the experiment with more coin tosses or by letting other people set up identical experiments and toss their own coins.

The point here is that if an hypothesis is correct, anyone and everyone will get the same results when the experiment is performed. If it isn't repeatable, you've got to doubt the conclusions even when you have complete confidence in the abilities and integrity of the original investigator.

If an hypothesis satisfies all these criteria, it can be accepted as a scientific theory. The scientific use of this term therefore differs from that used in general conversation; when writers discuss "wild-eyed theories" they are really speaking of untested hypotheses. Hypotheses may be true or false, but by definition theories describe real phenomena, and they make accurate predictions about the world. Examples of scientific theories include the theory of gravity and the theory of evolution. The "fact" of gravity is not in question, and the theory of gravity accounts for the available data. But this does not mean that theories cannot change over time. Newton's original theory of gravity, though used successfully for over two centuries, was profoundly modified and extended by Einstein. Similarly, the theory of evolution originally proposed by Charles Darwin has been greatly elaborated since it was first proposed in the middle of the last century. No one theory can tell the whole story, and all theories are continually undergoing modification and improvement as we learn more about our universe.

An Introduction to Diagnostics

Each of us has made simple diagnoses in our everyday experiences. When the car won't start, the kitchen faucet leaks, or the checkbook doesn't balance, most people will try to determine the nature of the problem. Sometimes the diagnosis is simple: the car battery is dead, the faucet is not completely turned off, or the amount of a check (or a deposit) was recorded incorrectly. Once the diagnosis is made, steps can be taken to remedy the situation.

Most people use similar observational skills to diagnose simple medical conditions. For example, imagine that you awaken with a headache, feeling weak and miserable. On arising you gaze at yourself in the bathroom mirror. Your face is flushed, your forehead is hot to the touch, and swallowing is painful. You know that these correspond to the general symptoms of the flu, and you also know that your lab partner missed Tuesday's class because of the flu. You diagnose yourself as sick with the flu, and you open the medicine cabinet in search of appropriate medication.

The steps taken in arriving at the conclusion "I've probably got the flu" were quite straightforward: (1) you made observations about your condition, (2) you compared your observations with available data, and (3) you determined the probable nature of the problem. Clinical diagnosis can be much more complicated, but these same steps are always required. In this section we will examine the basic principles of diagnosis. The goal is not to train you to be a clinician, but rather to demonstrate how these basic steps can be followed in a clinical setting.

Homeostasis and Normal Values EAP *p. 12*

The central principle of physiology is that physiological systems strive to maintain a relatively constant internal environment. Homeostatic control mechanisms are both flexible and dynamic. They operate to keep vital characteristics within the relatively narrow limits required for normal function. Examples include the mechanisms responsible for controlling levels of oxygen, carbon dioxide, glucose, pH, blood pressure, and other factors. As noted in Chapter 1 (EAP, *p. 12*), physiological values usually oscillate within a normal range of values. For this reason most laboratory reports indicate the values determined and the normal ranges for each value.

In many cases, the ranges can change over time (from infancy to old age) or from moment to moment, depending on the activity underway. For example, people often use 72 beats per minute (bpm) as a representative heart rate. That value is actually the average for young adults at rest. The real, measured values can be quite different. This becomes very apparent when you consider the normal *ranges*. The typical heart rate of resting young adults (60-80 bpm) differs from that of infants (70-170 bpm), and the adult heart rate during deep sleep (45-60 bpm) is very different from that found during heavy exercise (160-180 bpm).

HOMEOSTATIC FAILURE EAP *p. 14*

The ability to maintain homeostasis depends on two interacting factors: (1) the status of the physiological systems involved and (2) the nature of the stress imposed. Homeostasis is a balancing act, and each person is like a tightrope walker. Homeostatic systems must adapt to sudden or gradual changes in the environment, the arrival of pathogens, accidental injuries, and many other factors, just as a tightrope walker must make allowances for gusts of wind, frayed segments of the rope, and thrown popcorn.

The ability to maintain homeostatic balance varies with the age, general health, and genetic makeup of the individual. The geriatric patient or young infant with the flu is in much greater danger than an otherwise healthy young adult with the same viral infection. If homeostatic mechanisms cannot cope with a particular stress, physiological values will begin to drift outside of the normal range. This can ultimately affect all other systems, with potentially fatal results; after all, a person unable to maintain balance will eventually fall off the tightrope.

Consider a specific example. A young adult exercising heavily may have a heart rate of 180 bpm for several minutes. But such a heart rate can be disastrous for an older person already suffering from cardiovascular and respiratory problems. If the heart rate cannot be reduced, due to problems with the pacemaking or conducting systems of the heart, there will be damage to the cardiac muscle tissue, leading to decreased pumping efficiency and a fall in blood pressure.

This represents a serious threat to homeostasis. Other systems soon become involved, and the situation worsens. Filtration at the kidneys occurs at the normal range of arterial blood pressures. When the blood pressure becomes abnormally low, blood flow through peripheral tissues declines, and the kidneys stop working. Toxins then begin accumulating in the circulation. The reduced blood flow in other tissues soon leads to a generalized *hypoxia,* or low tissue oxygen level. Cells throughout the body then begin to suffer from oxygen starvation. The person is now in serious trouble, and unless steps are taken to correct the situation, survival is threatened.

HOMEOSTASIS AND DISEASE

Disease is the failure to maintain homeostatic conditions. The disease process may initially affect a tissue, an organ, or a system, but it will ultimately lead to changes in the function or structure of cells throughout the body. A disease can often be overcome through appropriate, automatic adjustments in physiological systems. In a case of the flu, the disease develops because the immune system cannot defeat the virus before it has infected cells of the respiratory passageways. For most people, the physiological adjustments made in response to the presence of this disease will lead to the elimination of the virus and the restoration of homeostasis. Some diseases cannot be easily overcome. In the case of the person with acute cardiovascular problems, some outside intervention must be provided to restore homeostasis and prevent fatal complications.

Diseases may result from:

- *Pathogens or parasites that invade the body:* Examples include the flu, mumps, measles, pinworms, and tapeworms.

- *Inherited genetic conditions that disrupt normal physiological mechanisms:* These conditions make normal homeostatic control difficult or impossible. Examples noted in later sections include the *lysosomal storage diseases* and *sickle cell anemia.*

- *The loss of normal regulatory control mechanisms:* For example, cancer involves the rapid, unregulated multiplication of abnormal cells. Many cancers have been linked to abnormalities in genes, responsible for controlling the rates of cell division. A variety of other diseases, called *autoimmune disorders,* result when regulatory mechanisms of the immune system fail, and normal tissues are attacked.

- *Degenerative changes in vital physiological systems:* Many systems become less adaptable and less efficient as part of the aging process. For example, there are significant reductions in bone mass, respiratory capacity, cardiac efficiency, and kidney filtration as part of the aging process. If these individuals are exposed to stresses that their weakened systems cannot tolerate, disease results.

- *Trauma, toxins, or other environmental hazards:* Accidents may damage internal organs, impairing their function. Toxins consumed in the diet or absorbed through the skin may disrupt normal metabolic activities.

- *Nutritional factors:* Diseases may result from diets inadequate in proteins, essential amino acids, essential fatty acids, vitamins, minerals, or water. Kwashiorkor, a protein deficiency disease, and scurvy, a disease caused by vitamin C deficiency, are two examples. Excessive consumption of high-calorie foods, fats, or fat-soluble vitamins can also cause disease.

Pathology is the study of disease, and *pathophysiology* is the study of functional changes caused by disease. Different diseases can often result in the same alteration of function and produce the same symptoms. For example, a patient with pale skin and complaining of a lack of energy and breathlessness may have (1) respiratory problems that prevent normal oxygen transfer to the blood, as in *emphysema,* or (2) cardiovascular problems that interfere with normal oxygen transport *(anemia)* or circulation (heart failure). Clinicians must ask questions and collect appropriate information to make a proper diagnosis. This often involves eliminating possible causes until a specific diagnosis is reached.

For example, if tests indicate that anemia is responsible for the these symptoms, the specific type of anemia must then be determined before treatment can begin. After all, the treatment for anemia due to a dietary iron deficiency will be very different from the treatment for anemia due to internal bleeding. Of course, you could not hope to identify the probable cause of the anemia unless you were already familiar with the physical and biochemical structure of red blood cells, and their role in the transport of oxygen. This brings us to a key concept: *All diagnostic procedures assume an understanding of normal anatomy and physiology.*

SYMPTOMS AND SIGNS

When disease processes affect normal functions, the alterations are called the *symptoms* and *signs* of the disease. An accurate diagnosis, or identification of the disease, is accomplished through the observation and evaluation of signs and symptoms.

A **symptom** is the patient's perception of a change in normal body function. Examples of symptoms include nausea, malaise, and pain. Symptoms are difficult to measure, and one must rely on asking appropriate questions. Examples of typical questions include:

"When did you first notice this symptom?"

"What does it feel like?"

"Does it come and go, or does it always feel the same?"

"Are there things you can do to make it feel better or worse?"

The answers provide information on the duration, sensations, recurrence, and potential triggering mechanisms for the symptoms important to the patient.

Pain, an important symptom of many illnesses, is often an indication of tissue injury. The flow chart in Figure A-2 demonstrates the types of pain and introduces important related terminology. Pain sensations and pathways are detailed in Chapter 10 of the text, and the control of pain is considered in related sections of the *Applications Manual* (p. 76).

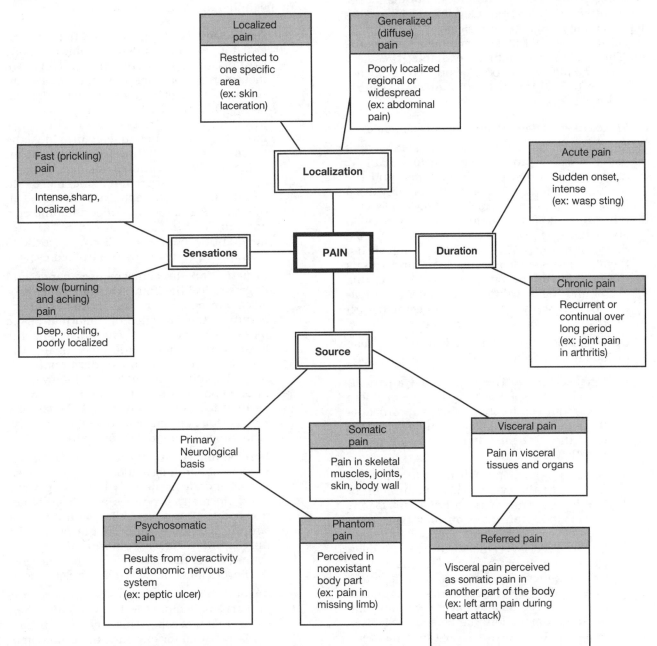

Figure A-2   Methods of Classifying and Describing Pain.

A **sign** is a physical manifestation of the disease. Unlike symptoms, signs can be measured and observed through sight, hearing, or touch. The yellow color of the skin caused by liver dysfunction or a detectable breast lump are signs of disease. An observable change due to a disturbance in the structure of tissue or cells is called a **lesion.** Lesions of the skin are considered in detail in a later section dealing with the integumentary system (p. 36).

Steps in Diagnosis

A person experiencing serious symptoms usually seeks professional help, and thereby becomes a patient. The consultant, whether a nurse, physician, or emergency medical technician, must determine the need for medical care based on observation and assessment of the patient's symptoms and signs. This is the process of diagnosis: the identification of a pathological process by its characteristic symptoms and signs.

Diagnosis is a lot like assembling a jigsaw puzzle. The more pieces (clues) available, the more complete the picture will be. The process of diagnosis is one of deduction and follows an orderly sequence of steps:

1. *The medical history of the patient is obtained:*

   The medical history is a concise summary of past medical disorders, general factors that may affect the function of body systems, and the health of the family. This information provides a framework for considering the individual's current problem. The examiner gains information about the person's concerns by asking specific questions and using good listening skills. Physical assessment begins here, and this is the time for unspoken questions like, "Is this person moving, speaking, and thinking normally?" The answers will later be integrated with the results of more precise observations.

   Other components of the medical history may include the following.

   - *Chief complaint:* The person, now a patient, is asked to specify the primary problem that requires attention. This is recorded as the *chief complaint.* An example would be the entry "Patient complains of pain in the right lower quadrant."

   - *Onset of symptoms:* When did the patient first notice the symptoms? The duration of the disease process is an important factor. For example, an infection may have been present for months, only gradually increasing in severity. This would be called a *chronic infection.* Chronic infections may be treated differently from an *acute infection* that produces sudden, intense symptoms. A disease process may have been underway for some time before the person recognizes that a problem exists. Over the initial peri-

od the individual experiences mild *subclinical symptoms* that are usually ignored.

   - *Review of systems:* The patient is asked questions that focus on the general status of each body system. This process may detect related problems or causative factors. For example, a chief complaint of headache pain may be *related* to visual problems (stars, spots, blurs, or blanks seen in the field of vision) or *caused* by visual problems (eyeglasses poorly fitted, or having the wrong prescription).

2. *A physical examination is performed:* The physical examination is a basic but vital part of the diagnostic process. The common techniques used in physical examination include *inspection* (vision), *palpation* (touch), *percussion* (tapping), and *auscultation* (listening).

   - **Inspection** is careful observation. A general inspection involves examining body proportions, posture, and patterns of movements. Local inspection is the examination of sites or regions of suspected disease. Of the four components of the physical exam, inspection is often the most important because it provides the largest amount of useful information. Many diagnostic conclusions can be made on the basis of inspection alone; most skin conditions, for example, are identified in this way. A number of endocrine problems and inherited metabolic disorders can produce subtle changes in body proportions that can be detected by the trained eye.

   - In **palpation** the clinician uses hands and fingers to feel the body. This procedure provides information on skin texture and temperature, the presence of abnormal tissue masses, the pattern of the pulse, and the location of tender spots. Once again, the procedure relies on an understanding of normal anatomy. A small, soft, lumpy mass in one spot is a salivary gland; in another location it could be a tumor. A tender spot is important in diagnosis only if the observer knows what organs lie beneath it.

   - **Percussion** is tapping with the fingers or hand to obtain information about the densities of underlying tissues. For example, the chest normally produces a hollow sound, because the lungs are filled with air. That sound changes in pneumonia, when the lungs contain large amounts of fluid. Of course, to get the clearest chest percussions, the fingers must be placed in the right spots.

   - **Auscultation** (aws-kul-Ā-shun; *auscultare,* to listen) is listening to body sounds, often using a stethoscope. This technique is particularly useful for checking the condition of the lungs during breathing. The wheezing sound heard in asthma is caused by constriction of the airways, and pneumonia produces a gurgling sound, indicating that fluid has accumulated in the lungs. Auscultation is also important in diagnosing heart conditions. Many cardiac problems affect the sound of the heartbeat or produce abnormal swirling sounds during blood flow.

Every examination also includes measurements of certain vital body functions, including body temperature, blood pressure, respiratory rate, and heart (pulse) rate. The results, called **vital signs,** are recorded on the patient's chart. As noted earlier, each of these values can vary over a normal range that differs according to the age, sex, and general condition of the individual. Table A-3 indicates the representative ranges for vital signs in infants, children, young adults, and persons over age 60.

3. *If necessary, diagnostic procedures are performed:* The physical examination alone may not provide enough information to permit a precise diagnosis. Diagnostic procedures can then be used to focus on abnormalities revealed by the physical examination. For example, if the chief complaint is knee pain after a fall, and the physical examination reveals swelling and localized, acute pain on palpation, the *preliminary diagnosis* may be a torn cartilage. MRI scan

Table A-3   Normal Range of Values for Resting Individuals in Different Age Groups

Vital Sign	Infant (3 months)	Child (10 years)	Adult
Blood Pressure (mm Hg)	90/50	125/60	95/60 to 140/90
Respiratory Rate (per minute)	30-50	18-30	8-18
Pulse Rate (per minute)	70-170	70-110	50-95

may be performed to ensure that there are not other problems, such as torn ligaments. With the information provided from the diagnostic procedure, the *final diagnosis* can be made with reasonable confidence. Diagnostic procedures thus extend, rather than replace, the physical examination.

There are two general categories of diagnostic procedures.

1. Tests performed on the individual, usually within a hospital facility. Information on representative tests of this type is summarized in Table A-4. These procedures allow the clinician to

Table A-4   Representative Diagnostic Tests, their Principles and Uses.

Procedure	Principle	Examples of Uses
Endoscopy	Insertion of fiber optic tubing into body orifice or through a small incision (laparoscopy and arthroscopy); permits visualization of body cavity or organ interior.	Allows direct visualization of internal structures and detection of abnormalities of surrounding soft tissue. Bronchoscopy: bronchus and lungs Laparoscopy: abdominopelvic organs, Cystoscopy: urinary bladder Esophagoscopy: esophagus, Gastroscopy: stomach Colonoscopy: colon, Arthroscopy: joint cavity
Standard X-rays	A beam of X-rays passes through the body and then strikes a photographic plate. Radiodense tissues block X-ray penetration, leaving unexposed (white) areas on film negative. (EAP, *p.21*)	Longbones: to detect fracture, tumor, growth patterns. Chest: to detect tumors, pneumonia, atelectasis, tuberculosis. Skull: to detect fractures, sinusitis, metastatic tumors.
Contrast X-rays	X-rays taken after infusion or ingestion of radiodense solutions. (EAP, *p.21*)	Barium Swallow (upper GI): Series of x-rays following the ingestion of barium, to detect abnormalities of esophagus, stomach, and duodenum. Barium Enema: series of x-rays following barium enema to detect abnormalities of colon. IV Pyelography: Series of x-rays following intravenous injection of radiopaque dye filtered by kidneys; reveals abnormalities of kidneys, ureters, and bladder; allows assessment of renal function. Mammogram: x-rays of each breast taken at different angles for early detection of breast cancer and other masses, such as cysts.

Procedure	Principle	Examples of Uses
Computerized Tomography (CT or CAT)	Produces cross-sectional images of area to be viewed; all sections together can produce a three-dimensional image for detailed examination. (EAP, p.22)	CT Scans of the head, abdominal region, (liver, pancreas, kidney), chest and spine are taken to assess organ size and position, to determine progression of a disease, and for detection of abnormal masses.
Nuclear Scans	Radioisotope ingested or injected into the body becomes concentrated in the organ to be viewed; gamma radiation camera records image on film. Area should appear uniformly shaded; dark or light areas suggest hyperactivity or hypoactivity of the organ.	Bone Scan: to detect tumors, infections, and degenerative diseases. Scans of the brain, heart, thyroid, liver, spleen, and kidney can be useful in assessing organ function and the extent of disease.
Radioactive Iodine Uptake Test (RAI)	Radioactive iodine is given orally and scans are taken at 3 different times to determine thyroid percentage uptake of radioiodine.	Aids in the determination of a hyperthyroid or hypothyroid condition.
Positron Emission Tomography (PET)	Radioisotopes are given by injection or inhalation; and gamma detectors absorb energy and transmit information to computers to generate cross-sectional images.	Used to measure metabolic activity of heart and brain, and to analyze blood flow through organs.
Magnetic Resonance Imaging (MRI)	A magnetic field is produced to align hydrogen protons, then exposed to radio waves that cause the aligned atoms to absorb energy. The energy is later emitted and captured to produce an image (EAP, p.22).	MRI gives excellent contrast of normal and abnormal tissue, tumor progression, demyelination obstructions in arteries, aneurysms, and determining the extent of organ disease.
Ultrasonography	A transducer contacting the skin or other body surface sends out sound waves and then picks up the echoes (EAP, p.22).	Used in obstetrics, to detect ectopic pregnancy, determine fetal size, check fetal rate of growth; upper abdominal ultrasound detects gallstones, visceral abnormalities, and measures kidneys.
Echocardiography		Used to assess the structure and function of the heart.
Electrocardiography (ECG)	Graphical recording of the electrical activity of the heart using electrodes on the skin surface.	Useful in detection of arrhythmias, such as premature ventricular contractions (PVCs) and fibrillation, and for assessing damage after myocar-dial infarction.
Electroencephalography (EEG)	Graphical recording of electrical activity in the brain through the use of electrodes on the surface of the scalp.	Analysis of brain wave frequency and amplitude aids in the diagnosis of tumors, seizure disorders, and strokes.
Electromyography (EMG)	Graphical recording of electrical activity resulting from skeletal muscle contraction using electrodes inserted into the muscles.	Determination of neural or muscular origin of muscle disorder; aids in the diagnosis of muscular dystrophy, myasthenia gravis, and other neuromuscular disorders.
Pulmonary Function Tests	Measurement of lung volumes and capacities by a spirometer or other device during different types of ventilations.	Aids in the differentiation between obstructive and restrictive lung diseases.
Pap Smears	Removal of cells for laboratory analysis.	Detection of precancerous cells or infections; most often used to assess mucosal cells of cervix.
Stress Testing	Monitoring of blood pressure, pulse rate, and ECG while exercising; may include intravenous injection of radioisotopes.	Aids in the determination of the extent of coronary artery disease, which may not be apparent while the individual is at rest.
Skin Tests	Injection of a substance under the skin, or placement of a substance on the skin surface to determine the response of the immune system.	Tuberculin test: injection of tuberculin protein under skin. Allergen test: injection of allergen or application of a patch containing allergen.

- Visualize internal structures (endoscopy, x-rays, scanning procedures, ultrasonography, mammography)
- Monitor physiological processes (EEG, ECG, PET, RAI, pulmonary function tests)
- Assess the patient's homeostatic responses (stress testing, skin tests)

2. Tests performed in a clinical laboratory on tissue samples, body fluids, or other materials collected from the patient. Table A-5 includes details on a representative sample of such tests.

THE PURPOSE OF DIAGNOSIS

Several hundred years ago, a physician would arrive at a final diagnosis and consider the job virtually done. Once the diagnosis was made, the patient and family would know what to expect. In effect, the physician was more of an oracle than a healer. Wounds could be closed, and limbs amputated, but there were few effective treatment options available. Therapy largely consisted of some combination of cupping and bleeding, often performed by barbers rather than surgeons.

Table A-5   Laboratory Tests Performed on Samples Taken from the Body

Blood tests: Serum, plasma or whole blood samples can be evaluated. Depending on the blood constituent or chemical being monitored, venous or arterial blood will be taken.

Laboratory Test	Significance
Complete Blood Count (CBC) in the number of red blood Hemoglobin (Hb, Hgb) Hematocrit (Hct)	Data from this test informs the practitioner RBC Count about a change cells; changes may indicate the presence of disease, hemorrhaging, or other problems. A CBC is usually performed during a normal physical exam to give the practitioner more information about the patient's general health. Changes may indicate blood loss, infections, or other problems.
RBC Indices: Mean Corpuscular Hemoglobin (MCH) Mean Corpuscular Hemoglobin Concentration (MCHC)	Provide information about the status of RBC production.
WBC Count Differential WBC Count	The white blood cell count reflects the body's immune system and the ability to fight infection. Increased WBC count could indicate the presence of infection.
Hemostasis Tests: Platelet Count Bleeding Time Factors Assay Plasma Fibrinogen Plasma Prothrombin Time (PT) Plasminogen	A decreased number of platelets could result in uncontrolled bleeding. Other constituents, such as fibrinogen, clotting factors, and prothrombin, also contribute to the clotting process and these can be assessed separately.
Serum Electrolytes Sodium Potassium Chloride Bicarbonate	Sodium, potassium, and chloride levels are important because these electrolytes function in nerve transmission, skeletal muscle contraction and cardiac rhythm. Abnormal levels of bicarbonate indicate problems with acid-base balance.
Iron (Serum)	Decreased levels cause iron deficiency anemia; increased levels may cause liver damage.
Arterial Blood Gases and pH pH P_{CO2} P_{O2}	Respiratory acidosis and alkalosis can be monitored with these values. Decreased oxygen levels occur in respiratory system dysfunction.
Hemoglobin electrophoresis: Hemoglobin A Hemoglobin F Hemoglobin S	Electrophoresis separates the differing types of hemoglobin for quantitative measurement. Abnormal levels of hemoglobin occur in many anemias. Abnormal types of hemoglobin occur in sickle cell anemias.
ABO and Rh typing	Blood typing is critical for correct matching of blood types prior to transfusion; Rh typing is important to assess during pregnancy to determine typing of mother and child.

Laboratory Test	*Significance*
Serum Cholesterol	Elevated cholesterol levels reflect the potential for atherosclerosis and coronary artery disease.
Serum Lipoproteins	Electrophoresis is used to separate the LDL fraction to determine ratio of HDL/LDL in the assessment of the potential of atherosclerotic disease.

Laboratory Test	*Significance*
Enzymes: Creatine Phosphokinase Isoenzymes (CPK-MM, CPK-MB, CPK-BB)	Abnormal levels of enzymes in the blood are usually due to cellular damage. CPK-MM is useful in the diagnosis of muscle disease; CPK-BB is used in the diagnosis of myocardial infarction.
Aspartate Aminotransferase (AST) Lactic Dehydrogenase (LDH)	AST levels are important to assess following a myocardial infarction or liver damage. Different isoenzymes of LDH can be useful in the detection of heart damage, liver problems, and pulmonary dysfunction.
Rheumatoid Factor	Measures presence of antibodies characteristic of *rheumatoid arthritis* and (less often) *systemic lupus erythematosus* and other autoimmune diseases.
Hormones	Increased or decreased levels reflect endocrine system disorders.
Blood Urea Nitrogen (BUN)	Used to assess kidney function.
Immunoglobin Electrophoresis (IgA, IgG, IgD, IgE, IgM)	Monitoring of infections and allergic response.
Alcohol	To determine level of intoxication or detect metabolic problems or poisoning.
Human Chorionic Gonadotropin (hCG)	To determine pregnancy.
Phenylketonuria	To detect a genetic disorder of amino acid metabolism, *phenylketonuria* (PKU).
Alpha Fetoprotein	To identify probability of fetal defects or presence of twins.
Glocose Tolerance Test	To detect hyperglycemia *(diabetes mellitus)* and other abnormalities.
Blood Culture	Presence of pathogen occurs in *septicemia* or other infectious disorders.

Urine tests: a single urine sample may be tested or urine may be collected over a period of time and tested (usually from two to twenty four hours). A routine urinalysis aids in the detection of kidney dysfunction as well as metabolic imbalances and other disorders. Presence of abnormal cellular constituents in urine indicates urinary system disorder, including infection or inflammation.

Creatine Clearance	Abnormal values indicate reduced renal function.
Urine Electrolytes Sodium Potassium	Abnormal levels reflect fluid or electrolyte imbalances.
Uric Acid	Increased levels occur with gout.
Human Chorionic Gonadotropin (hCG)	Determination of pregnancy.
Urine Culture	Presence of pathogen occurs in urinary tract infections.

Laboratory Test	Significance

Additional laboratory tests can be used to monitor other body fluids, excretory products, or tissues. Here are several examples:

Cerebrospinal Fluid:	Analysis of CSF pressure, color, sugar and protein content, and the presence of antibodies or pathogens.
Stool Sample:	Culturing of sample to identify microorganisms and determine their antibiotic sensitivities.
Semen Analysis:	Useful in diagnosis of male infertility.
Tissue Biopsy:	Removal of tissue for microscopic examination.

We have an incredible variety of treatment options today, and a final diagnosis is vital because it determines what treatment options will be selected. A modern physician with a new patient follows the *SOAP* protocol:

S is for *subjective:* The clinician obtains subjective information from the patient and completes the medical history.

O is for *objective:* The clinician now performs the physical examination and obtains objective information about the physical condition of the patient. This may include the use of diagnostic procedures.

A is for *assessment:* The clinician arrives at a diagnosis and, if necessary, reviews the literature on the condition. A clinician arrives at a diagnosis and reviews the literature on the condition. A preliminary conclusion concerning the *prognosis* (probable outcome) is made.

P is for *plan:* A treatment plan is designed. This can be very simple (take two aspirin) or very complex (radiation, chemotherapy, or surgery). If complex, one or more treatment options are usually prepared for review by the patient and, in many cases, the family. The options are discussed and the treatment plan finalized.

As you may have noticed, these are precisely the steps you followed at the beginning of this section: subjective (you felt ill), objective (flushed face, fever), assessment (flulike symptoms), and plan (take medicine). The SOAP protocol is both simple to remember and remarkably effective.

The primary goal of an introductory anatomy and physiology course is to provide you with the foundation for other, more specialized courses. In the sections of this manual dealing with body systems you will be introduced to clinical conditions that demonstrate the relationships between normal and pathological anatomy and physiology. The sections on diagnostic procedures and the Case Study exercises are intended to demonstrate how information can be extracted and organized to reach a reasonable tentative diagnosis. The goal is to acquaint you with the mechanics of the process. This will not enable you to make accurate diagnoses, for situations in the real world are much more complicated and variable than the examples provided here. Making a final diagnosis is usually a complex process that demands a far greater level of experience and training than this course can provide.

We will not discuss treatment plans for similar reasons; the treatment of serious diseases requires specialized training and competence in advanced biochemistry, pharmacology, microbiology, pathology, and other clinical disciplines. However, many of the discussions in later sections include information about the use of specific drugs and other therapeutic procedures in the treatment of disease. These are representative examples intended to show potential treatment strategies rather than to endorse specific protocols and therapies.

Applied Research Topics

The Application of Principles in Chemistry

Cells, tissues, organs, and organ systems are composed of chemicals. The survival of cells, tissues, organs, and systems depends on the control of chemical reactions, both within individual cells and in the extracellular fluids of the body. It is therefore not surprising that you cannot understand physiological principles without a familiarity with basic chemistry. As the understanding of physiological mechanisms has improved, physicians have become relatively adept at using chemical tests to diagnose disease. They have also developed ways of manipulating intracellular and extracellular chemical reactions to help restore homeostasis. In this section we will consider the practical application of some basic chemical principles introduced in Chapter 2.

THE MEDICAL IMPORTANCE OF RADIOISOTOPES
EAP *p. 27*

Many recent technological advances in medicine have involved the use of radioisotopes for diagnosis and the visualization of internal structures. This section will focus on two examples:

- *The use of radioactive tracers in diagnosis:* Radioisotopes can be attached to organic or inorganic molecules and injected into the body. Once within the body, these labeled compounds emit radiation energy that can be used to create images that provide information about tissue structure, tumorous growths, blocked or weakened blood vessels, or other abnormalities in the body.

- *The use of radioactive compounds to treat disease:* If a suitable radiation source can be accurately delivered to a target site, the radioactivity can be used to destroy abnormal cells or tissues.

℞ Radioisotopes and Clinical Testing

Radioisotopes are isotopes having unstable nuclei that emit subatomic particles in measurable amounts. **Alpha particles** are generally released by the nuclei of large radioactive atoms, such as uranium. Each alpha particle consists of a helium nucleus: two protons and two neutrons. **Beta particles** are electrons, more often released by radioisotopes of lighter atoms. **Gamma rays** are very-high-energy electromagnetic waves comparable to the X-rays used in clinical diagnosis. The **half-life** of any radioactive isotope is the time required for a 50 percent reduction in the amount of radiation it emits. The half-lives of radioisotopes range from fractions of a second to thousands of years.

Like X-rays, gamma rays, beta particles, and alpha particles can damage or destroy living tissues. The danger posed by radiation exposure varies, depending on the nature of the emission and the duration of exposure. But radiation also has a variety of beneficial uses in medical research and clinical diagnosis. Weakly radioactive isotopes with short half-lives can sometimes be used to check the structural and functional state of an organ without surgery.

Radioisotopes are useful because they can be incorporated into specific compounds normally found within the body. These labeled compounds, called **tracers,** can be introduced into the body and tracked by the radiation they release. After a labeled compound is swallowed, its uptake, distribution, and excretion can be determined by monitoring the radioactivity of samples taken from the digestive tract, body fluids, and waste products. For example, compounds labeled with radioisotopes of cobalt are used to monitor the intestinal absorption of vitamin B_{12}. Usually cobalt-58, a radioisotope with a half-life of 71 days, is used.

Radioisotopes can also be injected into the blood or other body fluids to provide information on circulatory anatomy and the anatomy and function of specific target organs. In **nuclear imaging** the radiation emitted by injected radioisotopes creates an image on a special photographic plate. Such a procedure may be used to identify regions where particular radioactive materials are concentrated or to check the circulation through vital organs. Radioisotopes can also produce pictures of specific organs, such as the liver, spleen, or thyroid, where labeled compounds are removed from the circulation.

The thyroid gland, shown in Figure A-3a, sits below the larynx (voicebox) on the anterior portion of the neck. The normal thyroid gland absorbs iodine, which is then used to produce thyroid hormones. As a result, the thyroid gland will actively absorb and concentrate radioactive iodine. The **thyroid scan** in Figure A-3b was taken following the injection of iodine-131, a radioisotope with an 8-day half-life. This procedure, called a *thyroid radioactive iodine uptake measurement*, or **RAIU,** can provide information on (1) the size and shape of the gland and (2) the amount of absorptive activity under way. Comparing the rate of iodine uptake with the level of circulating hormones makes it possible to evaluate the functional state of the gland.

Radioactive iodine is an obvious choice for imaging the thyroid gland. For most other tissues and organs a radioactive label must be attached to another compound. *Technetium,* a versatile label, is the primary radioisotope used in nuclear imaging today. The isotope, ^{99m}Tc, is artificially produced and has a half-life of 6 hours. This brief half-life significantly reduces the radiation exposure of the patient. Technetium is used in more than 80 percent of all scanning procedures. The nature of the technetium-labeled compound varies, depending on the identity of the target organ. Technetium scans are performed to examine the thyroid gland, spleen, liver, kidneys, digestive tract, bone marrow, and a variety of other organs.

PET (**P**ositron **E**mission **T**omography) scans utilize the same principles as standard radioisotope scans, but the analyses are performed by computer. The scans are much more sensitive, and the computers can reconstruct sections through the body that permit extremely precise localization. Among other things, this procedure can analyze blood flow through organs and assess the metabolic activity within specific portions of an organ as complex as the human brain.

Figure A-3c is a PET scan of the brain showing activity at a single moment in time. The scan is dynamic, however, and changing patterns of activity can be followed in real time. PET scans can be used to analyze normal brain function as well as to diagnose brain disorders. To date, the technique has served primarily as a research tool. Because the equipment is expensive and bulky, it is unlikely to be available anywhere except in large regional medical centers or universities. The research advantages of PET scans have diminished considerably since the

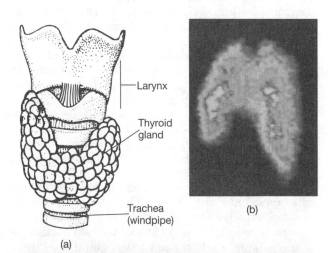

(a)

(b)

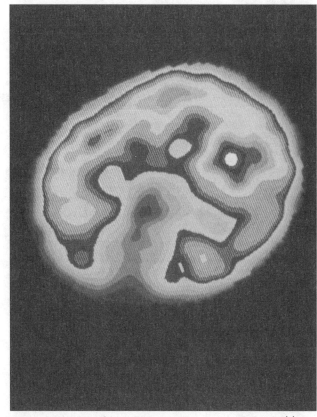

(c)

Figure A–3   Imaging techniques
(a) The position and contours of the normal thyroid gland as seen in dissection. (b) After it has been labeled with radioactive iodine, the thyroid can be examined by special imaging techniques. In this computer-enhanced image, different intensities indicate differing concentrations of the radioactive tracer. (c) A PET scan of the left cerebral hemisphere in lateral view. The light areas indicate regions of increased metabolic activity.

advent of real-time CT analysis *(cine-CT)* and the realization that MRI can be used to monitor small changes in blood flow and tissue activity without the use of radioactive tracers.

℞ Radiopharmaceuticals EAP *p. 27*

Nuclear medicine involving injected radioisotopes has been far more successful in producing useful images than in treating specific disorders. The problem is that relatively large doses of radiation must be used to destroy abnormal or cancerous tissues, and it is very difficult to control the distribution of these radioisotopes in the body with sufficient precision. As a result, both normal and abnormal tissues may be damaged by radiation exposure. For the same reason, it is difficult to control the radiation dosage administered to the target tissues. This is a problem because underexposure can have very little effect, whereas overexposure can cause the destruction of adjacent normal tissues.

Radioactive drugs, or *radiopharmaceuticals*, can be effective only if they are delivered precisely and selectively. One success story has been the treatment of *hyperthyroidism,* or thyroid oversecretion. As noted earlier, the thyroid gland selectively concentrates iodine. To treat hyperthyroidism, large doses of radioactive iodine (^{131}I) can be administered. The radiation released destroys the abnormal thyroid tissue and stops the excessive production of thyroid hormones. (The individual usually becomes *hypothyroid*—deficient in thyroid hormone—but this condition can be treated by taking thyroid hormones in tablet form.) This is now the preferred treatment method for hyperthyroid patients over 40 years of age.

A relatively new application of nuclear medicine involves attaching a radioactive isotope to a *monoclonal antibody (MoAb).* Antibodies are proteins produced in the body to provide a selective defense against foreign proteins, toxins, or pathogens. A substance that triggers antibody production is called an *antigen.* Monoclonal antibodies are produced by culturing immune cells that are sensitized to a particular antigen. (The process is detailed on p. 147.) The antibodies these cells manufacture can be extracted, labeled with radioactive materials, and concentrated. If injected into the body, the antibodies will bind to their target antigens, and in the process expose the surrounding tissues to radiation.

MoAbs specific for the antigens on certain types of tumor cells have already been approved by the FDA. When injected into the body, the radiolabeled MoAbs travel to the tumor site and attach to the specific antigen displayed on the cancer cell surface. Upon attachment, radiation is emitted from the target area. The amount of radiation emitted is low, however, and the procedure is used to produce images rather than to treat the disease. This technique is very sensitive and can detect small tumors for early diagnosis and treatment. Experiments continue, with the eventual goal of using radiolabeled MoAbs to destroy tumor cells.

📖 SOLUTIONS AND CONCENTRATIONS EAP *p. 33*

Physiologists and clinicians pay particular attention to ionic distributions across membranes, and the electrolyte composition of body fluids. Standard values for physiological tests are provided throughout the text, and summarized in Appendix V. Data must be analyzed from several different perspectives, and physiological values may be reported in several different ways. One method is to report the concentration of atoms, ions, or molecules in terms of weight per unit volume of solution. Although grams per liter (g/l) may be used, values are most often expressed in terms of grams (g), milligrams (mg), or micrograms (µg) per 100 ml. Since 100 ml is 0.1 liter, or 1 deciliter **(dl)**, the abbreviations most often used in this text are **g/dl** and **mg/dl.**

Osmolarity depends on the total number of individual atoms, ions, and molecules in solution, without regard to molecular weight, electrical charge or molecular identity. As a result, if fluid balance and osmolarity are being monitored, concentrations are usually reported in terms of moles per liter (M/l) or millimes per liter (mM/l or mmol/l), rather than in terms of g/dl or mg/dl. To convert from g/dl to M/l, multiply by 10 and divide by the atomic weight of the element. For example, a sample of plasma (blood with the cells removed) contains sodium ions at a concentration of roughly 0.32 g/dl (320 mg/dl). This value can be converted to M/l as follows:

$$\frac{\text{g/dl} \times 10}{\text{atomic weight}} = \frac{0.32 \times 10}{22.99}$$

$$= 0.140 \text{ M/l (140 mM/l)}$$

Moles or millimoles per liter can also be used to indicate the concentration of molecules in solution; the same conversion can be performed by substituting molecular weight for atomic weight in the above equation.

Because electrolyte concentrations have profound effects on living cells, it is often important to know how many positive and negative charges are present in a biological solution. In this case the important question is not just how many ions or molecules are present, but how many positive or negative charges they bear. For example, a single ion of calcium (Ca^{2+}) has twice the electrical charge of a single sodium atom (Na^+), although the two are identical in terms of their effects on osmolarity. One **equivalent (Eq)** represents a mole of positive or negative charges; physiological concentrations are often reported in terms of **milliequivalents per liter (mEq/l).** You should become familiar with both methods of expression, and fortunately the conversion from millimoles to millequivalents is relatively easy to perform. For **monovalent ions,** those with a +1 or –1 charge, millimole and millequivalent values are identical, and no calculation is needed. For **divalent ions,** with +2 or

–2 charges, the number of charges (mEq) is twice the number of ions (nM); if an ion had a +3 or –3 charge, the number of millequivalents would be 3 times the number of millimoles. To reverse the process, and convert mEq to mM, simply divide by the ionic valence (number of charges).

Table A-6 compares the different methods of reporting the concentration of major electrolytes in plasma in terms of weight, moles, and equivalents; the tables included in Appendix V provide data in terms currently accepted for clinical laboratory reports.

There is no doubt that physiologists and clinicians would benefit from the use of standardized reporting procedures; it can be very frustrating to consult three references and find that the first reports electrolyte concentrations in mg/dl, the second in mM/l, and the third in mEq/l. In 1984, the American Medical Association House of Delegates endorsed a plan to standardize clinical test results throught the use of metric **SI** (Systeme Internationale) units, with a target date of July 1, 1987 for the switchover. Unfortunately there was no mechanism for enforcing compliance, and the plan may not prove any more successful than the ill-fated attempt to drop the U.S. system of measurement in favor of the metric system.

The major problem is that the relationships to current normal values are difficult to remember. Electrolyte concentrations, now most often indicated in mEq/l, will be reported in SI units that represent mM/l. That means the values for sodium or potassium concentrations remain unchanged, but the normal values for calcium or magnesium are reduced by 50 percent. The situation becomes more confusing when metabolite concentrations are considered. Cholesterol and glucose concentrations are now reported in terms of mg/dl, but the SI units represent mM/l (mmol/l). However, total lipid concentrations, also currently listed as mg/dl, and total protein concentrations, now shown as g/dl, will be reported in terms of g/l. To be useful in a clinical setting, physicians must not only remember the definition of each SI unit, but convert and relearn the normal ranges. As a result, it appears unlikely that the conversion to SI units will be completed in the immediate future.

Table A-6   A Comparison of Methods for Reporting Concentrations of Solutes in the Blood

Solute	mg/dl	mM/l	mEq/l	SI Units	
Electrolytes					
Sodium (Na⁺)	320	140	140	140	mmol/l
Potassium (K⁺)	16.4	4.2	4.2	4.2	mmol/l
Calcium (Ca²⁺)	9.5	2.4	4.8	2.4	mmol/l
Chloride (Cl⁻)	354	100	100	100	mmol/l
Metabolites					
Glucose	90	5	nr	5	mmol/l
Lipids, total	600	nr	nr	0.6	g/l
Proteins, total	7 g/dl	nr	nr	70	g/l

Topics in Metabolism

Metabolism is the sum of all the biochemical reactions proceeding in the body. There are hundreds of thousands of reactions occurring in each cell, and at any given moment biochemical pathways may be producing phospholipids for the cell membrane or peptide hormones for secretion, while breaking down carbohydrates to generate ATP. This section will consider three aspects of metabolism:

1. Many disease processes are the result of a faulty biochemical pathway. For example, an enzyme may be missing or nonfunctional, or the necessary enzymatic substrates may be unavailable. *Phenylketonuria* and *albinism,* are examples of metabolic disorders that will be considered shortly.

2. Enzymes play a pivotal role in controlling metabolic processes in our cells. The mechanisms responsible for controlling enzymatic reactions are therefore important, and problems with enzymatic regulation can cause severe metabolic disorders.

3. Diet and nutrition have an obvious impact on metabolic operations within the body. A substantial research effort is under way to manipulate metabolic operations by dietary changes—the control of cholesterol in the diet is only one of several pertinent examples.

Ⓡ Artificial Sweeteners   EAP *p. 35*

Some people cannot tolerate sugar for medical reasons; others avoid it because recent dietary guidelines call for reduced sugar consumption, or in an effort to lose weight. Thus many people today are using artificial sweeteners in their foods and beverages.

Artificial sweeteners are organic molecules that can stimulate taste buds and provide a sweet taste to foods without adding substantial amounts of calories to the diet. These molecules have a much greater effect on the taste receptors than natural sweeteners, such as fructose or sucrose, so they can be used in minute quantities. For example, *saccharin* is about 300 times sweeter than sucrose. The popularity of this sweetener has declined since it was reported that saccharin may promote bladder cancer in rats. The risk is very small, however, and saccharin continues to be used. There are several other artificial sweeteners currently on the market, including *aspartame (NutraSweet®)*, *sucralose,* and *acesulfame potassium (Ace-K,* or *Surette ®)*. The market success of an artificial sweetener ultimately depends on its taste and its chemical properties. Stability in high temperatures (as in baking) and resistance to breakdown in an acidic pH (as in carbonated drinks) are important properties for any artificial sweetener.

Molecules of artificial sweeteners do not resemble those of natural sugars. Saccharin, acesulfame potassium, and sucralose cannot be broken down by the body and have no nutritional value. Aspartame consists of a pair of amino acids. Amino acids are the building blocks of proteins (as discussed later in this chapter), and they can be broken down in the body to provide energy. However, because aspartame is 200 times sweeter than sucrose, very small quantities are needed, so the sweetener adds few calories to a meal. Aspartame does not produce the bitter aftertaste sometimes attributed to saccharin, and thus is used in many diet drinks and low-calorie desserts.

Two new sweeteners, *thaumatin-1* and *monellin*, are proteins extracted from African berries. Thaumatin, roughly 100,000 times sweeter than sucrose, has been approved by the Food and Drug Administration for use in chewing gums. Another artificial sweetener, *cyclamate*, was banned in 1970 after experiments suggested that it caused bladder tumors in laboratory rats. These conclusions have been shown to be incorrect, so cyclamates may soon be reapproved. However, because this sweetener is only about 30 times sweeter than sucrose, it may not have much impact on the marketplace.

℞ Fat Substitutes
EAP *p. 36*

Although the average American diet is not as rich in fats as that of Eskimos, we still consume more fat than do people in many other parts of the world. Diets high in fat have been linked not only to heart disease but also to certain forms of cancer, and recent recommendations suggest that lowering the percentage of calories we derive from fat would benefit our health. This suggestion has led to an increased interest in the development of possible substitutes for fat.

Fat substitutes provide the texture, taste, and cooking properties of natural fats. One fat substitute, *Simplesse®* has been approved by the Food and Drug Administration; a second, *Trailblazer®*, is currently under review. Both are made from proteins of egg white and skim milk or whey. The heated proteins are treated to form small spherical masses that have the taste and texture of fats. Simplesse can be used in place of fats in any application other than baking; it is found in low-calorie "ice creams" under the trade name *Simple Pleasures®*. These fat substitutes can be broken down in the body, but they provide less energy than natural fats. For example, ice cream with Simplesse has half the calories of ice cream containing natural fats.

Several other fat substitutes currently being developed are derived from carbohydrates. One of these, *Olestra®*, is made by chemically combining sucrose and fatty acids. The resulting compounds cannot be used by the body, and so contribute no calories at all. Olestra has been approved as an ingredient in margarines and baked goods; its use as a shortening and cooking oil is under review. *Oatrim*, a fat substitute derived from soluble fiber and complex carbohydrates, is now used in muffins, cookies, no-fat cheeses, and lean hot dogs and luncheon meats.

Fat substitutes provide the texture of natural fats, but the flavor is usually quite different. For this reason efforts are underway to develop "designer fats" using fatty acids that provide fewer calories than the lipids normally found in the diet. For example, the fats in *SALATRIM®* contain long-chain fatty acids that are poorly absorbed and short-chain fatty acids that provide relatively few calories when broken down. On a gram-for-gram basis, the fats in SALATRIM provide only 60 percent of the calories of natural triglycerides.

Fat substitutes must be tested and approved by the Food and Drug Administration. To receive FDA approval, a substance must be proven harmless to laboratory animals when consumed at levels 100 times the expected dosage in humans. (This requirement is currently under review, as even otherwise harmless compounds can cause adverse effects at such high doses.)

The use of fat substitutes may also pose secondary metabolic problems. Fat-soluble vitamins (A, D, E, and K) are normally absorbed by the intestinal tract in company with dietary lipids. A drastic reduction in the lipid content of the diet may therefore lead to deficiencies of these vitamins.

⚕ Anomalies in Amino Acid Metabolism
EAP *p. 40*

1. *Phenylketonuria (PKU):* Patients diagnosed with *phenylketonuria (PKU)* lack the enzyme that converts the amino acid *phenylalanine* to another amino acid, *tyrosine.* Without this enzyme, phenylalanine accumulates in the blood and tissues. If this condition is not detected shortly after birth, mental retardation may occur. Newborn infants usually undergo a blood test for PKU 48 hours after nursing begins, because milk is a major source of phenylalanine. Abnormally high circulating levels of phenylalanine may indicate PKU. Once a diagnosis of PKU is made, the diet is controlled to avoid foods containing high levels of phenylalanine.

2. *Albinism: Albinism* is a genetic disorder that results in a lack of pigment production in the skin. The cause is a defective enzyme involved in the metabolism of the amino acid tyrosine. Because this enzyme is abnormal, the protein pigment *melanin* cannot be synthesized. The skin is white, and the hair and eyes are also affected. Among its other functions, melanin helps protect the skin from the effects of ultraviolet (UV) radiation. When outdoors, individuals with albinism must be careful to avoid skin damage from the UV radiation in sunlight.

Topics in Cellular Biology and Histology

Cells are the smallest living units in the body, but they are not the only forms of life. Discussions throughout the text and the *Applications Manual* assume that you are already familiar with the basic properties of cells and with the characteristics of potential *pathogens* (disease-causing organisms). This section begins with a review of the important distinctions among various types of pathogens, and then proceeds to a discussion of other topics at the cellular level of organization.

THE NATURE OF PATHOGENS

EAP *p. 47*

Chapter 3 of the text presented the structure of a "representative" cell. The cellular organization depicted in Figure 3-2 and described in that chapter is that of a *eukaryotic cell* (ū-kar-ē-OT-ik; *eu*, "true" + *karyon*, "nucleus"). The defining characteristic of eukaryotic cells is the presence of a nucleus. All eukaryotic cells have similar membranes, organelles, and methods of cell division. All multicellular animals and plants (as well as some single-celled organisms) are composed of eukaryotic cells.

The eukaryotic plan of organization is not the only one found in the living world, however. There are organisms that do not consist of eukaryotic cells. These organisms are of great interest to us because they include most of the pathogens that can cause human diseases.

Bacteria

Prokaryotic cells do not have nuclei or other membranous organelles. They do not have a cytoskeleton, and typically their cell membranes are surrounded by a semirigid cell wall made of carbohydrate and protein. Figure A-4a shows the structure of a representative **bacterium.**

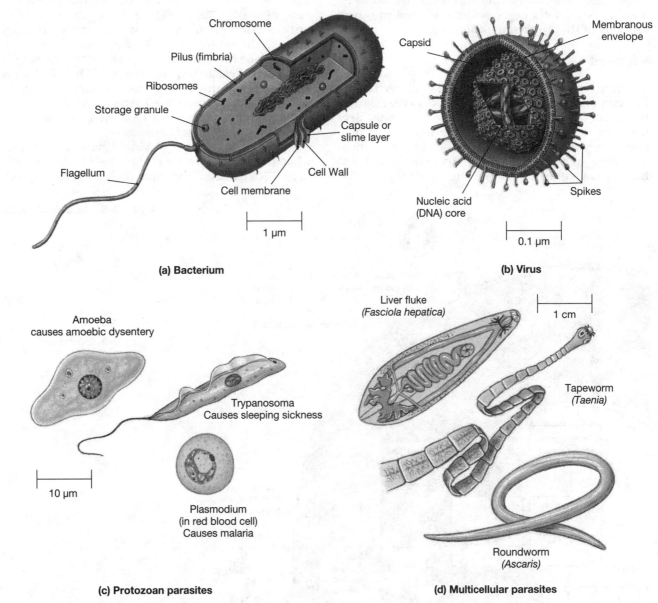

(a) Bacterium

(b) Virus

(c) Protozoan parasites

(d) Multicellular parasites

Figure A-4   Representative Pathogens.

(a) A bacterium, with prokaryotic characteristics indicated. Compare with Figure 3-2, (EAP *p.48),* which shows a representative eukaryotic cell. (b) A typical virus. Each virus has an inner chamber containing nucleic acid, surrounded by a protein capsid or an inner capsid and an outer membranous envelope. The herpes viruses are enveloped DNA viruses; they cause chickenpox, shingles, and herpes. (c) Protozoan pathogens. Protozoa are eukaryotic, single-celled organisms, common in soil and water. (d) Multicellular parasites. Several different groups of organisms are human parasites and many have complex life histories. Note: These illustrations are not drawn to a common scale.

Bacteria are usually less than 2 µm in diameter. Many bacteria are quite harmless, and many more—including some that live within our bodies—are actually beneficial to us in a variety of ways. Other bacteria are dangerous pathogens that will destroy body tissues if given the opportunity. These bacteria are dangerous because they absorb nutrients and release enzymes that damage cells and tissues. A few pathogenic bacteria also release toxic chemicals. Bacterial infections are responsible for many serious diseases, including tetanus, cholera, gangrene, pneumonia, meningitis, syphilis, gonorrhea, typhus, plague, tuberculosis, typhoid fever, and leprosy. These and other bacterial infections are considered in various chapters of the text, and in other sections of the *Applications Manual.*

Viruses

Another type of pathogen conforms neither to the prokaryotic nor the eukaryotic organizational plan. These tiny pathogens, called **viruses,** are not cells at all. In fact, when free in the environment, they do not show any of the characteristics of living organisms. They are classified as **infectious agents** because they can enter cells (either prokaryotic or eukaryotic) and replicate themselves.

Viruses consist of a core of nucleic acid (DNA or RNA) surrounded by a protein coat. (Some varieties have a membranous outer covering as well.) The structure of a representative virus is shown in Figure A-4b. Important viral diseases include influenza (flu), yellow fever, some leukemias, AIDS, hepatitis, polio, measles, mumps, rabies, and the common cold.

To enter a cell, a virus must first attach to the cell membrane. This attachment occurs at one of the normal membrane proteins. Once it has penetrated the cell membrane, the viral nucleic acid takes over the cell's metabolic machinery. In the case of a DNA virus (Figure A-5a), the viral DNA enters the cell nucleus, where transcription begins. The mRNA produced then enters the cytoplasm for translation, and the cell's ribosomes begin synthesizing viral proteins. The viral DNA replicates within the nucleus, "stealing" the cell's nucleotides. The replicated viral DNA and the new viral proteins then form new viruses that pass through the cell membrane.

In the case of an RNA virus, the story is somewhat more complicated (Figure A-5b). In the simplest RNA viruses, the viral RNA entering the cell

Figure A-5 Viral Replication

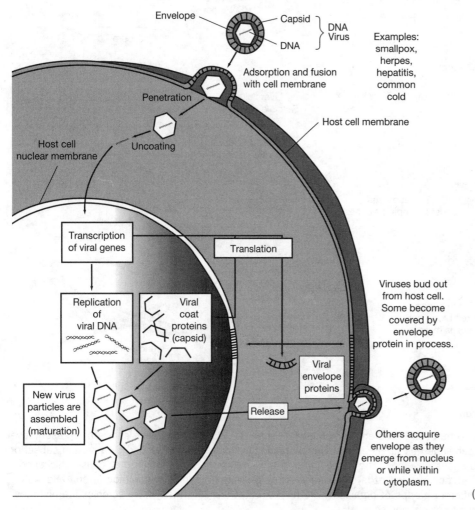

(a) Replication of a typical DNA virus.

functions as an mRNA strand that carries the information needed to direct the cell's ribosomes to synthesize viral proteins. These proteins include enzymes that perform the duplication of the viral RNA. When the cell is packed with new viruses, the cell membrane ruptures and the RNA viruses are released into the interstitial fluid.

In the *retroviruses,* a group that includes HIV (the virus responsible for AIDS), the replication process is more complex. These viruses carry an enzyme called *reverse transcriptase* that directs "reverse transcription:" the assembly of DNA based on the nucleotide sequence of an RNA strand. The DNA created in this way is then inserted into the infected cell's chromosomes. The viral genes are then activated, and the cell begins producing RNA through normal transcription. The RNA produced includes viral RNA, mRNA carrying the information for the synthesis of reverse transcriptase, and mRNA controlling the synthesis of viral proteins. These components then combine within the cytoplasm, which gradually becomes filled with viruses. These new RNA viruses are then shed at the cell surface.

Even if the host cell is not destroyed outright by these events, normal cell function is usually disrupted. In effect, the metabolic activity of the cell is diverted to create viral components, rather than performing tasks needed for cell maintenance and survival. Some viruses, however, can lie dormant within infected cells for long periods of time before initiating this process of replication.

Viruses are now becoming important as benefactors, as well as adversaries. In genetic engineering

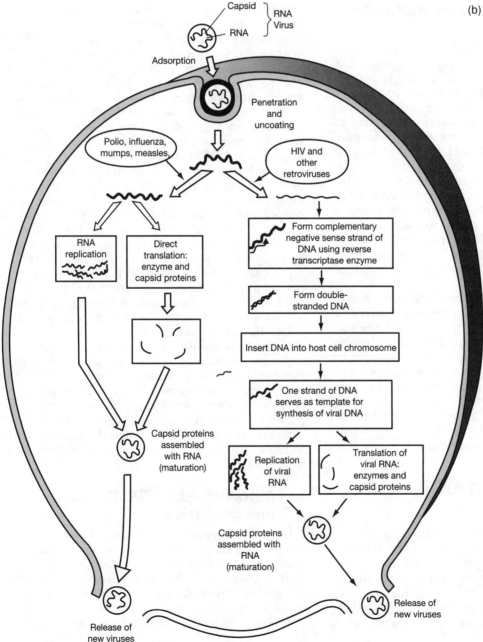

(b) Replication of RNA viruses.

procedures, viruses whose nucleic acid structure has been intentionally altered can be used to transfer copies of normal human genes into the cells of individuals with inherited enzymatic disorders. This was the method used to insert the gene for the enzyme missing in ADA patients (p. 000). Attempts are now planned to treat *cystic fibrosis (CF)* in the same way. Cystic fibrosis is a debilitating genetic defect whose most obvious—and potentially deadly—symptoms involve the respiratory system. The underlying problem is an abnormal gene that carries instructions for a chloride ion channel found in cell membranes throughout the body. Researchers have recently treated CF in laboratory animals by inserting the normal gene into a virus that infects cells lining the respiratory passageways. The virus could be given to human patients via an inhalant.

Unicellular Pathogens and Multicellular Parasites

Bacteria and viruses are the best known human pathogens, but there are eukaryotic pathogens as well. Examples of the most important types are included in Figure A-4c. **Protozoa** are unicellular eukaryotic organisms that are abundant in soil and water. They are responsible for a variety of serious human diseases, including *amoebic dysentery* and *malaria*. **Fungi** (singular *fungus*) are eukaryotic organisms that absorb organic materials from the remains of dead cells. Mushrooms are familiar examples of very large fungi. In a fungal infection, a microscopic fungus spreads through living tissues, killing cells and absorbing nutrients. Several relatively common skin conditions *(athlete's foot)* and a few more serious diseases *(histoplasmosis)* are the result of fungal infections.

Larger multicellular organisms (Figure A-4d), such as *flukes* or *nematodes*, generally referred to as *parasites*, can also invade the human body and cause diseases. These organisms, which range from microscopic flatworms to tapeworms a meter or more in length, usually cause weakness and discomfort, but do not *by themselves* kill their host. However, complications resulting from the parasitic infection, such as chronic bleeding or secondary infections by bacterial or viral pathogens, can ultimately prove fatal.

CELL STRUCTURE AND FUNCTION

Each cell in the body has a particular role to play in maintaining the integrity of the individual as a whole. Some conduct nerve impulses, while others manufacture hormones, build bones, or contract to produce body movements. When any of these cells malfunction, whether due to genetic abnormalities affecting enzyme function, a viral or bacterial infection, trauma, or cancer, homeostasis is threatened. The next section introduces clinical and practical applications of basic principles of cellular function, and discusses representative disorders resulting from problems at the cellular level of organization.

☤ Lysosomal Storage Diseases EAP *p. 60*

Problems with lysosomal enzyme production cause more than 30 storage diseases affecting children. In these conditions the lack of a specific lysosomal enzyme results in the buildup of materials normally removed and recycled by lysosomes. Eventually the cell cannot continue to function. Three important examples will be considered here: *Gaucher's disease, Tay-Sachs disease*, and *glycogen storage disease.*

Gaucher's disease is caused by the buildup of *cerebrosides*, glycolipids found in cell membranes. This is probably the most common type of lysosomal storage disease. There are two forms of this disease: (1) an infantile form, marked by severe neurological symptoms ending in death, and (2) a juvenile form, with enlargement of the spleen, anemia, pain, and relatively mild neurological symptoms. Gaucher's disease is most common among the Ashkenazi Jewish population, where it occurs at a frequency of approximately 1 in 1,000 births.

Tay-Sachs disease is another hereditary disorder caused by the inability to break down glycolipids. In this case the glycolipids are *gangliosides*, which are most abundant in neural tissue. Individuals with this condition develop seizures, blindness, dementia, and death, usually by age 3-4. Tay-Sachs disease is most common among the Ashkenazi Jewish population, where it occurs at a frequency of 0.3 per 1,000 births.

Glycogen storage disease (Type II) primarily affects skeletal muscle, cardiac muscle, and liver cells—the cells that synthesize and store glycogen. In this condition the cells are unable to mobilize glycogen normally, and large numbers of insoluble glycogen granules accumulate in the cytoplasm. These granules disrupt the organization of the cytoskeleton, interfering with transport operations and the synthesis of materials. In skeletal and heart muscle cells, the buildup leads to muscular weakness and potentially fatal heart problems.

📖 Mitochondrial DNA, Disease, and Evolution EAP *p. 61*

There are several inheritable disorders that result from abnormal mitochondrial activity. The mitochondria involved have defective enzymes that reduce their ability to generate ATP. Cells throughout the body may be affected, but symptoms involving muscle cells, neurons, and the receptor cells in the eye are most commonly seen because these cells have especially high energy demands. Disorders caused by defective mitochondria are called *mitochondrial cytopathies.* In several instances, the disorders have been linked to inherited abnormalities in mitochondrial DNA. In some cases, the problem appears in one population of cells only. For example, abnormal mitochondrial DNA has been found in the motor neurons whose degeneration is responsible for the condition of *Parkinson's disease,* a neurological disorder characterized by a shuffling gait and uncontrollable tremors.

More often, mitochondria throughout the body are involved. Examples of conditions caused by mitochondrial dysfunction include one class of epilepsies *(myoclonic epilepsy)* and a type of blindness *(Leber's hereditary optic neuropathy).* These are inherited conditions, but the pattern of inheritance is very unusual. Although men or women may have the disease, only affected women can pass the condition on to their children. The explanation for this pattern is that the disorder results from an abnormality in the DNA of mitochondria, not in the DNA of cell nuclei. All the mitochondria in the body are produced through the replication of mitochondria present in the fertilized ovum. Few if any of those mitochondria were provided by the father; most of the mitochondria of the sperm do not remain intact after fertilization takes place. As a result, children can usually inherit these conditions only from their mothers. This brings us to an interesting concept. Virtually all your mitochondria were inherited from your mother, and hers from her mother, and so on back through time. The same is true for every other human being. Now, it is known that over long periods of time small changes in DNA nucleotide sequences accumulate. Mitochondrial DNA, or mDNA, can therefore be used to estimate the degree of relationship between individuals. The greater the difference between the mDNA of two individuals, the more time has passed since the lifetime of their most recent common ancestor, and the more distant their relationship. On this basis it has been estimated that all human beings now alive shared a common female ancestor roughly 350,000 years ago. Appropriately enough, that individual has been called a "Mitochondrial Eve." The existence and history of Mitochondrial Eve remain controversial.

Topics in Molecular Biology

Molecular biology is the division of science that studies the synthesis, structure, and function of macromolecules important to life, such as proteins and nucleic acids. Deciphering the genetic code and relating the intricate structure of a protein to its particular functions are major goals of molecular biology. Research in this area has greatly enhanced our understanding of normal functions as well as disease processes.

The field of molecular biology has revolutionized the study of medicine by providing a clear biochemical basis for many complex diseases. For example, in *sickle cell anemia* red blood cells undergo changes in shape that result in blocked vessels and tissue damage due to oxygen starvation. It is now known that this condition results when an individual carries two copies of a defective gene that determines the structure of *hemoglobin,* the oxygen-binding protein found within red blood cells. The genetic defect is small, and changes just 2 of the 574 amino acids in this protein. That one change is enough to alter the functional properties of the hemoglobin molecule, leading to changes in the properties of the red blood cells. This type of disorder is often called a *molecular disease* because it results from abnormalities at the molecular level of organization.

Roughly 1,200 inherited disorders have now been identified, and researchers have located the defective genes responsible for cystic fibrosis, Duschenne's muscular dystrophy, and Tay-Sachs disease. Identifying the genetic defect is, of course, the vital first step toward an effective gene therapy or other treatment. The treatments that are now evolving make use of the principles of *genetic engineering.*

℞ GENETIC ENGINEERING AND GENE THERAPY   EAP *p. 65*

Once the mechanics of the genetic code were understood, everyone realized that it would be theoretically possible to change the genetic makeup of organisms—perhaps even of a human being. The popular term for activities related to this goal is *genetic engineering.*

What are some of the key problems confronting genetic engineers? Genes code for proteins; the makeup of each protein is determined by the sequence of codons (nucleotide triplets) in a stretch of DNA. A human cell has 46 chromosomes, 2 meters of DNA, and roughly 10^9 triplets. If all the DNA in the human body were extracted and strung together, the resulting strand would be long enough to make several hundred round trips between the earth and the sun. Simply finding a particular gene among the approximately 100,000 that each of us carries is an imposing task. Yet before a specific gene can be modified, its location must be determined with great precision. This involves preparing a map of the appropriate chromosome.

Mapping the Genome

Several techniques can be used to create a general map of the chromosomes. **Karyotyping** (KAR-ē-ō-tī-ping; *karyon,* nucleus + *typos,* a mark) is the determination of an individual's chromosome complement. Figure A-6a shows a set of normal human chromosomes. Each chromosome has characteristic banding patterns, and segments can be stained with special dyes. Unusual banding patterns can indicate structural abnormalities. These abnormalities are sometimes linked to specific inherited conditions, including a form of leukemia. *Down syndrome* (Figure A-6b) results from the presence of an extra chromosome, a copy of chromosome 21.

In December 1993, French researchers at the Centre d'Etude de Polymorphism Humaine (CEPH; "Center for the Study of Human Polymorphism") completed the first preliminary mapping of the entire human genome. This provided the landmarks and reference points needed to make more precise maps that indicate the locations of specific genes. More detailed maps have now been prepared for the Y chromosome and chromosome 21, the two smallest chromosomes, and work continues on the others.

Mapping is useful in itself, but it is only an intermediate step on the way to the ultimate goal: the determination of the nucleotide sequence of every gene in the human genome. As of 1994, a gene register maintained by Dr. Victor McKusick and other researchers at Johns Hopkins contained the locations of over 2,200 genes, an impressive number but a small percentage of the total. The sequencing process, which will take billions of dollars and another seven to ten years, will provide basic information about the location of genes on normal human chromosomes.

Gene Manipulation and Modification

Suppose that the location of a defective gene has been pinpointed. Before attempting to remedy the defect, one would have to determine the nature of the genetic abnormality. For example, the gene could be inactive or overactive or producing an abnormal protein. It could even be missing entirely.

Finally, it would be necessary to decide how to remedy the defect. Can the gene be turned on, turned off, modified, or replaced?

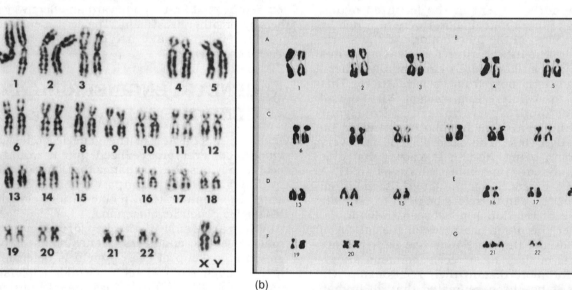

(a)   (b)

Figure A-6   Normal and Abnormal Karyotypes.

(a) A micrograph of the normal human chromosome set; the chromosomes have been arranged in this sequence to make comparisons easier. (b) The chromosomes of an individual with Down Syndrome. Note the extra copy of chromosome 21. (c) Down syndrome is associated with mental retardation, cardiovascular problems, and a variety of physical abnormalities. The boy (age 9) on the bicycle has the characteristic appearance of an individual with Down syndrome; compare his features with those of his normal sister.

(c)

What's the Problem? This can be a particularly difficult question to answer. Many of the 1,500 inheritable genetic disorders are classified according to general patterns of symptoms rather than any specific protein or enzyme deficiency. In some cases the approximate location of the gene has been determined, but the identity of the protein responsible for the clinical symptoms remains a mystery. It appears likely that there may be several different possible abnormalities in this protein that can result in different patterns of clinical disease.

What Can Be Done? If the gene is present but overproducing or underproducing, its activity might be controlled by introducing chemical repressors or inducers. Another approach relies upon **gene splicing** to produce a protein missing or present in inadequate quantities in the abnormal individual. Gene splicing (Figure A-7) begins with the localization of the gene, followed by its isolation. That gene is then "spliced" into the relatively simple DNA strand of a bacterium, creating recombinant DNA. Bacteria grow and reproduce rapidly under laboratory conditions, and before

long there is a colony of identical bacteria. All the members of the colony will carry the introduced gene and manufacture the corresponding protein. The protein can be extracted, concentrated, and administered to individuals whose diseases represent deficiencies in the activity of that particular gene. *Hemophilia*, a deficiency of blood clotting factors, and one form of diabetes (caused by an insulin deficiency) can be treated in this way.

Gene splicing is also used to obtain large quantities of proteins normally found in very small concentrations. Interferon, an antiviral protein, and human growth hormone are examples of compounds now being produced commercially using gene-splicing technology.

The most revolutionary strategies involve "fixing" abnormal cells by giving them copies of normal genes. In general, this method poses significant targeting problems, for the gene must be introduced into the right kind of cell. For example, placing liver enzymes in fingernails would not correct a metabolic disorder. But when the target cells can be removed and isolated, as in the case of bone marrow, the technique is

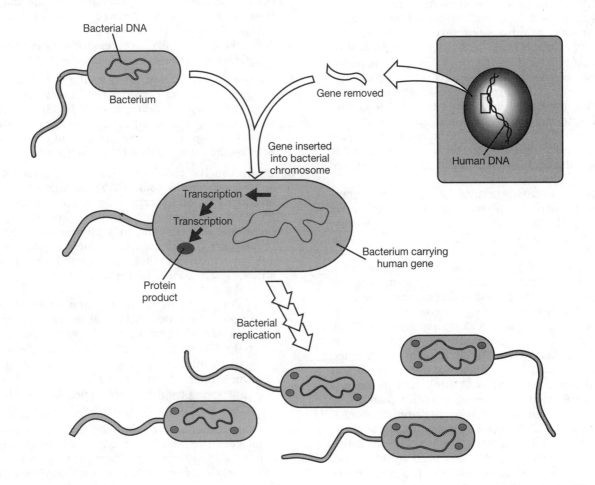

Figure A-7   Gene splicing.

A gene is removed from a human cell nucleus and attached to the DNA in a bacterium, where it directs production of a human protein. Bacterial replication creates a colony of bacteria that share the introduced gene and can yield large quantities of the protein product.

promising. Actual removal of a defective gene does not appear to be a practical approach, and the focus has been on adding genes that can take over normal functions.

In September 1990, the first gene therapy trials were initiated. The procedure was used to treat a 4-year-old girl afflicted with *adenosine deaminase deficiency (ADA)*. ADA is a rare condition that each year affects only about 20 children worldwide. Without this enzyme, toxic chemicals build up in cells of the immune system, and as these cells die, the body's defenses break down.

ADA results in a complex of symptoms known as *severe combined immunodeficiency disease*, or *SCID*. SCID can also be caused by other enzyme disorders affecting cells of the immune system. Symptoms include chronic respiratory infections, diarrhea, and a low resistance to viral or bacterial infections. Children with ADA usually die from infections that would pose no threat to normal children. A new drug called *PEG-ADA*, an altered form of the missing enzyme, can prolong life, but it does not cure the condition.

In this clinical trial, blood cells were collected, and cells of the immune system were removed. Short segments of DNA containing the normal gene for adenosine deaminase were then inserted into the nuclei of these cells, and the modified immune cells were returned to the body. Roughly a billion modified cells were reintroduced. Over time these modified cells divided to produce a large population of normal immune cells. The experiment was successful, and this protocol has been used to treat three other children with ADA deficiency. All have regained and retained apparently normal immune function, although the treatment must be repeated every few years.

This procedure attempts to relieve the symptoms of disease by inserting genes into defective somatic cells. They do not change the genetic structure of reproductive cells; because the eggs or sperm retain the original genetic pattern, the genetic defect will be passed to future generations. Researchers are much further away from practical methods of changing the genetic characteristics of reproductive cells. Mouse eggs fertilized outside the body have been treated and transplanted into the uterus of a second mouse for development. The gene added was one for a growth hormone obtained from a rat, and the large "supermouse" that resulted demonstrated that such manipulations can be performed. The possibilities for manipulating the characteristics of valuable animal stocks, such as cattle, sheep, or chickens, are quite exciting. The potential for altering the genetic characteristics of human beings is somewhat intimidating. Before any clinical variations on this theme are tested, our society will have to come to grips with a number of difficult ethical issues!

Just for a sense of the kinds of problems one might have to deal with, discuss the following questions with your friends and classmates.

Genetic Engineering—Questions to Think About

1. Your 51-year-old father has recently been diagnosed with a hereditary disorder affecting the brain. The prognosis is poor and there is currently no cure for the disorder. The physician advises that you be tested for the presence of the faulty gene.

   - Would you request the test for yourself (would you want to know if you had the defective gene)?
   - If you have the test done, who should have access to the results:

     Your insurance company?

     Your family, spouse, or fiancé?

     Your employer?

2. *Eugenics* is the control of the hereditary characteristics of individuals to improve the species. The science of eugenics became distorted through the work of scientists under Hitler's control. Prenatal testing now permits the diagnosis of a variety of inherited disorders before birth. This information could be used to "improve the species" by selectively terminating pregnancies. Is this advisable or ethical?

3. Many scientists today, such as the Nobel laureate James Watson, consider the human genome to be the blueprint for a human being. If an individual's genetic code affects every characteristic of that person, can people accused of crimes be held legally responsible for their actions?

4. Studies in the past have shown that men with an extra Y chromosome (XYY) are more violent and predisposed to crime than are men with the XY genotype. More recent studies have revealed no greater tendency toward violence among XYY individuals than among XY males. A prejudice against XYY males still exists due to the original study, now known to be seriously flawed.

   - What type of controls are needed to ensure that new information concerning genetic abnormalities is not released before it is confirmed? Not every person with a specific genotype will develop the same characteristics to the same degree. How can possible stereotypes be avoided when information is released?

5. You are an airline employer trying to offer your employees the least expensive health insurance available. The insurance company requires a blood test on each new potential employee to determine genetic abnormalities. Somehow you learn that a potential candidate for a job as a pilot has a genetic predisposition for a heart attack. Would this information affect your decision to hire that individual?

✠ CANCER

EAP *p.68*

Twenty-five percent of all Americans develop cancer at some point in their lives. It has been estimated that 70-80 percent of these cases involve chemical exposure, environmental factors, or both, and almost 40 percent of these are due to a single stimulus: cigarette smoke. During 1994, roughly 538,000 Americans were killed by some form of cancer, making this Public Health Enemy Number 2, second only to heart disease.

Cancer Causes

A relatively small number of cancers are actually inherited; 18 types have been identified to date, including two forms of leukemia. Most cancers develop through the interaction of genetic and environmental factors, and it is difficult to separate the two completely.

Genetic Factors: Two related genetic factors are involved in the development of cancer: *hereditary predisposition* and *oncogene activation.*

An individual born with genes that increase the likelihood of cancer is said to have a hereditary predisposition for the disease. Under these conditions a cancer is not guaranteed, but it is a lot more likely. The inherited genes usually affect tissue abilities to metabolize toxins, control mitosis and growth, perform repairs after injury, or identify and destroy abnormal tissue cells. As a result, body cells become more sensitive to local or environmental factors that would have little effect on normal tissues.

Cancers may also result from somatic mutations that modify genes involved with cell growth, differentiation, or mitosis. As a result, an ordinary cell is converted into a cancer cell. The modified genes are called **oncogenes** (ON-kō-jēns); the normal genes are called **proto-oncogenes**. Oncogene activation occurs by alteration of normal somatic genes. Because these mutations do not affect reproductive cells, the cancers caused by active oncogenes are not inherited.

A proto-oncogene, like other genes, has a regulatory component that turns the gene "on" and "off" and a structural component that contains the triplets that determine protein structure. Mutations in either portion of the gene may convert it to an active oncogene. A small mutation can accomplish this; changing 1 nucleotide out of a chain of 5,000 can convert a normal proto-oncogene to an active oncogene. In some cases, a viral infection can trigger activation of an oncogene. For example, one of the papilloma viruses appears to be responsible for many cases of cancer of the cervix.

More than 50 proto-oncogenes have been identified. In addition, a group of anticancer genes has been discovered. These genes, called *tumor-suppressing genes (TSG),* or *anti-oncogenes,* suppress division and growth in normal cells. Mutations that alter TSGs make oncogene activation more likely. TSG mutation has been suggested as important in promoting several cancers, including several blood cell cancers, breast cancer, and ovarian cancer. Examples of important suppressor genes include the genes *p53* and *p16.* Mutations affecting the p53 gene are responsible for the majority of cancers of the colon, breast, and liver. Abnormal p16 gene activity may be involved in as many as half of all cancer cases.

Environmental Factors: Many cancers can be directly or indirectly attributed to environmental factors called *carcinogens* (kar-SIN-o-jens). Carcinogens stimulate the conversion of a normal cell to a cancer cell. Some carcinogens are *mutagens* (MŪ-ta-jens)—that is, they damage DNA strands and sometimes cause chromosomal breakage. Radiation is an example of a mutagen that has carcinogenic effects.

There are many different chemical carcinogens in the environment. Plants manufacture poisons that protect them from insects and other predators, and although their carcinogenic activities are often relatively weak, many common spices, vegetables, and beverages contain compounds that can be carcinogenic if consumed in large quantities. Animal tissues may also store or concentrate toxins, and hazardous compounds of many kinds can be swallowed in contaminated food. A variety of laboratory and industrial chemicals, such as coal tar derivatives and synthetic pesticides, have been shown to be carcinogenic. Cosmic radiation, X-rays, UV radiation, and other radiation sources can also cause cancer. It has been estimated that 70-80 percent of all cancers are the result of chemical and/or environmental factors, and almost half (40 percent) are due to a single stimulus: cigarette smoke.

Specific carcinogens will affect only those cells capable of responding to that particular physical or chemical stimulus. The responses vary because differentiation produces cell types with specific sensitivities. For example, benzene can produce a cancer of the blood, cigarette smoke a lung cancer, and vinyl chloride a liver cancer. Very few stimuli can produce cancers throughout the body; radiation exposure is a notable exception. In general, cells undergoing mitosis are most likely to be vulnerable to chemical or radiational carcinogens. As a result, the cancer rates are highest in epithelial tissues, where stem cell divisions occur rapidly and relatively low in nervous and muscle tissues, where divisions do not normally occur.

Cancer Formation and Growth

Physicians who specialize in the identification and treatment of cancers are called **oncologists** (on-KOL-o-jists; *onkos,* "mass"). Pathologists and oncologists classify cancers according to their cellular appearance and their sites of origin. Over a hundred kinds have been described, but broad categories are usually used to indicate the location of the primary tumor. Table A-7 summarizes information concerning benign and malignant tumors (cancers) associated with the major tissues of the body.

Cancer develops in a series of steps diagrammed in Figure A-8. Initially the cancer cells are restricted to a single location, called the **primary tumor** or **primary neoplasm.** All the cells in the tumor are usually the daughter cells of a single malignant cell. At first the growth of the primary tumor simply distorts the tissue, and the basic tissue organization remains intact. Metastasis begins as tumor cells "break out" of the primary tumor and invade the surrounding tissue. They may then enter the lymphatic system and accumulate in nearby lymph nodes. When this invasion is followed by penetration of nearby blood vessels, the cancer cells begin circulating throughout the body. The growth of blood vessels into the tumor is a vital step in the development and spread of the cancer. Without those vessels, the growth of the cancer cells will be limited by the availability of oxygen and nutrients. A peptide called *antiangiogenesis factor* can prevent the growth of blood vessels and slow the growth of cancers. This peptide, produced in normal human cartilage, can be extracted in large quantities from sharks, whose skeletons are entirely cartilaginous. Sharks are now being collected to obtain antiangiogenesis factor for use in experimental cancer therapies.

Cancer cells within the circulatory system, responding to cues that are as yet unknown, ultimately migrate out of the blood vessels to establish **secondary tumors** at other sites. These tumors are extremely active metabolically, and their presence stimulates the growth of blood vessels into the area. The increased circulatory supply provides additional nutrients and further accelerates tumor growth and metastasis. Death may occur as a result of compression of vital organs, because nonfunctional cancer cells have killed or replaced the normal cells in vital organs, or because the cancer cells have starved normal tissues of essential nutrients.

Cancer Detection and Incidence

A statistical profile of cancer incidence and survival rates has been included as Table A-8. Interestingly enough, the picture changes when you look at data from other countries. For example, bladder cancer is

Table A-7   Benign and Malignant Tumors in the Major Tissue Types

Tissue	Description
Epithelia	
Carcinoma	Any cancer of epithelial origin
Adenocarcinoma	Cancers of glandular epithelia
Angiosarcomas	Cancers of endothelial cells
Mesotheliomas	Cancers of mesothelial cells
Connective tissues	
Fibromas	Benign tumors of fibroblast origin
Lipomas	Benign tumors of adipose tissue
Liposarcomas	Cancers of adipose tissue
Leukemias, Lymphomas	Cancers of blood-forming tissues
Chondromas	Benign tumors in cartilage
Chondrosarcomas	Cancers of cartilage
Osteomas	Benign tumors in bone
Osteosarcomas	Cancers of bone
Muscle tissues	
Myxomas	Benign muscle tumors
Myosarcomas	Cancers of skeletal muscle tissue
Cardiac Sarcomas	Cancers of cardiac muscle tissue
Leiomyomas	Benign tumors of smooth muscle tissue
Leiomyosarcomas	Cancers of smooth muscle tissue
Neural tissues	
Gliomas, Neuromas	Cancers of neuroglial origin

common in Egypt, stomach cancer in Japan, and liver cancer in Africa. A combination of genetic, dietary, and environmental factors are thought to be responsible for these differences.

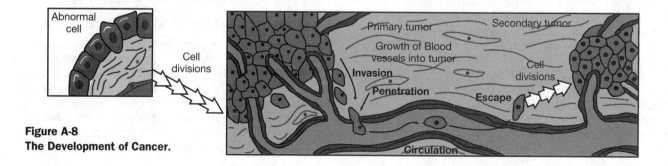

Figure A-8
The Development of Cancer.

Clinical Staging and Tumor Grading

Detection of a cancer often begins during a routine physical examination, when the physician detects an abnormal lump or growth. A tumor or neoplasm is defined as a "new growth" resulting from uncontrolled cell division. A tumor may be malignant or benign, metastasizing rapidly or spreading very slowly; only malignant tumors are called cancers.

Many laboratory and diagnostic tests are necessary for the correct diagnosis of cancer. Information is usually obtained through examination of a tissue sample, or *biopsy*, often supplemented by medical imaging and blood studies. A biopsy is one of the most significant diagnostic procedures because it permits a direct look at the tumor cells. Not only do malignant cells have an abnormally high mitotic rate, but they are structurally distinct from normal body cells.

If the tissue appears cancerous, other important questions must be answered, including:

- What is the measurable size of the primary tumor?

- Has the tumor invaded surrounding tissues?

- Has the cancer already metastasized to develop secondary tumors?

- Are any regional lymph nodes affected?

The answers to these questions are combined with observations from the physical exam, the biopsy results, and information from any imaging procedures to develop an accurate *diagnosis* (an identification of the disease) and *prognosis* (the probable outcome of the disorder).

In an attempt to develop a standard system, national and international cancer organizations have developed the *TNM* system for staging cancers. The letters refer to *tumor* (T) size and invasion, *lymph node* (N) involvement, and degree of *metastasis* (M).

- Tumor size is graded on a scale of 0 to 4, with the largest dimensions and greatest amount of invasion categorized as T4. (T0 indicates the absence of a primary tumor.)

- Lymph node involvement is graded on a scale of 0 to 3. A designation of N0 indicates that no lymph nodes have been invaded by cancer cells. A classification of N1–3 indicates the involvement of increasing numbers of lymph nodes.

A classification of N1 indicates involvement of a single lymph node less than 3 cm in diameter.

A classification of N2 includes one medium-sized node (3–6 cm) or multiple nodes, all smaller than 6 cm.

A classification of N3 indicates the presence of a single lymph node larger than 6 cm in diameter, whether or not other nodes are involved.

This categorization is important because lymph nodes filter the tissue fluids from nearby capillary beds. The fluid, called *lymph*, then returns to the general circulation. Once cancer cells have entered the lymphatic system, they can spread very quickly throughout the body.

Table A–8 Cancer Incidence and Survival Rates in the United States

Site	Estimated New cases (1996)	Estimated Deaths (1996)	Five-Year Survival Rates Diagnosis Date 1970-73	Five-Year Survival Rates Diagnosis Date 1986-91
DIGESTIVE TRACT				
Esophagus	12,300	11,200	4%	11%
Stomach	22,800	14,000	13%	19%
Colon and rectum	133,500	54,900	47%	60%
RESPIRATORY TRACT				
Lung and bronchus	177,000	158,700	10%	14%
URINARY TRACT				
Kidney and Other Urinary Structures	30,600	12,000	46%	59%
Bladder	52,900	11,700	61%	82%
REPRODUCTIVE SYSTEM				
Breast	185,700	44,560	68%	84%
Ovary	26,700	14,800	36%	44%
Testis	7,400	370	72%	95%
Prostate gland	317,100	41,400	63%	87%
NERVOUS SYSTEM	17,900	13,300	20%	28%
CIRCULATORY SYSTEM	102,200	56,210	22%	41%
SKIN (MELANOMA ONLY)	38,300	7,300	68%	87%

Data courtesy of the American Cancer Society.

- Metastasis is graded using a scale of 0 to 1. M0 indicates that there is no evidence of metastasis, whereas M1 indicates that the cancer cells have produced secondary tumors in other portions of the body.

This grading system provides a general overview of the progression of the disease. For example, a tumor classified as T1N1M0 obviously has a better prognosis than T4N2M1. The latter tumor will be much more difficult to treat. The grading system alone does not provide all the information needed to plan a treatment, however, because different types of cancer progress in different ways, and the therapies must vary as a result. Thus *leukemia,* a cancer of the blood forming tissues, will be treated differently than colon cancer. Specific treatments will be considered in discussions dealing with cancers affecting individual body systems. The next section provides a general overview of the strategies used to treat cancer.

Cancer Treatment

It is unfortunate that the media tend to describe cancer as though it were one disease rather than many. This simplistic perspective fosters the belief that some dietary change, air ionizer, or wonder drug will be found that can prevent the affliction. There is no single, universally effective cure for cancer; there are too many separate causes, possible mechanisms, and individual differences.

The goal of cancer treatment is to achieve **remission.** A tumor in remission either ceases to grow or decreases in size. Basically the treatment of malignant tumors must accomplish one of the following to produce remission:

1. *Surgical removal or destruction of individual tumors:* Tumors containing malignant cells can be surgically removed or destroyed by radiation, heat, or freezing. These techniques are very effective, if the treatment is undertaken before extensive metastasis has occurred. For this reason early detection is important in improving survival rates for all forms of cancer.

2. *Killing metastatic cells throughout the body:* This is much more difficult and potentially dangerous, because healthy tissues are likely to be damaged at the same time. At present the most widely approved treatments are *chemotherapy* and radiation.

Chemotherapy involves the administration of drugs that will either kill the cancerous tissues or prevent mitotic divisions. These drugs often affect stem cells in normal tissues, and the side effects are usually unpleasant. For example, because chemotherapy slows the regeneration and maintenance of epithelia of the skin and digestive tract, patients lose their hair and experience nausea and vomiting. Several different drugs are often administered simultaneously, or in sequence, because over time cancer cells can develop a resistance to a single drug. Chemotherapy is often used in the treatment of many kinds of metastatic cancer.

Massive doses of radiation are sometimes used to treat advanced cases of *lymphoma,* a cancer of the immune system. In this rather drastic procedure enough radiation is administered to kill all the blood-forming cells in the body. After treatment, new blood cells must be provided by a bone marrow transplant. Later sections dealing with the lymphatic system contain additional information about marrow transplants, lymphomas, and other cancers of the blood.

An understanding of molecular mechanisms and cell biology is leading to new approaches that may revolutionize cancer treatment. One approach focuses attention on the fact that cancer cells are ignored by the immune system. In **immunotherapy** chemicals are administered that help the immune system recognize and attack the cancer cells. More elaborate experimental procedures involve the creation of customized antibodies using the gene-splicing techniques discussed on p. 00. The resulting antibodies are specifically designed to attack the tumor cells in one patient; although this technique shows promise, it remains difficult, costly, and very labor intensive.

A second new procedure builds upon the first. In **boron neutron capture therapy (BNCT)** antibodies made to attack cancer cells are labeled with an isotope of boron (B). After these antibodies are administered, the patient is irradiated with neutrons. These neutrons do not damage normal tissues. However, the boron atoms absorb neutrons and release alpha particles (2 neutrons + 2 protons). This radiation kills the cancer cells quite effectively; because the cancer cells absorb the radiation, other, normal tissues are unaffected.

Cancer and Survival

Advances in chemotherapy, radiation procedures, and molecular biology have produced significant improvements in the survival rates for several types of cancer. However, the improved survival rates indicated in Table A-8 not only reflect advances in therapy, but in early detection. Much of the credit goes to increased public awareness and concern about cancer. In general, the odds of survival increase markedly if the cancer is detected early, especially before it undergoes metastasis. Despite the variety of possible cancers, the American Cancer Society has identified seven "warning signs" that mean it's time to consult a physician. These are presented in Table A-9.

Table A–9   Seven Warning Signs of Cancer

Change in bowel or bladder habits

A sore that does not heal

Unusual bleeding or discharge

Thickening or lump in breast or elsewhere

Indigestion or difficulty in swallowing

Obvious change in wart or mole

Nagging cough or hoarseness

"Jaws" and the Fight against Cancer

EAP p. 85

Cartilage cells survive despite the fact that they are crowded together in an avascular matrix. In other tissues, when cells are crowded and active, blood vessels grow into the area and improve oxygen and nutrient delivery. Cartilage secreting a chemical that blocks the growth of blood vessels. This compound has been named **antiangiogenesis factor** (*anti-*, against + *angeion*, vessel + *gennan*, to produce).

One reason cancers can grow so explosively is that blood vessels branch into the developing tumor, delivering supplies to the renegade cells. This growth could theoretically be prevented by antiangiogenesis factor, but the quantities produced in normal human cartilage are extremely small.

Sharks are highly successful marine predators. Although large, their skeletons are cartilaginous, rather than bony. Sharks are now being collected to obtain the antiangiogenesis factor from their cartilages. Because they contain so much cartilage, substantial quantities can be extracted from a single animal.

Tissue Structure and Disease

Pathologists (pa-THOL-o-jists) are physicians who specialize in the study of disease processes. Diagnosis, rather than treatment, is usually the main focus of their activities. In their analyses, pathologists integrate anatomical and histological observations to determine the nature and severity of the disease. Because disease processes affect the histological organization of tissues and organs, tissue samples, or **biopsies,** often play a key role in their diagnoses.

Figure A-9 diagrams the histological changes induced by one relatively common irritating stimulus, cigarette smoke. The first abnormality to be observed is **dysplasia** (dis-PLĀ-zē-uh), a change in the normal shape, size, and organization of tissue cells. It is usually a response to chronic irritation or inflammation, and the changes are reversible. The normal trachea (windpipe) is lined by a pseudo-stratified, ciliated, columnar epithelium. The cilia move a mucous layer that traps foreign particles and moistens incoming air. The drying and chemical effects of smoking first paralyze the cilia, halting the movement of mucus (Figure A-9a) As mucus builds up, the individual coughs to dislodge it (the well-known "smoker's cough").

Epithelia and connective tissues may undergo more radical changes in structure, caused by the division and differentiation of stem cells. **Metaplasia** (me-tuh-PLĀ-zē-uh) is a structural change that dramatically alters the character of the tissue. In our example, heavy smoking first paralyzes the cilia, and over time the epithelial cells lose their cilia altogether. As metaplasia occurs, the epithelial cells produced by stem cell divisions no longer differentiate into ciliated columnar cells. Instead, they form a stratified squamous epithelium that provides a greater resistance to drying and chemical irritation (Figure A-9b). This epithelium protects the underlying tissues more effectively, but it completely eliminates the moisturization and cleaning properties of the epithelium. The cigarette smoke will now have an even greater effect on more delicate portions of the respiratory tract. Fortunately, metaplasia is reversible, and the epithelium gradually returns to normal once the individual quits smoking.

During **anaplasia** (a-nuh-PLĀ-zē-uh) tissue organization breaks down. Tissue cells change size and shape, often becoming unusually large or abnormally small. In anaplasia (Figure A-9c), which occurs in smokers developing one form of lung cancer, the cells divide more frequently, but not all divisions proceed in the normal way, and many of the tumor cells have abnormal chromosomes. Unlike dysplasia and metaplasia, anaplasia is irreversible.

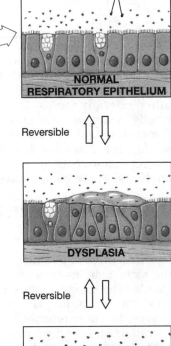

(a) The cilia of respiratory epithelial cells are damaged and paralyzed by exposure to cigarette smoke. These changes cause the local buildup of mucus and reduce the effectiveness of the epithelium in protecting deeper, more delicate portions of the respiratory tract.

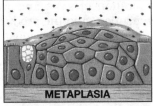

(b) In metaplasia, a tissue changes its structure. In this case the stressed respiratory surface converts to a stratified epithelium that protects underlying connective tissues but does nothing for other areas of the respiratory tract.

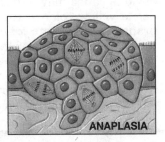

(c) In anaplasia, the tissue cells become tumor cells; anaplasia produces a cancerous tumor.

FIGURE A-9

Changes in a Tissue under Stress

The Body Systems: Clinical and Applied Topics

This section relates aspects of the normal anatomy and physiology of each body system to specific clinical conditions, diagnostic procedures, and other relevant topics. Each body system will be briefly reviewed, and specific disorders affecting the system will be approached from a diagnostic standpoint. Case studies and other exercises following major sections will give you the chance to apply the concepts you have learned.

The Integumentary System

The structures of the integumentary system include the skin, hair, nails, and several types of exocrine glands. The integumentary system has a variety of functions, including the protection of underlying tissues, the maintenance of body temperature, the excretion of salts and water in sweat, cutaneous sensation, and the production of vitamin D_3.

The skin is the most visible organ of the body. As a result, abnormalities are easily recognized. A bruise, for example, often creates a swollen and discolored area where the walls of blood vessels have been damaged. Skin color, skin tone, and the overall condition of the skin often accompany illness or disease. These changes can assist in diagnosis. For example, extensive bruising without obvious cause may indicate a blood clotting disorder; a yellow color in the skin and mucous membranes may indicate *jaundice*, a sign that usually indicates some type of liver disorder. The general condition of the skin may also be significant. For example, color changes or changes in skin flexibility, elasticity, dryness, or sensitivity often appear following the malfunctions of other organ systems.

EXAMINATION OF THE SKIN

When examining a patient, dermatologists use a combination of investigative interviews ("What have you done?" or "How does it feel?") and physical examination to arrive at a diagnosis. The condition of the skin is carefully observed. Notes are made concerning the presence of **lesions,** which are changes in skin structure caused by disease processes. These lesions are also called **skin signs,** because they are measurable, visible abnormalities of the skin surface. Figure A-10 diagrams the most common skin signs and related disorders.

The distribution of lesions may be an important clue to the source of the problem. For example, in *shingles* there are painful vesicular eruptions on the skin that follow the path of peripheral sensory nerves. A ring of slightly raised scaly (papular) lesions is typical of fungal infections that may affect the trunk, scalp, and nails. Examples of skin disorders caused by infection or allergic reactions are included in Table A-10, with descriptions of the related lesions. Skin lesions caused by trauma are considered in a separate section entitled "Classification of Wounds" (p. 43).

This table considers signs on the skin surface, but signs involving the accessory organs of the skin can also be important. For example,

- Nails have a characteristic shape that can change due to an underlying disorder. An example is *clubbing* of the nails, often a sign of *emphysema* or *congestive heart failure.* In these conditions the fingertips broaden, and the nails become distinctively curved.

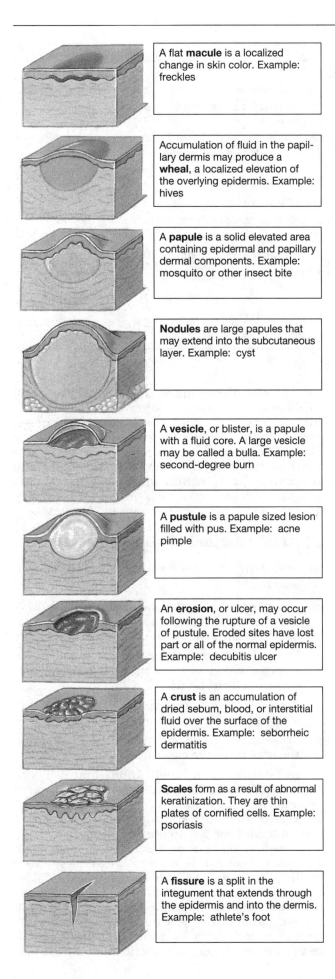

A flat **macule** is a localized change in skin color. Example: freckles

Accumulation of fluid in the papillary dermis may produce a **wheal**, a localized elevation of the overlying epidermis. Example: hives

A **papule** is a solid elevated area containing epidermal and papillary dermal components. Example: mosquito or other insect bite

Nodules are large papules that may extend into the subcutaneous layer. Example: cyst

A **vesicle**, or blister, is a papule with a fluid core. A large vesicle may be called a bulla. Example: second-degree burn

A **pustule** is a papule sized lesion filled with pus. Example: acne pimple

An **erosion**, or ulcer, may occur following the rupture of a vesicle of pustule. Eroded sites have lost part or all of the normal epidermis. Example: decubitis ulcer

A **crust** is an accumulation of dried sebum, blood, or interstitial fluid over the surface of the epidermis. Example: seborrheic dermatitis

Scales form as a result of abnormal keratinization. They are thin plates of cornified cells. Example: psoriasis

A **fissure** is a split in the integument that extends through the epidermis and into the dermis. Example: athlete's foot

Figure A-10   Skin Signs

- The condition of the hair can be an indicator of the overall health of the individual. For example, depigmentation and coarseness of hair occurs in protein deficiency disease called *kwashiorkor.*

Diagnosing Skin Conditions

A single vascular lesion, such as a hematoma, may have multiple causes. This is one of the challenges facing dermatologists—the signs may be apparent, but not the underlying causes. Making matters more difficult, many different skin disorders produce the same uncomfortable sensations. For example, **pruritis** (prōō-RĪ-tus), an irritating itching sensation, is an extremely common symptom associated with a variety of skin conditions. Questions concerning medical history, medications, possible sources of infection, and other signs, such as bleeding at the gums, can be the key to making an accurate diagnosis.

Pain is another common symptom of many skin disorders. Although pain is unwelcome, cutaneous sensation is an important function of the integumentary system. This is dramatically demonstrated in the condition of *leprosy,* or *Hansen's disease.* Hansen's disease is caused by bacteria that has an affinity for cooler regions of the body. The bacteria destroy cutaneous nerve endings sensitive to touch, pain, hot, and cold. Damage to the extremities then occurs and accumulates because the individual is no longer aware of painful stimuli. Hansen's disease is considered in more detail in a later section (p. 72).

Examples of diagnostic tests that may prove useful include:

- Scrapings of affected tissue, a process often performed to check for fungal infections

- Culturing of bacteria removed from a lesion to aid in identification and determine drug sensitivity

- Biopsy of affected tissue to view cellular structure

- Skin tests: Various types of disorders can be detected through use of a *skin test.* In a skin test a localized area of the skin is exposed to an inactivated pathogen, a portion of a pathogen, or a substance capable of producing an allergic reaction in sensitive individuals. Exposure may be via injection or surface application. For example, in a tuberculosis skin test a small quantity of tuberculosis antigens is injected *intradermally (intra,* within). If the individual has been infected in the past, or currently has tuberculosis, there will be erythema and swelling at the injection site 24-72 hours later.

Table A-10   Skin Signs of Various Disorders

Cause	Examples	Resulting Skin Lesion
Viral infections	Chicken pox	Lesions begin as macules and papules but develop into vesicles
Viral infections	Measles (rubeola)	A maculopapular rash that begins at the face and neck and spreads to the trunk and extremities
Viral infections	Erythema infectiosum (Fifth's disease)	A maculopapular rash that begins on the cheeks (slapped cheek appearance) and spreads to the extremities
Viral infections	Herpes	Raised vesicles that heal with a crust
Bacterial infections	Impetigo	Vesiculopustular lesions with exudate and yellow crusting
Fungal infections	Ringworm	An annulus (ring) of scaly papular lesions with central clearing
Parasitic infections	Scabies	Linear burrows with a small papule at one end
Parasitic infections	Lice (pediculosis)	Dermatitis: excoriation (scratches) due to pruritis (itching)
Allergies to medication	Penicillin	Wheals (urticaria or hives)
Food allergies	Eggs, certain fruits	Wheals
Environmental allergies	Poison ivy	Lesions ranging from vesicles to bulla formation

Patch testing is used to check sensitivity to *allergens,* environmental agents that can cause allergic reactions. In a patch test the allergen is applied to the surface of the skin. If erythema, swelling, and/or itching develop, the individual is sensitive to that allergen.

✝ Disorders of Keratin Production

EAP *p. 99*

Not all skin signs are the result of infectious, traumatic, or allergic conditions. Excessive production of keratin is called **hyperkeratosis** (hī-per-ker-a-TŌ-sis). The most obvious effects are easily observed as calluses and corns. Calluses are thickened patches that appear on already thick-skinned areas, such as the palms of the hands or the heels of the foot, in response to chronic abrasion and distortion. Corns are more localized areas of excessive keratin production, and they form in areas of thin skin on or between the toes.

In **psoriasis** (so-RĪ-a-sis) the stratum germinativum becomes unusually active, causing hyperkeratosis in specific areas, including the scalp, elbows, palms, soles, groin, and nails. Normally an individual stem cell divides once every 20 days, but in psoriasis it may divide every day and a half. Keratinization is abnormal and often incomplete by the time the outer layers are shed. The affected areas appear to be covered with small silvery scales that continually flake away. Psoriasis develops in 20-30 percent of the individuals with an inherited tendency for the condition. Roughly 5 percent of the general U.S. population has psoriasis to some degree, often triggered by stress and anxiety. Most cases are painless and treatable.

Xerosis (ze-RŌ-sis), or "dry skin," is a common complaint of the elderly and people who live in arid climates. Under these conditions cell membranes in the outer layers of the skin gradually deteriorate, and the stratum corneum becomes more a collection of scales than a single sheet. The scaly surface is much more permeable than an intact layer of keratin, and the rate of insensible perspiration increases. In persons afflicted with severe xerosis, the rate of insensible perspiration may increase by up to 75 times.

▯ Transdermal Medications   EAP *p. 99*

Several drugs are now routinely administered transdermally:

- Transdermal *scopolamine,* a drug that affects the nervous system, is used to control the nausea associated with motion sickness.

- Transdermal *nitroglycerin* can be used to improve blood flow within heart muscle and prevent a heart attack.

- Transdermal *estrogens* may be administered to women to reduce symptoms of menopause.

- Transdermal *nicotine* can be used to suppress the urge to smoke cigarettes.

In addition, pain medications and drugs to control high blood pressure may be administered via transdermal patches.

DMSO (dimethyl-sulfoxide) is a transdermal drug intended for the treatment of injuries to the muscles and joints of domesticated animals, such as horses or cows. It is a solvent that rapidly crosses the skin, and drugs dissolved in DMSO will be carried into the body at the same time. DMSO has not been tested and approved for the treatment of human patients in the United States, either for joint or muscle injuries or as a transdermal solvent. However, it can be prescribed in Canada and

Europe. The long-term risks associated with it are unknown; reported short-term side effects include nausea, vomiting, cramps, and chills.

☤ Skin Cancers and the Ozone Hole

EAP *p. 100*

Almost everyone has several benign tumors of the skin; freckles and moles are examples. Skin cancers are the most common form of cancer, and the most common skin cancers are caused by prolonged exposure to the ultraviolet radiation in sunlight.

A **basal cell carcinoma** is a malignant cancer that originates in the germinativum (basal) layer. This is the most common skin cancer, and roughly two-thirds of these cancers appear in areas subjected to chronic UV exposure.

Squamous cell carcinomas are less common, but almost totally restricted to areas of sun-exposed skin. Metastasis seldom occurs in squamous cell carcinomas and virtually never in basal cell carcinomas, and most people survive these cancers. The usual treatment involves surgical removal of the tumor, and 95 percent of patients survive five years or more after treatment. (This statistic, the five-year survival rate, is a common method of reporting long-term prognosis.)

Compared with these common and seldom life-threatening cancers, **malignant melanomas** (mel-a-NŌ-mas) are extremely dangerous. In this condition, cancerous melanocytes grow rapidly and metastasize through the lymphatic system. The outlook for long-term survival changes dramatically, depending on when the condition is diagnosed. If localized, the five-year survival rate is 99 percent; if widespread, the survival rate drops to 14 percent.

To detect melanoma at an early stage, it is essential to know what to look for when examining your skin. The key points can be remembered most easily using the mnemonic ABCD.

- A is for *asymmetry:* melanomas tend to be irregular in shape. Often they are raised; they may ooze or bleed.

- B is for *border:* usually unclear, sometimes notched.

- C is for *color:* usually mottled, with many different colors (tan, brown, black, red, pink, white, and/or blue).

- D is for *diameter:* a growth more than about 5 mm (0.2 in.) in diameter, or roughly the area covered by the eraser on a pencil, is dangerous.

A new experimental treatment for melanoma uses genetic engineering technology to manufacture antibodies that target the MSH (melanocyt stimulating hormone) receptors on the surfaces of melanocytes. Melanocytes coated with these antibodies are then recognized and attacked by cells of the immune system.

Fair-skinned individuals who live in the tropics are most susceptible to all forms of skin cancer, because their melanocytes are unable to shield them from the ultraviolet radiation. Sun damage can be prevented by avoiding exposure to the sun during the middle hours of the day, and by using a sunblock (not a tanning oil)—a practice that also delays the cosmetic problems of sagging and wrinkling. *Everyone* who expects to be out in the sun for any length of time should choose a broad-spectrum sunblock with a sun protection factor (SPF) of at least 15; blonds, redheads, and people with very fair skin are better off with a sun protection factor of 20 to 30. (One should also remember the risks before spending hours in a tanning salon or tanning bed.)

The use of sun screens now becomes even more important as the ozone gas in the upper atmosphere is destroyed by our industrial emissions. Ozone absorbs UV before it reaches the earth's surface, and in doing so, it assists the melanocytes in preventing skin cancer. Australia, which is most affected by the depletion of ozone near the south pole (the "ozone hole"), is already reporting an increased incidence of skin cancers.

Tumors in the Dermis

Tumors seldom develop in the dermis, and those that do appear are usually benign. Two forms of **hemangiomas** may appear among dermal blood vessels during development. Viewed from the surface, these form prominent **birthmarks.** A **capillary hemangioma** involves capillaries of the papillary layer. It usually enlarges after birth, but subsequently fades and disappears. **Cavernous hemangiomas,** or "port-wine stains," affect larger vessels in the dermis, and such birthmarks usually last a lifetime.

📋 Abnormal Skin Pigmentation EAP *p. 100*

Several diseases that have primary impacts on other systems may have secondary effects on skin color and pigmentation. Because the skin is easily observed, these color changes can be useful in diagnosis. For example:

- In *jaundice* (JAWN-dis) the liver is unable to excrete bile, and a yellowish pigment accumulates in body fluids. In advanced stages, the skin and whites of the eyes turn yellow.

- Some tumors affecting the pituitary gland result in the secretion of large amounts of *melanocyte-stimulating hormone* (MSH). This hormone causes a darkening of the skin, as if the individual has an extremely deep bronze tan.

- In *Addison's disease* the pituitary gland secretes large quantities of *ACTH,* a hormone that is structurally similar to MSH. The result of ACTH on the skin coloration is also similar to that of MSH.

- In *vitiligo* (vi-ti-LĪ-gō) individuals lose their melanocytes. The condition develops in about 1 percent of the population, and the incidence increases among individuals with thyroid gland disorders, *Addison's disease,* and several other disorders. It is suspected that this disorder develops when the immune defenses malfunction, and antibodies attack normal melanocytes. The primary problem with vitiligo is cosmetic, especially for individuals with darkly pigmented skin. Michael Jackson is said to suffer from vitiligo.

Color changes affecting the integument as a whole are discussed in the text. Examples include *cyanosis,* and *Addison's disease* Dilation of the blood vessels in the dermis of a light-skinned individual can produce a color change known as **erythema** (*erythros,* "red"). Erythema may affect the entire body surface or it may be localized to a particular region. For example, a transient erythema is produced on the cheeks when some individuals become embarrassed. More permanent localized vascular changes in the skin usually reflect either an increase or decrease in local vascularity, damage to blood vessels, or problems with the clotting system. Table A-11 introduces several major types of vascular lesions. Less frequently, localized color changes may accompany tumor or cancer formation; these processes will be considered separately in the discussion of skin cancers.

Dermatitis

EAP *p. 101*

Because of the abundance of sensory receptors in the skin, regional infection or inflammation can be very painful. **Dermatitis** (der-muh-TĪ-tis) is an inflammation of the skin that primarily involves the papillary layer. In typical dermatitis, inflammation begins in a portion of the skin exposed to infection or irritated by chemicals, radiation, or mechanical stimuli. Dermatitis may cause no physical discomfort, or it may produce an annoying itch, as in poison ivy. Other forms of this condition can be quite painful, and the inflammation may spread rapidly across the entire integument.

There are many forms of dermatitis, some of them quite common:

- **Contact dermatitis** usually occurs in response to strong chemical irritants. It produces an itchy rash that may spread to other areas; poison ivy is an example.
- *Eczema* (EK-se-muh) is a dermatitis that can be triggered by temperature changes, fungus, chemical irritants, greases, detergents, or stress. Hereditary or environmental factors or both can encourage the development of eczema.
- *Diaper rash* is a localized dermatitis caused by a combination of moisture, irritating chemicals from fecal or urinary wastes, and flourishing microorganisms.
- *Urticaria* (ur-ti-KAR-ē-uh), also known as *hives,* is an extensive allergic response to a food, drugs, an insect bite, infection, stress, or other stimulus.

DISORDERS OF THE ACCESSORY ORGANS OF THE SKIN

There are many disorders that affect accessory organs of the skin, especially the hair and the exocrine glands. We will consider only three relatively common examples, *baldness, hirsutism,* and *acne.*

Baldness and Hirsutism

EAP *p. 102*

Two factors interact to determine baldness. A bald individual has a genetic susceptibility triggered by large quantities of male sex hormones. Many women carry the genetic background for baldness, but unless major hormonal abnormalities develop, as in certain endocrine tumors, nothing happens.

Male pattern baldness affects the top of the head first, only later reducing the hair density along the sides. Thus hair follicles can be removed from the sides and implanted on the top or front of the head, temporarily delaying a receding hairline. This procedure is rather expensive (thousands of dollars) and not every transplant is successful.

Alopecia areata (al-ō-PĒ-shē-ah ar-ē-A-ta) is a localized hair loss that can affect either sex. The cause is not known, and the severity of hair loss varies from case to case. This condition is associated with several disorders of the immune system; it has also been suggested that periods of stress may promote alopecia areata in individuals already genetically prone to baldness.

Table A-11   Examples of Vascular Lesions

Lesion	Features	Some Possible Causes
Ecchymosis	Reddish purple, blue, or yellow bruising related to trauma	Blood clotting disorder if bruising is abnormal; some vitamin deficiencies; thrombocytopenia; increased tendency to bruise is normal with aging or sundamaged skin
Hematoma	Pooling of blood from a broken vessel forming a mass; associated with pain and swelling	
Petechiae	Small red to purple pinpoint dots appearing in clusters	Leukemia; septicemia (toxins in blood), thrombocytopenia
Erythema	Red flushed color of skin due to dilation of blood vessels in the skin	Extensive: drug reactions; Localized: burns, dermatitis

Hairs are dead, keratinized structures, and no amount of oiling, shampooing, or dousing with kelp extracts, vitamins, or nutrients will influence the follicle buried in the dermis. Skin conditions that affect follicles can contribute to hair loss; temporary baldness can also result from exposure to radiation or to many of the toxic (poisonous) drugs used in cancer therapy.

Untested treatments for baldness were banned by the Federal Drug Administration in 1984. *Minoxidil*, a drug originally marketed for the control of blood pressure, appears to stimulate inactive follicles when rubbed onto the scalp. It is now available on a prescription basis. Treatment involves applying a 2 percent solution to the scalp twice daily; after 4 months, over one-third of patients reported satisfactory results. It is most effective in preventing progression of early hair loss.

Hirsutism (HER-soot-izm; *hirsutus,* "bristly") refers to the growth of hair on women in patterns usually characteristic of men. Because considerable overlap exists between the two sexes in terms of normal hair distribution, and there are significant racial and genetic differences, the precise definition is more often a matter of personal taste than objective analysis. Age and sexual hormones may play a role, for hairiness increases late in pregnancy, and menopause produces a change in body hair patterns.

Severe hirsuitism is often associated with abnormal androgen (male sex hormone) production, either in the ovaries or in other endocrine organs. Unwanted follicles can be permanently "turned off" by plucking a growing hair and removing the papilla. Electrocautery, which destroys the follicle with a jolt of electricity, requires the services of a professional, but the results are more reliable. Patients may also be treated with drugs that reduce or prevent androgen stimulation of the follicles.

☤ Acne   EAP *p. 103*

Individuals with a genetic tendency toward acne have larger than average sebaceous glands, and when the ducts become blocked the secretions accumulate. Inflammation develops, and bacterial infection may occur. The condition usually surfaces at puberty, as sexual hormone production accelerates and stimulates the sebaceous glands. Their secretory output may be further encouraged by anxiety, stress, physical exertion, certain foods, and drugs.

The visible signs of acne are called **comedos** (ko-MĒ-doz). "Whiteheads" contain accumulated, stagnant secretions. "Blackheads" contain more solid material that has been invaded by bacteria. Although neither condition indicates the presence of dirt in the pores, washing may help to keep superficial oiliness down.

Acne usually fades after sex hormone concentrations stabilize. Topical (applied) antibiotics, vitamin A derivatives, such as *Retin-A,* or peeling agents may help reduce inflammation and mini-

mize scarring. In cases of severe acne, the most effective treatment usually involves the discouragement of bacteria by the administration of antibiotic drugs; because oral antibiotic therapy has risks, including the development of antibiotic-resistant bacteria, this therapy is not used unless other treatment methods have failed. Truly dramatic improvements in severe cases have been obtained with the prescription drug *Accutane.* This compound is structurally similar to vitamin A, and it reduces oil gland activity on a long-term basis. A number of minor side effects, including dry skin rashes, have been reported; these apparently disappear when the treatment ends. However, the use of Accutane during the first month of pregnancy carries a high (25 times normal) risk of inducing birth defects.

INJURIES AND TREATMENT OF SKIN CONDITIONS

Topical (applied on the skin) **anti-inflammatory drugs,** such as the steroid **hydrocortisone,** can be used to reduce the redness and itching that accompanies a variety of skin conditions. Some systemic (injected or swallowed) drugs may also be helpful; aspirin is a familiar systemic drug with anti-inflammatory properties. Isolated growths, such as skin tumors or warts, may be surgically removed. Alternatively, abnormal cells may be destroyed by electrical currents (**electrosurgery**) or by freezing (**cryosurgery**). Ultraviolet radiation may help conditions such as acne or psoriasis, whereas using a sunscreen or sunblock is important in controlling sun-sensitive outbreaks.

☤ Injury and Repair   EAP *p. 104*

The skin can regenerate effectively even after considerable damage has occurred, because stem cells persist in both the epithelial and connective tissue components. Germinative cell divisions replace epidermal cells, and mesenchymal cell divisions replace lost dermal cells. This process can be slow, and when large surface areas are involved problems of infection and fluid loss complicate the situation. The relative speed and effectiveness of skin repair vary depending on the type of wound involved. A slender, straight cut, or *incision,* may heal relatively quickly compared with a deep scrape, or *abrasion,* which involves a much greater area.

Figure A-11 shows stages in the regeneration of the skin after an injury. When damage extends through the epidermis and into the dermis, bleeding usually occurs. The blood clot, or **scab,** that forms at the surface temporarily restores the integrity of the epidermis and restricts the entry of additional microorganisms. Cells of the stratum germinativum undergo rapid divisions and begin to migrate along the sides of the wound in an attempt to replace the missing epidermal cells.

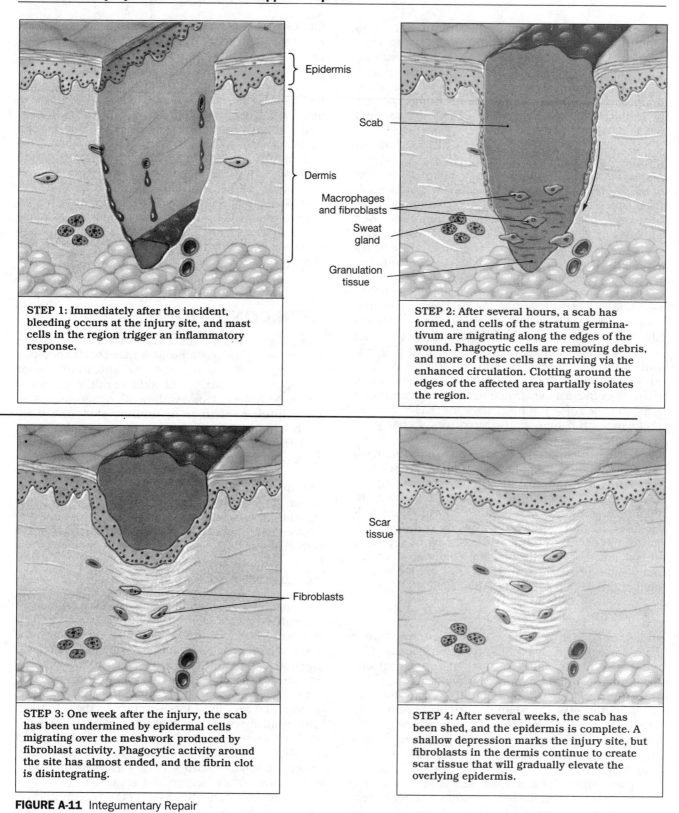

STEP 1: Immediately after the incident, bleeding occurs at the injury site, and mast cells in the region trigger an inflammatory response.

STEP 2: After several hours, a scab has formed, and cells of the stratum germinativum are migrating along the edges of the wound. Phagocytic cells are removing debris, and more of these cells are arriving via the enhanced circulation. Clotting around the edges of the affected area partially isolates the region.

STEP 3: One week after the injury, the scab has been undermined by epidermal cells migrating over the meshwork produced by fibroblast activity. Phagocytic activity around the site has almost ended, and the fibrin clot is disintegrating.

STEP 4: After several weeks, the scab has been shed, and the epidermis is complete. A shallow depression marks the injury site, but fibroblasts in the dermis continue to create scar tissue that will gradually elevate the overlying epidermis.

FIGURE A-11 Integumentary Repair

If the wound covers an extensive area, or involves a region covered by thin skin, dermal repairs must be under way before epithelial cells can cover the surface. Fibroblast and mesenchymal cell divisions produce mobile cells that invade the deeper areas of injury. Endothelial cells of damaged blood vessels also begin to divide, and capillaries follow the fibroblasts, providing a circulatory supply. The combination of blood clot, fibroblasts, and an extensive capillary network is called **granulation tissue.** Over time, the clot dissolves and the number of capillaries declines. Fibroblast activity leads to the appearance of collagen fibers and typical ground substance.

These repairs do not restore the integument to its original condition, however, for the region will contain an abnormally large number of collagen fibers and relatively few blood vessels. Severely damaged hair follicles, sebaceous or sweat glands, muscle cells, and nerves are seldom repaired, and they too are replaced by fibrous tissue. The formation of this rather inflexible, fibrous, noncellular scar tissue can be considered a practical limit to the healing process.

It is not known what regulates the extent of scar tissue formation, and the process is highly variable. For example:

- Surgical procedures performed on a fetus do not leave scars, perhaps because damaged fetal tissues do not produce the same types of growth factors that adult tissues do.

- In some adult individuals, most often those with dark skin, scar tissue formation may continue beyond the requirements of tissue repair. The result is a flattened mass of scar tissue that begins at the injury site and grows into the surrounding dermis. This thickened area of scar tissue, called a **keloid** (KĒ-loyd), is covered by a shiny, smooth epidermal surface. Keloids most often develop on the upper back, shoulders, anterior chest, and earlobes. They are harmless, and some aboriginal cultures intentionally produce keloids as a form of body decoration.

Skin repairs proceed most rapidly in young, healthy individuals. For example, barring infection, it takes 3–4 weeks to complete the repairs to a blister site in a young adult. The same repairs at age 65–75 take 6–8 weeks. However, this is just one example of the changes that occur in the integumentary system as a result of the aging process.

A Classification of Wounds   EAP *p. 104*

Injuries, or **trauma,** involving the integument are very common, and a number of terms are used to describe them. Traumatic injuries usually affect all components of the integument, and each type of wound presents a different series of problems to clinicians attempting to limit damage and promote healing.

An **open wound** is an injury producing a break in the epithelium. The major categories of open wounds have been illustrated in Figure A-12. **Abrasions** are the result of scraping against a solid object. Bleeding may be slight, but a considerable area may be open to invasion by microorganisms. **Incisions** are linear cuts produced by sharp objects. Bleeding may be severe if deep vessels are damaged. The bleeding may help to flush the wound, and closing the incision with bandages or stitches can limit the area open to infection while healing is underway. A **laceration** is a jagged, irregular tear in the surface produced by solid impact or an irregular object. Tissue damage is more extensive, and repositioning the opposing sides of the injury may be difficult. Despite the bleeding that usually occurs, lacerations are prone to infection. **Punctures** result from slender, pointed objects piercing the epithelium. Little bleeding results, and any microbes delivered under the epithelium in the process are likely to find conditions to their liking. In an **avulsion,** chunks of tissue are torn away by the brute force of an auto acci-

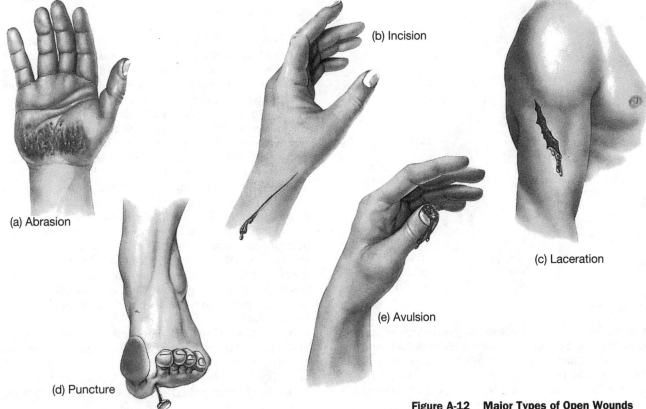

(a) Abrasion

(b) Incision

(c) Laceration

(d) Puncture

(e) Avulsion

Figure A-12   Major Types of Open Wounds

dent, explosion, or other incident. Bleeding may be considerable, and even more serious internal damage may be present.

Closed wounds may affect any internal tissue, but because the epithelium is intact, the likelihood of infection is reduced. A **contusion** is a bruise causing bleeding in the dermis. "Black and blue" marks are familiar examples of contusions that are taken lightly; contusions of the head, such as "black eyes," may be harmless or a sign of dangerous intracranial bleeding. In general, closed wounds affecting internal organs and organ systems are serious threats to life.

☤ Burns and Grafts   EAP *p.104*

Burns result from exposure of the skin to heat, radiation, electrical shock, or strong chemical agents. The severity of the burn reflects the depth of penetration and the total area affected.

First- and second-degree burns are also called **partial-thickness burns** because damage is restricted to the superficial layers of the skin. Accessory structures such as hair follicles and glands are usually unaffected. **Full-thickness burns,** or **third-degree burns,** destroy the epidermis and dermis, extending into subcutaneous tissues. These burns are actually less painful than second-degree burns, because sensory nerves are destroyed along with accessory structures, blood vessels, and other dermal components. Extensive third-degree burns cannot repair themselves, and the site remains exposed to potential infection.

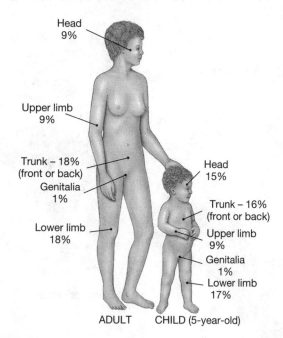

Figure A-13

A Quick Method for Estimating the Percentage of Surface Area Affected by Burns in Adults and Small Children. The method of estimation is called the "rule of nines" because of the surface area proportion in the adult. This rule must be modified in children because their proportions are quite different.

Roughly 10,000 people die from burns each year in the United States. The larger the area burned, the more significant the effects on integumentary function. Figure A-13 presents a standard reference for calculating the percentage of total surface area involved. Burns that cover more than 20 percent of the skin surface represent serious threats to life because they affect the following functions:

- *Fluid and electrolyte balance.* Even areas with partial-thickness burns lose their effectiveness as barriers to fluid and electrolyte losses. In full-thickness burns, the rate of fluid loss through the skin may reach five times the normal level.

- *Thermoregulation.* Increased fluid loss means increased evaporative cooling. More energy must be expended to keep body temperature within acceptable limits.

- *Protection from attack.* The epidermal surface, damp from uncontrolled fluid losses, encourages bacterial growth. If the skin is broken at a blister or the site of a third-degree burn, infection is likely. Widespread bacterial infection, or **sepsis** (*septikos*, rotting), is the leading cause of death in burn victims.

Effective treatment of full-thickness burns focuses on these procedures:

1. Replacing lost fluids and electrolytes.
2. Providing sufficient nutrients to meet increased metabolic demands for thermoregulation and healing.
3. Preventing infection by cleaning and covering the burn while administering antibiotic drugs.
4. Assisting tissue repairs.

Because full-thickness burns cannot heal unaided, surgical procedures are necessary to encourage healing. In a **skin graft,** areas of intact skin are transplanted to cover the burn site. A **split-thickness graft** takes a shaving of the epidermis and superficial portions of the dermis. A **full-thickness graft** involves the epidermis and both layers of the dermis.

With the development of fluid replacement therapies, infection control methods, and grafting techniques, the recovery rate for severe burns has improved dramatically. At present, young patients with burns over 80 percent of the body have an approximately 50 percent chance of recovery.

Recent advances in cell culture techniques may improve survival rates further. It is now possible to remove a small section of undamaged epidermis and grow it under controlled laboratory conditions. Over time, the germinative cell divisions produce large sheets of epidermal cells that can then be used to cover the burn area. From initial samples the size of postage stamps, square yards of epidermis have been grown and transplanted onto body surfaces. Although questions remain concerning the strength and flexibility of the repairs, skin cultivation represents a substantial advance in the treatment of serious burns.

℞ Synthetic Skin   EAP *p. 99*

Traditional skin grafts involve covering areas of complete skin loss or destruction with pieces of undamaged skin from other areas of the body. If the damaged area is large, there may not be enough normal skin available for grafting. Epidermal culturing can produce a new epithelial layer to cover a burn site. The cells in a 3-square centimeter epidermal sample from the victim can be cultured in a controlled environment that contains epidermal growth factors, fibroblast growth factors, and other stimulatory chemicals. The number of cells doubles in the first 18 hours, and the cells are then separated, and culturing continues. After three days in culture, the number of cells has increased by more than 16 times. After a week, the number has increased by over 300 times. The cells are now in small clusters, and they have begun forming layers. Germinative cells are on the bottom, attached to the glass of the culturing vessel, and the more superficial layers of cells roughly resemble those of normal epidermis, although keratin production does not occur. This artificially produced epidermis can then be transplanted to cover an injury site. The larger the area that must be covered, the longer the culturing process continues. After three or four weeks of culturing, the cells obtained in the original sample can provide enough epidermis to cover the entire body surface of a normal adult.

The major problem that reduces grafting success is the contamination of the wound site by bacteria while the epidermis is being cultured. This is prevented by the use of a skin *allograft*. Skin from a frozen cadaver is removed and placed over the wound as a temporary method of sealing the surface. Before the immune system of the patient attacks the graft, it will be partially or completely removed to provide a binding site for the epidermal transplant. After grafting, the complete reorganization and repair of the dermis and epidermis at the injury site takes approximately five years.

A second new procedure provides a model for dermal repairs that takes the place of normal tissue. A special synthetic skin is used. The imitation has a plastic (silastic) "epidermis" and a dermis composed of collagen fibers and ground cartilage. The collagen fibers are taken from cow skin and the cartilage from sharks. (The use of shark cartilage represents a second contribution to medicine by these predators; the first, *antiangiogenesis factor*, was noted on p. 32.) Over time, fibroblasts migrate among the collagen fibers and gradually replace the model framework with their own. The silastic epidermis is intended only as a temporary cover that will be replaced by either skin grafts or a cultured epidermal layer.

CRITICAL THINKING QUESTIONS

1-1. Charlie is badly burned in an accident with fireworks on the Fourth of July. When he reaches the emergency room, the examining physician determines the severity of the incident as a third-degree burn. The physician would likely order
  a. IV (intravenous) fluids and electrolytes
  b. antibiotics
  c. a high-nutrient diet
  d. all of the above

1-2. John, who has sailed boats professionally for years, visits his dermatologist. John's wife has noticed a mole on the back of his neck with suspicious characteristics. The dermatologist makes note of the irregular border of the nevus (mole), the unusual red and white coloration, and the surrounding inflammation. John's wife reports the mole has undergone these changes in the last month. The dermatologist recommends immediate biopsy of the nevus, suspecting
  a. basal cell carcinoma
  b. melanoma
  c. squamous cell carcinoma
  d. hemangioma

1-3. Carrie, a vegetarian, tells her physician about a yellow discoloration of the palms of her hands and the soles of her feet. Further inspection revels that her forehead and the area around her nose are also yellow in color. The physician determines that Carrie has not been exposed to hepatitis, and she has normal liver function tests. What might be the problem?

1-4. Sam likes to work on his motorcycle on weekends, and he spends a great deal of time cleaning engine parts with organic solvents. He doesn't wear gloves while working, and notices that after he is finished his hands feel painfully dry for several hours. What is the probable cause?

1-5. During pregnancy it is not uncommon for skin and hair color to become darker, especially around the genitals, nipples, and face. What is the probable cause of these changes?

The Skeletal System

The skeletal framework of the body is composed of 206 or more bones and the associated tendons, ligaments, and cartilages. The skeletal system has a variety of important functions, including the support of soft tissues, blood cell production, mineral and lipid storage, and, through its relationships with the muscular system, the support and movement of the body as a whole. Skeletal system disorders can thus affect many other systems. The skeletal system is in turn influenced by the activities of other systems. For example, weakness or paralysis of skeletal muscles will lead to a weakening of the associated bones.

Although the bones you study in the lab may seem rigid and permanent structures, the living skeleton is dynamic and undergoing continual remodeling. The remodeling process involves bone deposition by osteoblasts and bone reabsorption by osteoclasts. The net result of the remodeling varies depending on

1. *The age of the individual:* During development, bone deposition occurs faster than bone resorption, and the skeleton grows larger. At maturity, bone deposition and resorption are in balance; as the aging process continues, the rate of bone deposition declines and the bones become less massive. This gradual weakening, called *osteopenia,* begins at age 30-40 and may ultimately progress to *osteoporosis* (p. 49).

2. *The applied physical stresses:* Heavily stressed bones become thicker and stronger, and lightly stressed bones become thinner and weaker. Skeletal weakness can therefore result from muscular disorders, such as *myasthenia gravis* (p. 60) or the *muscular dystrophies* (p. 59), and conditions that affect CNS motor neurons, such as spinal cord injuries (p. 68), *demyelination disorders* (p. 65), or *multiple sclerosis* (p. 65).

3. *Circulating hormone levels:* Changing levels of growth hormone, androgens and estrogens, thyroid hormones, parathyroid hormone, and calcitonin increase or decrease the rate of mineral deposition in bone. As a result, many disorders of the endocrine system will have an impact on the skeletal system. For example,

- Conditions affecting the skin, liver, or kidneys can interfere with calcitriol production.

- Thyroid or parathyroid disorders can alter thyroid hormone, parathyroid hormone, or calcitonin levels.

- Pituitary gland disorders or liver disorders can affect GH or somatomedin production.

- Reproductive system disorders can alter circulating levels of androgens or estrogens.

Many of these conditions will be detailed in the section dealing with the endocrine system (pp. 80–88). Figure A-14 provides an overview

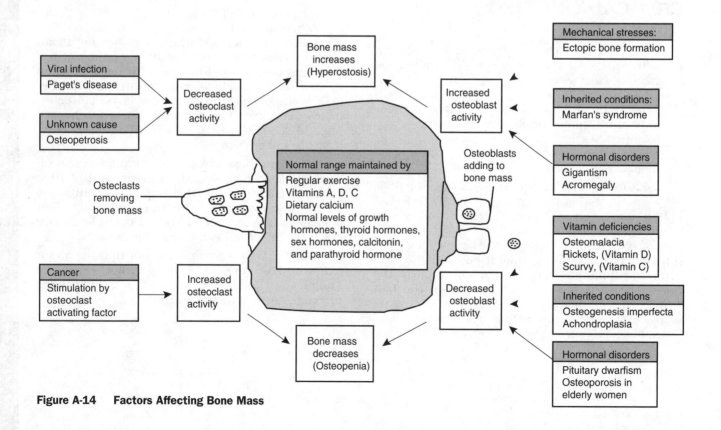

Figure A-14   Factors Affecting Bone Mass

of factors affecting the balance between bone deposition and bone resorption.

4. *Rates of calcium and phosphate absorption and excretion:* For bone mass to remain constant, the rate of calcium and phosphate excretion, primarily at the kidneys, must be balanced by the rate of calcium and phosphate absorption at the digestive tract. Dietary calcium deficiencies or problems at the digestive tract affecting calcium and phosphate absorption will thus have a direct effect on the skeletal system.

5. *Genetic and environmental factors:* Genetic or environmental factors may affect the structure of bone or the remodeling process. There are a number of inherited abnormalities of skeletal development, such as *Marfan's syndrome* and *achondroplasia* (p. 48). When bone fails to form embryonically in certain areas, underlying tissues can be exposed and associated function can be altered. This occurs in a *cleft palate* (a condition in which the palatine processes of the maxil-

lary bones do not fuse) and in *spina bifida* (p. 53). Environmental stresses can alter the shape and contours of developing bones. For example, some cultures use lashed boards to shape an infant's or child's skull to a form considered fashionable. Environmental forces can also result in the formation of bone in unusual locations. These *heterotopic bones* may develop in a variety of connective tissues exposed to chronic friction, pressure, or mechanical stress. For example, cowboys in the nineteenth century sometimes developed heterotopic bones in the dermis of the thigh, from friction with the saddle.

Figure A-15 diagrams the relationships between the major classes of skeletal disorders affecting bones (Figure A-15a) and joints (Figure A-15b). Some of these conditions are the result of conditions that primarily affect the skeletal system *(osteosarcoma, osteomyelitis)* and others result from problems originating in other systems *(acromegaly, rickets).* These conditions affect the structure and function of the bones of the skeleton.

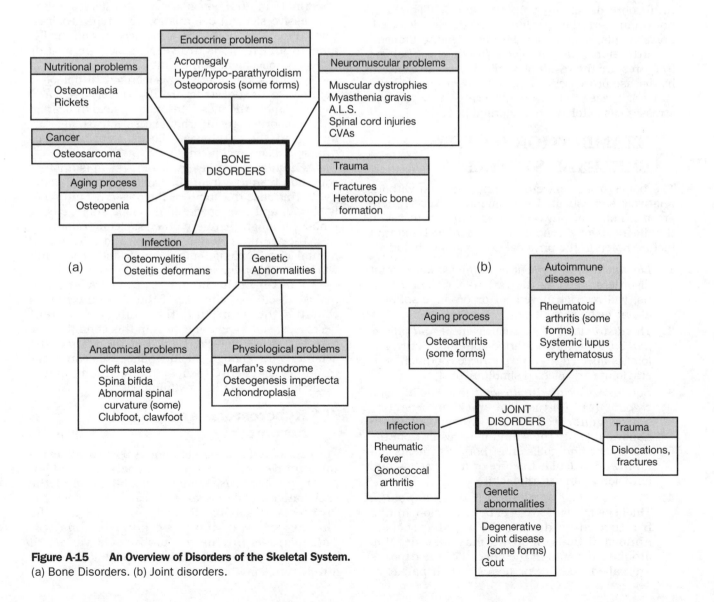

Figure A-15   An Overview of Disorders of the Skeletal System.
(a) Bone Disorders. (b) Joint disorders.

Traumatic injuries, such as fractures or dislocations, and infections also affect the cartilages, tendons, and ligaments associated with the bones involved. A somewhat different array of conditions affect the soft tissues of the bone marrow. Areas of red bone marrow contain the stem cells for red blood cells, white blood cells, and platelets. The bone marrow becomes abnormal in diseases of the blood that are characterized by blood cell overproduction (*leukemia, polycythemia,* p. 96, 91) or underproduction (several *anemias,* p. 92, 95).

THE SYMPTOMS OF BONE AND JOINT DISORDERS

A common symptom of a skeletal system disorder is pain. Bone pain and joint pain are common symptoms associated with many bone disorders. As a result, the presence of pain does not provide much help in identifying a specific bone or joint disorder. Chronic, aching bone or joint pain may be tolerated, and a person often will not seek medical assistance until more definitive symptoms appear. This may not occur until the condition is relatively advanced. For example, a symptom that may require immediate attention is a *pathologic fracture.* Pathologic fractures are the result of weakening of the skeleton by disease processes, such as *osteosarcoma* (a bone cancer). These fractures may be caused by physical stresses easily tolerated by normal bones.

EXAMINATION OF THE SKELETAL SYSTEM

The bones of the skeleton cannot be seen without relatively sophisticated equipment. However, there are a number of physical signs that can assist in the diagnosis of a bone or joint disorder. Important factors noted in the physical examination include

1. *Limitation of movement or stiffness:* Many joint disorders, such as the various forms of arthritis, will restrict movement or produce stiffness at one or more joints.

2. *The distribution of joint involvement and inflammation:* In a *monoarthritic* condition, only one joint is affected. In a *polyarthritic* condition, several joints are affected simultaneously.

3. *Sounds associated with joint movement:* Bony crepitus (KREP-i-tus) is a crackling or grating sound generated during movement of an abnormal joint. The sound may result from the movement and collision of bone fragments following an articular fracture or from friction and abrasion at an arthritic joint.

4. *The presence of abnormal bone deposits:* Thickened, raised areas of bone develop around fracture sites during the repair process. Abnormal bone deposits may also develop around the joints in the fingers. These deposits are called *nodules* or *nodes.* When palpated, nodules are solid and painless. Nodules, which can restrict movement, often form at the interphalangeal joints of the fingers in arthritis.

5. *Abnormal posture:* Bone disorders that affect the spinal column can result in abnormal posture. This is most apparent when the condition alters the normal spinal curvature. Examples include *kyphosis, lordosis,* and *scoliosis* (p. 51). A condition involving an intervertebral joint, such as a herniated disc, will also produce abnormal posture and movement.

Inherited Abnormalities in Skeletal Development   EAP *p. 114*

There are several inherited conditions that result in abnormal bone formation. Three examples are *osteogenesis imperfecta, Marfan's syndrome,* and *achondroplasia.*

Osteogenesis imperfecta (im-per-FEK-ta) is an inherited condition, appearing in 1 individual in about 20,000, that affects the organization of collagen fibers. Osteoblast function is impaired, growth is abnormal, and the bones are very fragile, leading to progressive skeletal deformation and repeated fractures. Fibroblast activity is also affected, and the ligaments and tendons are very "loose," permitting excessive movement at the joints.

Marfan's syndrome is also linked to defective connective tissue structure. Extremely long and slender limbs, the most obvious physical indication of this disorder, result from excessive cartilage formation at the epiphyseal plates. (Marfan's syndrome is discussed further on p. 000.)

Achondroplasia (a-kon-drō-PLA-sē-a) is another condition resulting from abnormal epiphyseal activity. In this case the epiphyseal plates grow unusually slowly, and the individual develops short, stocky limbs. Although there are other skeletal abnormalities, the trunk is normal in size, and sexual and mental development remain unaffected. The adult will be an *achondroplastic dwarf.*

In **osteomalacia** (os-tē-ō-ma-LA-shē-ah; *malakia,* softness) the size of the skeletal elements remains the same, but their mineral content decreases, softening the bones. In this condition the osteoblasts are working hard, but the matrix isn't accumulating enough calcium salts. This can occur in adults or children whose diet contains inadequate levels of calcium or vitamin D_3.

Hyperostosis and Acromegaly   EAP *p. 114*

The excessive formation of bone is termed **hyperostosis** (hī-per-os-TŌ-sis). In **osteopetrosis** (os-tē-ō-pe-TRŌ-sis; *petros,* stone) the total mass of the skeleton gradually increases because of a decrease in osteoclast activity. Remodeling stops, and the shapes of the bones gradually change. Osteopetrosis in children produces a variety of skeletal deformities. The primary cause for this relatively rare condition is unknown.

In **acromegaly** (*akron*, extremity + *megale*, great) an excessive amount of growth hormone is released after puberty, when most of the epiphyseal plates have already closed. Cartilages and small bones respond to the hormone, however, resulting in abnormal growth at the hands, feet, lower jaw, skull, and clavicle. Figure A-16 shows a typical acromegalic individual.

Ⓡ Stimulation of Bone Growth and Repair EAP *p. 116*

Despite the considerable capacity for bone repair, every fracture does not heal as expected. A *delayed union* is one that proceeds more slowly than anticipated. *Nonunion* may occur as a result of complicating infection, continued movement, or other factors preventing complete callus formation.

There are several techniques for inducing bone repair. Surgical bone grafting is the most common treatment for nonunion. This method immobilizes the bone fragments and provides a bony model for the repair process. Dead bone or bone fragments can be used; alternatively, living bone from another site, such as the iliac crest or part of a rib, can be inserted. As an alternative to bone grafting, surgeons can insert a shaped patch, made by mixing crushed bone and water. Bone transplants from cadavers have been used, but unless thoroughly sterilized, blood-borne diseases, such as AIDS, may be transmitted. The calcium carbonate skeleton of tropical corals has been sterilized and used as another alternative.

Another approach involves the stimulation of osteoblast activity by strong electrical fields at the injury site. This procedure has been used to promote bone growth after fractures have refused to heal normally. Wires may be inserted into the skin, implanted in the adjacent bone, or wrapped around a cast. The overall success rate of about 80 percent is truly impressive.

One experimental method of inducing bone repair involves mixing bone marrow cells into a soft matrix of bone collagen and ceramic. This combination is used like a putty at the fracture site. Mesenchymal cells in the marrow divide, producing chondrocytes that create a cartilaginous patch that is later converted to bone by periosteal cells. A second experimental procedure uses a genetically engineered protein to stimulate the conversion of osteoprogenitor cells into active osteoblasts. Although results in animal experimentation have been promising, neither technique has yet been approved for human trials.

⚕ Osteoporosis and Age-Related Skeletal Abnormalities EAP *p. 118*

Osteoporosis (os-tē-ō-por-Ō-sis; *porosus*, porous) is a condition that produces a reduction in bone mass sufficient to compromise normal function. The distinction between the "normal" osteopenia of aging and the clinical condition of osteoporosis is therefore a matter of degree. Current estimates indicate that 29

Figure A-16   A Person with Acromegaly.

percent of women between the ages of 45 and 79 can be considered osteoporotic. The increase in incidence after menopause has been linked to decreases in the production of estrogens (female sex hormones). The incidence of osteoporosis in men of the same age is estimated at 18 percent.

The excessive fragility of the bones frequently leads to breakage, and subsequent healing is impaired. Vertebrae may collapse, distorting the vertebral articulations and putting pressure on spinal nerves. Therapies that boost estrogen levels, dietary changes to elevate calcium levels in the blood, and exercise that stresses bones and stimulates osteoblast activity appear to slow but not completely prevent the development of osteoporosis.

Osteoporosis can also develop as a secondary effect of many cancers. Cancers of the bone marrow, breast, or other tissues release a chemical known as osteoclast-activating factor. This compound increases both the number and activity of osteoclasts and produces a severe osteoporosis.

Infectious diseases that affect the skeletal system become more common in older individuals. In part this reflects the higher incidence of fractures, combined with slower healing and reduction in immune defenses.

Osteomyelitis (os-t-ē-ō-mī-e-Lī-tis; *myelos*, marrow) is a painful infection of a bone most often caused by bacteria. This condition, most common in people over 50 years of age, can lead to dangerous systemic infections. A virus appears to be responsible for **Paget's disease,** also known as

osteitis deformans (ost-ē-Ī-tis de-FOR-mans). This condition may affect up to 10 percent of the population over 70. Osteoclast activity accelerates, producing areas of acute osteoporosis, and osteoblasts produce abnormal matrix proteins. The result is a gradual deformation of the skeleton.

INDIVIDUAL VARIATION IN THE SKELETAL SYSTEM   EAP *p. 120*

A comprehensive study of a human skeleton can reveal important information about the individual. For example, there are characteristic racial differences in portions of the skeleton, especially the skull and pelvis, and the development of various ridges and general bone mass can permit an estimation of muscular development and body weight. Details such as the condition of the teeth or the presence of healed fractures can provide information about the individual's medical history. Two important details, sex and age, can be determined or closely estimated on the basis of measurements indicated in Tables A-12 and A-13. Table A-12 identifies characteristic differences between the skeletons of males and females, but not every skeleton shows every feature in classic detail. Many differences, including markings on the skull, cranial capacity, and general skeletal features, reflect differences in average body size, muscle mass, and muscular strength. The general changes in the skeletal system that take place with age are summarized in Table A-13. Note how these changes begin at age 1 and continue throughout life. For example, fusion of the epiphyseal plates begins about age 3, and degenerative changes in the normal skeletal system, such as a reduction in mineral content in the bony matrix, typically do not begin until age 30–45.

An understanding of individual variation and of the normal timing of skeletal development is important in clinical diagnosis and treatment. Several professions focus on specific aspects of skeletal form and function. Each specialist has a different perspective, with its own techniques, traditions, and biases. For example, a person with back pain may consult an *orthopedist*, an *osteopath*, or a *chiropractor*.

Septal Defects and Sinus Problems
EAP *p. 124*

Flushing the nasal epithelium with mucus produced in the paranasal sinusesoften succeeds in removing a mild irritant. Buta viral or bacterial infection produces an inflammation of the mucous membrane of the nasalcavity. As swelling occurs, the communicating passageways narrow. Drainage of mucous slows, congestion increases, and the victim experiences headaches and a feeling of pressure within the facial bones. This condition of sinus inflammation and congestion is called **sinusitis.** The maxillary sinuses are often involved. Because gravity does little to assist mucus drainage from these sinuses, the effectiveness of the flushing

---

TABLE A-12   Sexual Differences in the Human Skeleton

Region/Feature	Male (as compared to female)	Female (as compared to male)
SKULL		
General appearance	Heavier, rougher	Lighter, smoother
Forehead	Sloping	More vertical
Sinuses	Larger	Smaller
Cranium	About 10% larger (average)	About 10% smaller
Mandible	Larger, robust	Lighter, smaller
Teeth	Larger	Smaller
PELVIS		
General appearance	Narrow, robust, heavy, rough	Broader, lighter, smoother
Pelvic inlet	Heart-shaped	Oval to round
Iliac fossa	Relatively deep	Relatively shallow
Ilium	Extends farther above sacral articulation	More vertical; less extension above sacroiliac joint
Angle inferior to **pubic symphysis**	Under 90°	100° or more
Acetabulum	Directed laterally	Faces slightly anteriorly as well as laterally
Obturator foramen	Oval	Triangular
Ischial spine	Points medially	Points posteriorly
Sacrum	Long, narrow triangle with pronounced sacral curvature	Broad, short triangle with less curvature
Coccyx	Points anteriorly	Points inferiorly
OTHER SKELETAL ELEMENTS		
Bone weight	Heavier	Lighter
Bone markings	More prominent	Less prominent

action is reduced and pressure on the sinus walls typically increases.

Temporary sinus problems may accompany allergies or the exposure of the mucous epithelium to chemical irritants or invading microorganisms. Chronic sinusitis may occur as the result of a **deviated** (nasal) **septum.** In this condition the nasal septum has a bend in it, most often at the junction between the bony and cartilaginous regions. Septal deviation often blocks drainage of one or more sinuses, producing chronic cycles of infection and inflammation. A deviated septum can result from developmental abnormalities or injuries to the nose, and the condition can usually be corrected or improved by surgery.

⚕ TMJ Syndrome   EAP p. 125

The temporomandibular joint permits a considerable degree of mandibular movement. The connective tissue sheath, or *capsule*, that surrounds the joint is relatively loose, and the opposing bone surfaces are separated by a fibrocartilage pad. In **TMJ sydrome,** or *myofascial pain syndrome,* the mandible is pulled slightly out of alignment, usually by spasms in one of the jaw muscles. The individual experiences (1) facial pain that radiates around the ear on the affected side and (2) an inability to open the mouth fully.

TMJ syndrome is a repeating cycle of: muscle spasm → misalignment → pain → muscle spasm. It has been linked to unconscious behaviors, such as grinding of the teeth during sleep, and emotional stress. Treatment focuses on breaking the cycle of pain and muscle spasm and, when necessary, providing emotional support. The application of heat to the affected joint, coupled with the use of anti-inflammatory drugs or local anesthetics or both, may be helpful. If teeth grinding is suspected, special mouth guards may be worn at night.

⚕ Kyphosis, Lordosis, Scoliosis   EAP p. 126

In **kyphosis** (kī-FŌ-sis), the normal thoracic curvature becomes exaggerated posteriorly, producing a "roundback" appearance. This can be caused by (1) osteoporosis or a compression fracture affecting the anterior portions of vertebral bodies, (2) chronic contractions in muscles that insert on the vertebrae, or (3) abnormal vertebral growth. In **lordosis** (lōr-DŌ-sis), or "swayback," the abdomen and buttocks protrude because of an anterior exaggeration of the lumbar curvature.

Scoliosis (skō-lē-Ō-sis) involves an abnormal lateral curvature. This lateral deviation may occur in one or more of the movable vertebrae. Scoliosis is the most common distortion of the spinal curvature. Scoliosis may result from developmental

6

TABLE A-13   Age-Related Changes in the Skeleton

Region/Feature	Event(s)	Age (Years)
GENERAL SKELETON		
Bony matrix	Reduction in mineral content	Values differ for males versus females between ages 45 and 65; similar reductions occur in both sexes after age 65.
Markings	Reduction in size, roughness	Gradual reduction with increasing age and decreasing muscular strength and mass.
SKULL		
Fontanels	Closure	Completed by age 2
Metopic suture	Fusion	2–8
Occipital bone	Fusion of ossification centers	1–4
Styloid process	Fusion with temporal bone	12–16
Hyoid bone	Complete ossification and fusion	25–30
Teeth	Loss of "baby teeth"; appearance of secondary dentition; eruption of posterior molars	Detailed in Chapter 17 (digestive system).
Mandible	Loss of teeth; reduction in bone mass; change in angle at mandibular notch	Accelerates in later years (60+).
VERTEBRAE		
Curvature	Appearance of major curves	Described in Figure 7-17.
Intervertebral discs	Reduction in size, percentage contribution to height	Accelerates in later years (60+).
LONG BONES		
Epiphyseal plates	Fusion	Ranges vary, but general analysis permits determination of approximate age.
PECTORAL AND PELVIC GIRDLES		
Epiphyses	Fusion	Overlapping ranges somewhat narrower than the above, including 14–16, 16–18, 22–25 years.

problems, such as incomplete vertebral formation, or from muscular paralysis affecting one side of the back. In four out of five cases it is impossible to determine the structural or functional cause of the abnormal spinal curvature. Scoliosis usually appears in girls during adolescence, when periods of growth are most rapid. Treatment consists of a combination of exercises, braces, and sometimes surgical modifications of the affected vertebrae. Early detection greatly improves the chances for successful treatment.

⚕ Problems with the Intervertebral Discs
EAP p. 128

An intervertebral disc compressed beyond its normal limits may become temporarily or permanently damaged. If the posterior ligaments are weakened, as often occurs with advancing age, the compressed nucleus pulposus may distort the annulus fibrosus, forcing it partway into the vertebral canal. This condition is often called a *slipped disc* (Figure A-17), although the disc does not actually slip. The most common sites for disc problems are at C_5–C_6, L_4–L_5, and between L_5 and S_1. A disc problem can occur at any age as the result of an accidental injury, such as a hard fall or a "whiplash" injury to the neck, but with advanced age, the supporting ligaments may become so weak that the problem may occur without warning or apparent cause.

If the nucleus pulposus breaks through the annulus fibrosis, it often protrudes into the vertebral canal. This condition is called a **herniated disc.** When a disc herniates, sensory nerves are distorted, and the protruding mass can also compress the nerve

roots passing through the adjacent intervertebral foramen. The result is severe backache, an abnormal posture (abnormal vertebral flexion), abnormal sensory function, often a burning or tingling sensation from the lower back and lower limbs, and in some cases a partial loss of control over skeletal muscles innervated by the compressed nerve fibers. The location of the injured disc can usually be determined by noting the distribution of abnormal sensations. For example, someone with a herniated disc at L_4–L_5 will experience pain in the hip, groin, the posterior and lateral surfaces of the thigh, the lateral surface of the calf, and the top of the foot; a herniation at L_5–S_1 produces pain in the buttocks, the posterior thigh, the posterior calf, and the sole of the foot.

Most lumbar disc problems can be successfully treated with some combination of rest, back braces, analgesic (pain-killing) drugs, and physical therapy. Surgery to relieve the symptoms is required in only about 10 percent of cases involving lumbar disc herniation. The primary method of treatment involves removing the offending disc, and, if necessary, fusing the vertebral bodies together to prevent relative movement. Accessing the disc requires removal of the nearest vertebral arch by shaving away the laminae. For this reason the procedure is known as a *laminectomy* (la-mi-NEK-to-mē).

In cases where the herniated portion of the disc does not extend well into the vertebral foramen, portions of the disc may be removed with a suction cutter guided to the site by radiological imaging. Although this procedure is faster and easier than a laminectomy, relatively few herniated discs fall within this category.

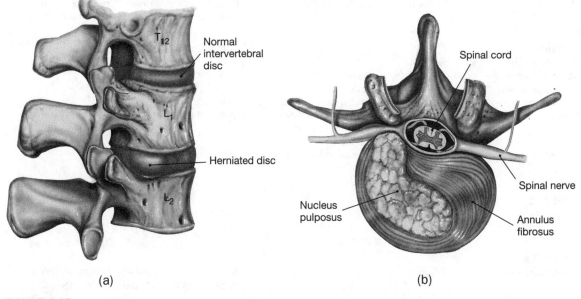

(a)   (b)

FIGURE A-17

Damage to the Intervertebral Discs. (a) Lateral view of the lumbar region of the spinal column, showing a herniated intervertebral disc. (b) Sectional view through a herniated disc, showing release of the nucleus pulposus and its effect on the spinal cord and adjacent nerves.

⚕ Spina Bifida
EAP *p. 128*

Spina bifida (SPĪ-na BI-fi-da) results when the vertebral laminae fail to unite during development. The neural arch is incomplete, and the membranes that line the dorsal body cavity bulge outward. In mild cases, most often involving the sacral and lumbar regions, the condition may pass unnoticed. This condition, called *spina bifida occulta*, can be detected only through X-ray or other scanning procedures. Spina bifida occulta is very common, and it may affect 10 percent of the U.S. population. In some cases, the protective layers, or *meninges*, that surround the spinal cord may bulge through the open vertebral laminae. The result is a prominent fluid-filled sac, or *meningocele*, beneath the skin of the back.

The vertebral abnormalities in more severe forms of spina bifida are often associated with abnormal development of the spinal cord and associated nerves. Symptoms may range from mild problems with balance and movement to a general sensory and motor paralysis of the lower body.

⚕ Problems with the Ankle and Foot
EAP *p. 135*

The ankle and foot are subjected to a variety of stresses during normal daily activities. In a *sprain*, a ligament is stretched to the point where some of the collagen fibers are torn. The ligament remains functional, and the structure of the joint is not affected. The most common cause of a **sprained ankle** is a forceful inversion of the foot that stretches the lateral ligament. An ice pack is usually required to reduce swelling, and with rest and support the ankle should heal in about three weeks.

In more serious incidents, the entire ligament may be torn apart, or the connection between the ligament and the malleolus may be so strong that the bone breaks before the ligament. In general, a broken bone heals more quickly and effectively than does a torn ligament. A dislocation often accompanies such injuries.

In a **dancer's fracture** the proximal portion of the fifth metatarsal is broken. This usually occurs while the body weight is being supported by the longitudinal arch. A sudden shift in weight from the medial portion of the arch to the lateral, less elastic border breaks the fifth metatarsal close to its distal articulation.

Individuals with abnormal arch development are more likely to suffer metatarsal injuries. Someone with flat feet loses or never develops the longitudinal arch. "Fallen arches" may develop as tendons and ligaments stretch and become less elastic. Obese individuals or those who must constantly stand or walk on the job are likely candidates. Children have very mobile articulations and elastic ligaments, so they often have flexible flat feet. Their feet look flat only while they are standing, and the arch appears when they stand on their toes or sit down. This condition usually disappears as growth continues.

Clawfeet are also produced by muscular abnormalities. In this case the median longitudinal arch becomes exaggerated because the plantar flexors are overpowering the dorsiflexors. Muscle cramps or nerve paralysis may be responsible; the condition tends to develop in adults, and it gets progressively worse with age.

Congenital talipes equinovarus ("*clubfoot*") results from an inherited developmental abnormality that affects 2 in 1000 births. Boys are affected roughly twice as often as girls. One or both feet may be involved, and the condition may be mild, moderate, or severe. The underlying problem is abnormal muscle development that distorts growing bones and joints. Usually the tibia, ankle, and foot are affect-ed, and the feet are turned medially and inverted. The longitudinal arch is exaggerated, and if both feet are involved, the soles face one another. Prompt treat-ment with casts or other supports in infancy helps to alleviate the problem, and fewer than half of the cases require surgery. Kristi Yamaguchi, Olympic GoldMedalist in figure skating, was born with this condition.

⚕ Rheumatism, Arthritis, and Synovial Function
EAP p. 138

Rheumatism (RŌŌ-ma-tizm) is a general term that indicates pain and stiffness affecting the skeletal system, the muscular system, or both. There are several major forms of rheumatism. **Arthritis** (ar-THRĪ-tis) includes all the rheumatic diseases that affect synovial joints. Arthritis always involves damage to the articular cartilages, but the specific cause may vary. For example, arthritis can result from bacterial or viral infection, injury to the joint, metabolic problems, autoimmune disorders, or severe physical stresses.

Proper synovial function depends on healthy articular cartilages. When an articular cartilage has been damaged, the matrix begins to break down, and the exposed cartilage changes from a slick, smooth gliding surface to a rough feltwork of bristly collagen fibers. This feltwork drastically increases friction, damaging the cartilage further. Eventually the central area of the articular cartilage may completely disappear, exposing the underlying bone.

Fibroblasts are attracted to areas of friction, and they begin tying the opposing bones together with a network of collagen fibers. This network may later be converted to bone, locking the articulating elements into position. Such a bony fusion, called **ankylosis** (an-kē-LŌ-sis), eliminates the friction, but only by the drastic remedy of making movement impossible.

The diseases of arthritis are usually considered as either **degenerative** or **inflammatory** in nature. Degenerative diseases begin at the articular cartilages, and modification of the underlying bone and inflammation of the joint occur secondarily. Inflammatory diseases start with the inflammation of synovial tissues, and damage later spreads to the articular surfaces. We will consider a single example of each type.

6

Osteoarthritis (os-tē-ō-ar-THRĪ-tis), also known as **degenerative arthritis** or **degenerative joint disease** (DJD), usually affects older individuals. In the U.S. population, 25 percent of women and 15 percent of men over 60 years of age show signs of this disease. The condition seems to result from cumulative wear and tear on the joint surfaces. Some individuals, however, may have a genetic predisposition to develop osteoarthritis, for researchers have recently isolated a gene linked to the disease. This gene codes for an abnormal form of collagen that differs from the normal protein in only 1 of its 1,000 amino acids.

Rheumatoid arthritis is an inflammatory condition that affects roughly 2.5 percent of the adult population. The cause is uncertain, although allergies, bacteria, viruses, and genetic factors have all been proposed. The synovial membrane becomes swollen and inflamed, a condition known as **synovitis** (sī-nō-VĪ-tis). The cartilaginous matrix begins to break down, and the process accelerates as dying cartilage cells release lysosomal enzymes.

Advanced stages of inflammatory and degenerative forms of arthritis produce an inflammation that spreads into the surrounding area. Ankylosis, common in the past when complete rest was routinely prescribed for arthritis patients, is rarely seen today. Regular exercise, physical therapy, and drugs that reduce inflammation, such as aspirin, can slow the progress of the disease. Surgical procedures can realign or redesign the affected joint, and in extreme cases involving the hip, knee, elbow, or shoulder, the defective joint can be replaced by an artificial one. Joint replacement has the advantage of eliminating the pain and restoring full range of motion. Prosthetic (artificial) joints, such as those shown in Figure A-18, are weaker than natural ones, but elderly people seldom stress them to their limits.

⚡ Hip Fractures, Aging, and Professional Athletes EAP *p. 145*

Today there are two very different groups of people suffering hip fractures: (1) individuals over age 60, whose bones have been weakened by osteoporosis, and (2) young, healthy professional athletes, who subject their hips to extreme forces. When the injury is severe, the vascular supply to the joint is damaged. As a result, two problems gradually develop:

1. *Avascular necrosis:* The mineral deposits in the bone of the pelvis and femur are turned over very rapidly, and osteocytes have high energy demands. A reduction in blood flow first injures and then kills them. When bone maintenance stops in the affected region, the matrix begins to break down. This process is called *avascular necrosis.*

2. *Degeneration of articular cartilages:* The chondrocytes in the articular cartilages absorb nutrients from the synovial fluid, which circulates around

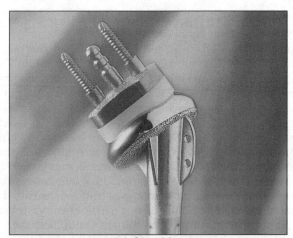

(a) Shoulder

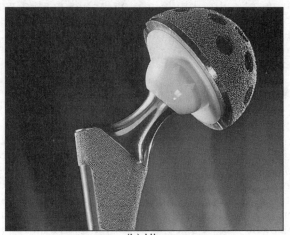

(b) Hip

(c) Knee

Figure A-18   Artificial Joints.

the joint cavity as the bones change position. A fracture of the femoral neck is usually followed by joint immobility and poor circulation to the synovial membrane. The combination results in a gradual deterioration of the articular cartilages of the femur and acetabulum.

In recent years the frequency of hip fractures has increased dramatically among young, healthy professional athletes, the best-known example being Bo Jackson. At age 28 he was playing professional football with the LA Raiders and professional baseball for the Kansas City Royals—and starring in both sports. He was also the centerpiece for successful advertising campaigns for sporting goods, using the "Bo knows" slogan. But on January 13, 1991, things changed dramatically when Bo was tackled near the sidelines in an NFL playoff game. The combination of pressure and twisting applied by the tackler and the tremendous power of Bo's thigh muscles produced a fracture-dislocation of the hip. Roughly 15 percent of the inferior acetabular fossa was broken away. The femur was not broken, but it was dislocated. Although he experienced severe pain, the immediate damage to the femur and hip was sufficiently limited.

The initial optimism began to fade when it became apparent that the complications of the injury were more damaging than the injury itself. The dislocation tore blood vessels in the capsule and along the femoral neck, where the capsular fibers attach. The result was avascular necrosis and the degeneration of the articular cartilages at the hip.

After more than a year of rehabilitation, Bo Jackson had surgery to replace the damaged joint with an artificial hip. In this "total hip" procedure, the damaged portion of the femur was removed, and an artificial femoral head and neck is attached by a spike that extends into the marrow cavity of the shaft. Special cement may be used to anchor it in place and to attach a new articular surface to the acetabulum. After this procedure, Bo reentered professional baseball as a designated hitter and occasional first baseman for the Chicago White Sox, and was later traded to the California Angels. He is a bit slower, and his batting average has slipped, but everyone is hoping for the best.

Joint replacement eliminates the pain and restores full range of motion. However, prosthetic (artificial) joints, such as those shown in Figure A-19b, are weaker that natural ones. They are usually implanted in older people who seldom stress them to their limits. Whether an artificial hip will be able to tolerate the stresses of professional sports is an open question. It is also an important question, because the frequency of such injuries is increasing. By early 1994, hip fractures and the associated potential for avascular necrosis had sidelined Marvin Jones of the Jets and Mike Sherrard of the Giants, perhaps permanently.

CRITICAL THINKING QUESTIONS

2-1. Diane, a 65-year-old woman, is brought into the emergency room by her daughter, Mary. Diane has had severe back pain since she fell accidentally yesterday. Her daughter insists it was a minor fall and is confused about the severity of the pain, and tenderness over the thoracic spine. X-ray studies of the thoracic vertebrae reveal fractures of T_{10}, T_{11}, and T_{12}. The X-rays also reveal a decreased bone density in all the vertebrae. Laboratory tests are within normal limits. The physician diagnoses the fracture as a vertebral fracture most likely due to:

    a. osteomalacia   c. osteoporosis
    b. osteomyelitis   d. Paget's disease

2-2. Beth, a dentist, has a patient complaining of a toothache on the lower right side of his mouth. Examination shows severe tooth decay in the first mandibular molar, with facial swelling along the mandibular angle. Beth suspects an infection and takes an X-ray that shows a reduction in bone density around the root of the decayed tooth. What is the likely diagnosis?

    a. osteomalacia   c. fracture of the mandible
    b. osteomyelitis   d. tumor of the mandible

2-3. While studying a 2,000-year old campsite, an archeologist discovers several adult male and female skeletons. Chemical analysis of the bones of both sexes indicates that osteopenia had been occurring for several years before their deaths. The male skeletons all showed heavy thickenings and prominences in the bones of the upper limbs. The bones of the lower limbs were of normal size, and the markings indicated that only minimal muscle stress had been applied during life. How old do you think these individuals were when they died, and what does the skeletal information tell you about the lifestyle of the males?

2-4. Lori suffers from a severe case of temporomandibular joint syndrome. What symptoms would you expect to observe as a result of this condition?

2-5. Two patients sustain hip fractures. In one case, a pin is inserted into the joint and the injury heals well, In another, the fracture fails to heal. Identify the types of fractures that are probably involved. Why did the second patient's fracture not heal, and what steps could be taken to restore normal function?

The Muscular System

The muscular system includes over 700 skeletal muscles that are directly or indirectly attached to the skeleton by tendons or aponeuroses. The muscular system produces movement, as the contractions of skeletal muscles pull on the attached bones. Muscular activity does not always result in movement, however; it can also be important in stabilizing skeletal elements and preventing movement. Skeletal muscles are also important in guarding entrances or exits of internal passageways, such as those of the digestive, respiratory, urinary, or reproductive systems, and in generating heat to maintain our stable body temperatures.

Skeletal muscles contract only under the command of the nervous system. For this reason, clinical observation of muscular activity may provide direct information about the muscular system, and indirect information about the nervous system. Assessment of facial expressions, posture, speech, and gait can be an important part of the physical examination. Classical signs of muscle disorders include the following:

- *Gower's sign* is a distinctive method of standing from a sitting or lying position on the floor. This method is used by children with *muscular dystrophy* (p. 59). When a young child with this disorder moves from a sitting position to a standing position, he or she pushes the trunk off the floor with the hands, and then moves the hands to the knees. The hands are then used as braces to force the body into the standing position. This extra support is necessary because the pelvic muscles are too weak to swing the weight of the trunk over the legs.

- *Ptosis* is a drooping of the upper eyelid. It may be seen in *myasthenia gravis* (p. 60), *botulism* (p. 60), *myotonic dystrophy* (p. 60), or following damage to the cranial nerve innervating the *levator palpabrae superioris*, a muscle of the eyelid.

- A abnormal dense region, or *muscle mass*, may sometimes be seen or felt within a skeletal muscle. A muscle mass may result from torn muscle tissue, a hematoma, a parasitic infection, such as *trichinosis* (p. 58), or bone deposition, as in *myositis ossificans*.

- Abnormal contractions may indicate problems with the muscle tissue or its innervation. *Muscle spasticity* exists when a muscle has excessive muscle tone. A *muscle spasm* is a sudden, strong, and painful involuntary contraction.

- *Muscle flaccidity* exists when the relaxed skeletal muscle appears soft and relaxed and its contractions are very weak.

- *Muscle atrophy* is skeletal muscle deterioration, or *wasting*, due to disuse, immobility, or interference with the normal muscle innervation.

- Abnormal patterns of muscle movement, such as *tics, choreiform movements,* or *tremors,* and

muscular paralysis are usually caused by nervous system disorders. These movements will be described further in sections dealing with abnormal nervous system function.

SIGNS AND SYMPTOMS OF MUSCULAR SYSTEM DISORDERS

Two common symptoms of muscular disorders are *pain* and *weakness* in the affected skeletal muscles. The potential causes of muscle pain include:

1. *Muscle trauma:* Examples of traumatic injuries to a skeletal muscle would include a laceration, a deep bruise or crushing injury, a muscle tear, or a damaged tendon.

2. *Muscle infection:* Skeletal muscles may be infected by viruses, as in some forms of myositis, or colonized by parasitic worms, such as those responsible for *trichinosis* (p. 58). These infections usually produce pain that is restricted to the involved muscles. Diffuse muscle pain may develop in the course of other infectious diseases, such as influenza or measles.

3. *Related problems with the skeletal system:* Muscle pain may result from skeletal problems, such as arthritis (p. 53) or a sprained ligament near the point of muscle origin or insertion.

4. *Problems with the nervous system:* Muscle pain may be experienced due to inflammation of sensory neurons or stimulation of pain pathways in the CNS.

Muscle strength can be evaluated by applying an opposite force against a specific action. For example, the examiner might exert a gentle extending force while asking the patient to flex the arm. Because the muscular and nervous systems are so closely interrelated, a single symptom, such as muscle weakness, can have a variety of different causes (Figure A-19). Muscle weakness may also develop as a consequence of a condition that affects the entire body, such as anemia or acute starvation.

Figure A-20 presents a simple diagnostic chart for representative muscle disorders.

✝ Necrotizing Fasciitis EAP *p. 154*

Several bacteria produce enzymes such as *hyaluronidase* or *cysteine protease*. Hyaluronidase breaks dwn hyaluronic acid and disassembles the associated proteoglycans. Cystein protease breaks down conective tissue proteins. These bacteria are dangerous because they can spread rapidly by liquifying the matrix and dissolving the intercellular cement that holds epithelial cells together. The *Streptococci* are an example of a group of bacteria that secrete both of these enzymes. *Streptococcus A* bacteria are involved in many human diseases,

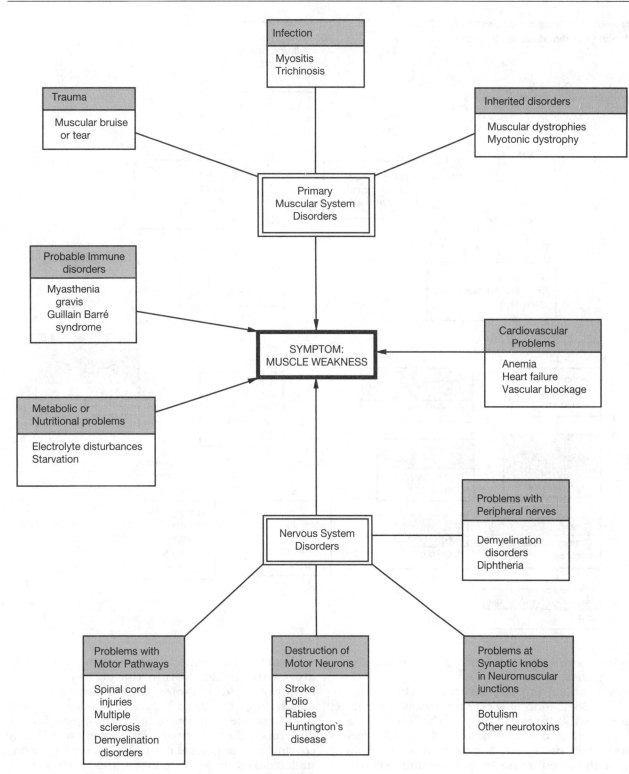

Figure A-19   Some Possible Causes of Muscle Weakness

most notably "strep throat," a pharyngeal infection. In most cases the immune response is sufficient to contain and ultimately defeat these bacteria before extensive tissue damage has occurred.

However, in 1994 tabloid newspapers had a field day recounting stories of "killer bugs" and "flesh-eating bacteria" that terrorized residents of the city of Gloucester, England. The details were horrific—minor cuts become major open wounds, with interior connective tissues dissolving. Although in fact there were only 7 reported cases, 5 of the victims died. The pathogen responsible was a strain of *Streptococcus A* that overpowered immune defenses and swiftly invaded and destroyed soft tissues. More over, the pathogens eroded their way along the fascial wrapping that covers skeletal muscles and other organs. The term for this condition is **necrotizing fasciitis.** In some cases the

Figure A-20   A Simple Diagnostic Flow Chart for Disorders of the Muscular System.

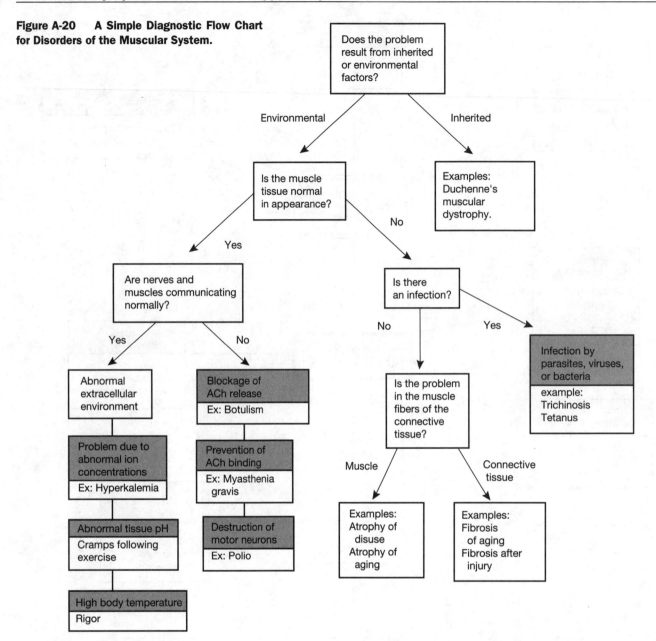

muscle tissue was also destroyed, a condition called *myositis.*

The problem is not restricted to the United Kingdom. Some form of soft tissue invasion occurs roughly 75-150 times annually in the U.S.; this is 5-10 percent of the *Streptococcus A* wound infections reported each year. At present it is uncertain whether the recent surge in myositis and necrotizing faciitis reflects increased awareness of the condition or the appearance of a new strain of strep bacteria.

✠ Trichinosis   EAP *p. 154*

Trichinosis (trik-i-NŌ-sis; *trichos,* hair + *nosos,* disease) results from infection by a parasitic nematode worm, *Trichinella spiralis.* Symptoms include diarrhea, weakness, and muscle pain. The muscular symptoms are caused by the invasion of

skeletal muscle tissue by larval worms, which create small pockets within the perimysium and endomysium. Muscles of the tongue, eyes, diaphragm, chest, and leg are most often affected.

Larvae are common in the flesh of pigs, horses, dogs, and other mammals. The larvae are killed by cooking, and people are most often exposed by eating undercooked pork. Once eaten, the larvae mature within the intestinal tract, where they mate and produce eggs. The new generation of larvae then migrates through the body tissues to reach the muscles, where they complete their early development. The migration and subsequent settling produce a generalized achiness, muscle and joint pain, and swelling in infected tissues. An estimated 1.5 million Americans carry *Trichinella* around in their muscles, and up to 300,000 new infections occur each year. The mortality rate for people who have symptoms severe enough to require treatment is approximately 1 percent.

✝ Fibromyalgia and Chronic Fatigue Syndrome

EAP *p. 154*

Fibromyalgia (-*algia*, pain) is an inflammatory disorder that has been recognized only in the last 10 years. Although first described in the early 1800s, the condition is still somewhat controversial because the reported symptoms cannot be linked to any anatomical or physiological abnormalities. However, physicians now recognize a distinctive pattern of symptoms that warrant consideration as a clinical entity.

Fibromyalgia may be the most common musculoskeletal disorder affecting women under 40 years of age. There may be 3-6 million cases in the United States today. Symptoms include chronic aches, pain, and stiffness, and multiple tender points at specific, characteristic locations. The four most common tender points are (1) just below the kneecap, (2, 3) distal to the medial and lateral epicondyles of the humerus, and (4) the junction between the second rib and the cartilage attaching it to the sternum. An additional clinical criterion is that the pains and stiffness cannot be explained by other mechanisms. Patients suffering from this condition invariably report chronic fatigue; they feel tired on awakening and often complain of awakening repeatedly during the night.

Most of these symptoms could be attributed to other problems. For example, chronic depression can lead to fatigue and poor-quality sleep. As a result, the pattern of tender points is really the diagnostic key to fibromyalgia. This symptom clearly distinguishes fibromyalgia from **chronic fatigue syndrome** (CFS). The current symptoms accepted as a definition of CFS include (1) sudden onset, usually following a viral infection, (2) disabling fatigue, (3) muscle weakness and pain, (4) sleep disturbance, (5) fever, and (6) enlargement of cervical lymph nodes.

Attempts to link either fibromyalgia or CFS to a viral infection or to some physical or psychological trauma have not been successful, and the cause remains mysterious. Treatment at present is limited to relieving symptoms when possible. For example, anti-inflammatory medications may help relieve pain, drugs can be used to promote sleep, and exercise programs may help maintain normal range of motion.

✝ The Muscular Dystrophies

EAP *p. 157*

The **muscular dystrophies** (DIS-trō-fē-z) are inherited diseases that produce progressive muscle weakness and deterioration. One of the most common and best understood conditions is **Duchenne's muscular dystrophy (DMD).**

This form of muscular dystrophy appears in childhood, often between the ages of 3 and 7. The condition generally affects only males. A progressive muscular weakness develops, and the individual usually dies before age 20 because of respiratory paralysis. Skeletal muscles are primarily affected, although for some reason the facial muscles continue to function normally. In later stages of the disease, the facial muscles and cardiac muscle tissue may also become involved.

The skeletal muscle fibers in a DMD patient are structurally different from those of normal individuals. Abnormal membrane permeability, cholesterol content, rates of protein synthesis, and enzyme composition have been reported. DMD sufferers also lack a protein, called *dystrophin,* found in normal muscle fibers. It is attached to the inner surface of the sarcolemma near the triads. Although the functions of this protein remain uncertain, it is suspected that dystrophin plays a role in the regulation of calcium ion channels in the sarcolemma. Researchers have recently identified and cloned the gene for dystrophin. Rats with DMD have been cured by insertion of this gene into their muscle fibers, a technique that may eventually be used to treat human patients.

The inheritance of DMD is sex-linked; women carrying the defective genes are unaffected, but each of their male children will have a 50 percent chance of developing DMD. Now that the specific location of the gene has been identified, it is possible to determine whether or not a woman is carrying the defective gene. It is also possible to use an innovative prenatal test to determine if a fetus has this condition. In this procedure, a small sample of fluid is collected from the membranous sac that surrounds the fetus. This fluid contains fetal cells, called *amniocytes,* that are collected and cultivated in the laboratory. Researchers then insert a gene, called *MyoD,* that triggers their differentiation into skeletal muscle fibers. These cells can then be tested not only for the signs of muscular dystrophy, but for indications of other inherited muscular disorders.

MYOTONIC DYSTROPHY. Myotonic dystrophy is a form of muscular dystrophy that occurs in the United States at an incidence of 13.5 per 100,000 population. Symptoms may develop in infancy, but more often develop after puberty. As with other forms of muscular dystrophy, adults developing myotonic dystrophy experience a gradual reduction in muscle strength and control. Problems with other systems, especially cardiovascular and digestive systems, often develop. There is no effective treatment.

The inheritance of myotonic dystrophy is unusual because the children of an individual with myotonic dystrophy often develop more severe symptoms. The increased severity of the condition appears to be related to the presence of multiple copies of a specific gene on chromosome 19. For some reason, the nucleotide sequence of that gene gets repeated several times, and the number can increase from generation to generation. This has been called a "genetic stutter." The greater the number of copies, the more severe the symptoms. It is not known why the stutter develops, nor how the genetic duplication affects the severity of the condition.

⚕ Botulism
EAP *p. 159*

Botulinus (bot-yōō-LĪ-nus) **toxin** prevents the release of ACh at the synaptic terminal. It thus produces a severe and potentially fatal paralysis of skeletal muscles. A case of botulinus poisoning is called **botulism.**[1] The toxin is produced by a bacterium, *Clostridium botulinum,* that does not need oxygen to grow and reproduce. Because the organism can live quite happily in a sealed can or jar, most cases of botulism are linked to improper canning or storing procedures, followed by failure to cook the food adequately before eating. Canned tuna or beets, smoked fish, and cold soups have most often been involved with cases of botulism. Boiling for a half-hour destroys both the toxin and the bacteria.

Symptoms usually begin 12-36 hours after eating a contaminated meal. The initial symptoms are often disturbances in vision, such as seeing double or a painful sensitivity to bright lights. These symptoms are followed by other sensory and motor problems, including blurred speech and an inability to stand or walk. Roughly half of botulism patients experience intense nausea and vomiting. These symptoms persist for a variable period (days to weeks), followed by a gradual recovery; some patients are still recovering after a year.

The major risk of botulinus poisoning is respiratory paralysis and death by suffocation. Treatment is supportive: bed rest, observation, and, if necessary, use of a mechanical respirator. In severe cases drugs that promote the release of ACh, such as *guanidine hydrochloride,* may be administered. The overall mortality rate in the United States is about 10 percent.

⚕ Myasthenia Gravis
EAP *p. 162*

Myasthenia gravis (mī-as-THĒ-nē-a-GRA-vis) is characterized by a general muscular weakness that is often most pronounced in the muscles of the arms, head, and chest. The first symptom is usually a weakness of the eye muscles and drooping eyelids. Facial muscles are often weak as well, and the individual develops a peculiar smile known as the "myasthenic snarl." As the disease progresses, pharyngeal weakness leads to problems with chewing and swallowing, and it becomes difficult to hold the head upright.

The muscles of the upper chest and upper extremities are next to be affected. All the voluntary muscles of the body may ultimately be involved. Severe myasthenia gravis produces respiratory paralysis, with a mortality rate of 5-10 percent. However, the disease does not always progress to such a life-threatening stage. For example, roughly 20 percent of patients experience eye problems with no other symptoms.

The conditions results from a decrease in the number of ACh receptors on the motor end plate.

[1]This disorder was described 200 years ago by German physicians treating patients poisoned by dining on contaminated sausages. *Botulus* is the Latin word for sausage.

Before the remaining receptors can be stimulated enough to trigger a strong contraction, the ACh molecules are destroyed by cholinesterase. As a result, muscular weakness develops.

The primary cause of myasthenia gravis appears to be a malfunction of the immune system. The body attacks the ACh receptors of the motor end plate as if they were foreign proteins. For unknown reasons, women are affected twice as often as men. Estimates of the incidence of this disease in the United States range from 2 to 10 cases per 100,000 population.

One approach to therapy involves the administration of drugs, such as *Neostigmine,* that are termed **cholinesterase inhibitors.** These compounds, as their name implies, are enzyme inhibitors—they tie up the active sites at which cholinesterase normally binds ACh. With cholinesterase activity reduced, the concentration of ACh at the synapse can rise enough to stimulate the surviving receptors and produce muscle contraction.

⚕ Polio
EAP *p. 164*

Because skeletal muscles depend on their motor neurons for stimulation, disorders that affect the nervous system can have an indirect affect on the muscular system. The *poliovirus* is a virus that does not produce clinical symptoms in roughly 95 percent of infected individuals. The virus may produce variable symptoms in the remaining 5 percent. Some individuals develop a nonspecific illness resembling the flu. A second group of individuals develop a brief *meningitis* (p. 67), an inflammation of the protective membranes surrounding the CNS. In the third group of people, the virus attacks somatic motor neurons in the CNS.

In this form of the disease, the individual develops a fever 7-14 days after infection. The fever subsides, and then recurs roughly a week later, accompanied by muscle pain, cramping, and flaccid paralysis of one or more limbs. Respiratory paralysis may also occur, and the mortality rate for this form of polio is 2-5 percent for children and 15-30 percent of adults. If the individual survives, some degree of recovery usually occurs over a period of up to six months.

For as yet unknown reasons, the survivors of paralytic polio may develop progressive muscular weakness 20-30 years after the initial infection. This *postpolio syndrome* is characterized by fatigue, muscle pain and weakness, and, in some cases, muscular atrophy. There is no treatment for this condition, although rest seems to help.

Polio has been almost completely eliminated from the U.S. population through a successful immunization program. In 1954 there were 18,000 new cases in the United States; there were 8 in 1976, and none in 1992. The World Health Organization now reports that polio has been eradicated from the entire western hemisphere, and the virus may be eliminated worldwide by the year

2000. Unfortunately, many parents today refuse to immunize their children against the poliovirus on the assumption that the disease has been "conquered." This is a mistake because (1) there is still no cure for polio, (2) the virus remains in the environment in many areas of the world, and (3) approximately 38 percent of children ages 1-4 have not been immunized. A major epidemic could therefore develop very quickly if the virus were brought into the United States from another part of the world.

✝ Hernias   EAP *p. 177*

When the abdominal muscles contract forcefully, pressure in the abdominopelvic cavity can increase dramatically, and those pressures are applied to internal organs. If the individual exhales at the same time, the pressure is relieved, because the diaphragm can move upward as the lungs collapse. But during vigorous isometric exercises or when lifting a weight while holding one's breath, pressure in the abdominopelvic cavity can rise to 106 kg/cm^2 (1,500 lb/in.^{2}), roughly 100 times normal pressures. Pressures this high can cause a variety of problems, among them the development of a *hernia*.

A **hernia** develops when a visceral organ protrudes through an abnormal opening. There are many types of hernias; we will consider only *inguinal* (groin) *hernias* and *diaphragmatic hernias* here.

Late in the development of the male, the testes descend into the scrotum by passing through the abdominal wall at the **inguinal canals.** In the adult male, the permatic ducts and associated blood vessels penetrate the abdominal musculature at the **inguinal canals** on their way to the abdominal reproductive organs. In an inguinal hernia (Figure A-21) the inguinal canal enlarges, and the abdominal contents such as a portion of the intestine (or more rarely the bladder) are forced into the inguinal canal. If the herniated structures become trapped or twisted within the inguinal sac, surgery may be required to prevent serious complications. Inguinal hernias are not always caused by unusually high abdominal pressures. Injuries to the abdomen, or inherited weakness or dispensability of the canal, may have the same effect.

The esophagus and major blood vessels pass through an opening in the diaphragm, the muscle that separates the thoracic and abdominopelvic cavities. In a **diaphragmatic hernia,** also called a *hiatal hernia* (hī-Ā-tal; *hiatus*, a gap or opening), abdominal organs slide into the thoracic cavity, most often through the *esophageal hiatus*, the opening used by the esophagus. The severity of the condition will depend on the location and size of the herniated organ(s). Hiatal hernias are actually very common, and most go unnoticed. Radiologists see them in about 30 percent of patients examined with barium contrast techniques. When clinical complications develop, they usually occur because abdominal organs that have pushed into the thoracic cavity are exerting pressure on structure or organs there. As is the case with inguinal hernias, a diaphragmatic hernia may result from congenital factors or from an injury that weakens or tears the diaphragmatic muscle.

🏃 Sports Injuries   EAP *p.182*

Sports injuries affect amateurs and professionals alike. At the amateur level, a 5-year study of college football players indicated that 73.5 percent experienced mild injuries, 21.5 percent moderate injuries, and 11.6 percent severe injuries during their playing careers. Contact sports are not the only activities that show a significant injury rate; a study of 1650 joggers running at least 27 miles per week reported 1819 injuries in a single year.

Muscles and bones respond to increased use by enlarging and strengthening. Poorly conditioned individuals are therefore more likely to subject their bones and muscles to intolerable stresses than are people in good condition. Training is also important in minimizing the use of antagonistic muscle groups and keeping joint movements within the intended ranges of motion. Planned warm-up exercises before athletic events stimulate circulation, improve muscular performance and control, and help prevent injuries to muscles, joints, and ligaments. Stretching exercises stimulate muscle circulation and help keep ligaments and joint capsules supple. Such conditioning extends the range of motion and prevents sprains and strains when sudden loads are applied.

Dietary planning can also be important in preventing injuries to muscles during endurance events, such as marathon running. Emphasis has often been placed on the importance of carbohydrates, leading to the practice of "carbohydrate loading" before a marathon. But muscles also utilize amino acids extensively while operating within aerobic limits, and an adequate diet must include both carbohydrates and proteins.

Improved playing conditions, equipment, and regulations also play a role in reducing the incidence of sports injuries. Jogging shoes, ankle or knee braces,

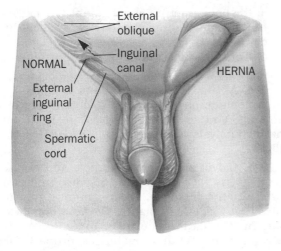

Figure A-21   Inguinal Hernia

External oblique

NORMAL

Inguinal canal

HERNIA

External inguinal ring

Spermatic cord

helmets, and body padding are examples of equipment that can be effective. The substantial penalties now earned for "personal fouls" in contact sports have reduced the numbers of neck and knee injuries.

Several injuries common to those engaged in active sports may also affect nonathletes, although the primary causes may differ considerably. A partial listing of activity-related conditions would include the following:

- *Bone bruise:* Bleeding within the periosteum of a bone
- *Bursitis:* Inflammation of the bursae around one or more joints
- *Muscle cramps:* Prolonged, involuntary, and painful muscular contractions
- *Sprains:* Tears or breaks in ligaments or tendons
- *Strains:* Tears in muscles
- *Stress fractures:* Cracks or breaks in bones subjected to repeated stresses or trauma
- *Tendinitis:* Inflammation of the connective tissue surrounding a tendon

Many of these conditions have been discussed in previous chapters.

Finally, many sports injuries would be prevented if people who engage in regular exercise would use common sense and recognize their personal limitations. It can be argued that some athletic events, such as the "ultramarathon," place such excessive stresses on the cardiovascular, muscular, respiratory, and urinary systems that they cannot be recommended, even for athletes in peak condition.

Carpal Tunnel Syndrome   EAP *p.186*

Tenosynovitis is the inflammation of a tendon sheath. Carpal tunnel syndrome results from tenosynovitis of the tendon sheath surrounding the flexor tendons of the palm. The inflammation leads to compression of the *median nerve,* a mixed (sensory and motor) nerve that innervates the palm. Symptoms include pain, especially on palmar flexion, a tingling sensation or numbness on the palm, and weakness in the abductor pollicis. This condition is fairly common and often strikes those engaged in repetitive hand movements, such as typing, working at a computer keyboard, or playing the piano. Treatment involves administration of anti-inflammatory drugs such as aspirin, injection of anti-inflammatory agents, such as *glucocorticoids* (steroid hormones produced by the adrenal cortex), and use of a splint to prevent wrist flexion and stabilize the region.

Carpal tunnel syndrome is an example of a *cumulative trauma disorder,* or *overuse syndrome.* These disorders are caused by repetitive movements of the arms, hands, and fingers. These musculoskeletal problems now account for over 50 percent of all work-related injuries in the United States.

CRITICAL THINKING QUESTIONS

3-1. A patient experiencing a severe hyperkalemia could have the following related problems:

a. the potassium ion concentration of the interstitial fluid would be less than normal
b. the membrane potential of nerves and muscles would be more negative
c. unresponsive skeletal muscles and cardiac arrest
d. muscle weakness and increased strength of twitch contractions
e. all of the above

3-2. Making hospital rounds, Dr. R., an anesthesiologist, meets with a first semester anatomy and physiology student named CeCe who is scheduled for surgery the next day. Having just finished the unit on skeletal muscles and the nervous system, CeCe is eager to learn about the anesthesia which will be used during the surgery. Dr. R. explains he will be using a drug, *tubocurarine chloride,* that competes with acetylcholine and blocks the action of this neurotransmitter at the neuromuscular junction. What effect will this have on CeCe's skeletal muscles?

a. produce paralysis of all the skeletal muscles
b. cause tetany of the skeletal muscles

c. increase the force and strength of muscle contractions

CeCe answers this question correctly but becomes immediately concerned about this effect on a select group of skeletal muscles. What is CeCe concerned about?

3-3. Tom broke his right leg in a football game. After six weeks in a cast, the cast is finally removed, and when he takes his first few steps, he loses his balance and falls. What is the most likely explanation?

a. the bone fracture is not completely healed
b. the right leg muscles have atrophied due to disuse
c. Tom has an undiagnosed neuromuscular disorder

3-4. A sample of muscle tissue is taken from a champion tennis player and a nonathlete of the same age and sex. Both samples are subjected to enzyme analysis. How would you expect the two to differ?

3-5. Calvin steps into a pothole and twists his ankle. He is in a great deal of pain, and cannot stand. In the hospital, the examining physician notes that Calvin can plantar flex and dorsiflex the foot, but he cannot perform inversion without extreme pain. What muscle has probably been injured?

NOTES

The Nervous System

The nervous system is a highly complex and interconnected network of neurons and supporting neuroglia. Neural tissue is extremely delicate, and the characteristics of the extracellular environment must be kept within narrow homeostatic limits. When homeostatic regulatory mechanisms break down, under the stress of environmental factors, infection, or trauma, symptoms of neurological disorders appear.

There are literally hundreds of different disorders of the nervous system. A *neurological examination* attempts to trace the source of the problem through evaluation of the sensory, motor, behavioral and cognitive functions of the nervous system. Figure A-22 introduces several major categories of nervous system disorders. Many of these examples will be discussed in the sections that follow.

THE SYMPTOMS OF NEUROLOGICAL DISORDERS

The nervous system has varied and complex functions, and the systems of neurological disorders are equally diverse. However, there are a few symptoms that accompany many different disorders.

- *Headache:* The majority of headaches (roughly 90 percent) are *tension headaches* due to muscle tension or *migraine headaches* that have both neurological and circulatory origins Neither of these conditions is life threatening.

- *Muscle weakness:* Muscle weakness can have an underlying neurologic basis, as noted in the section on muscle disorders (see Figure A-20, p. 58). The examiner must determine the origin of the symptom. Myopathies (muscle disease) must be differentiated from neurologic diseases such as demyelinating disorders, neuromuscular junction dysfunction, and peripheral nerve damage.

- *Paresthesias:* Loss of feeling, numbness, or tingling sensations may develop following damage to (1) a sensory nerve (cranial or spinal nerve) or (2) sensory pathways inside the CNS. The effects may be temporary or permanent. For example, a *pressure palsy* (p. 73) may last a few minutes, whereas the paresthesia that develops distal to an area of severe spinal cord damage (p. 68) will probably be permanent.

Figure A-22   Nervous System Disorders

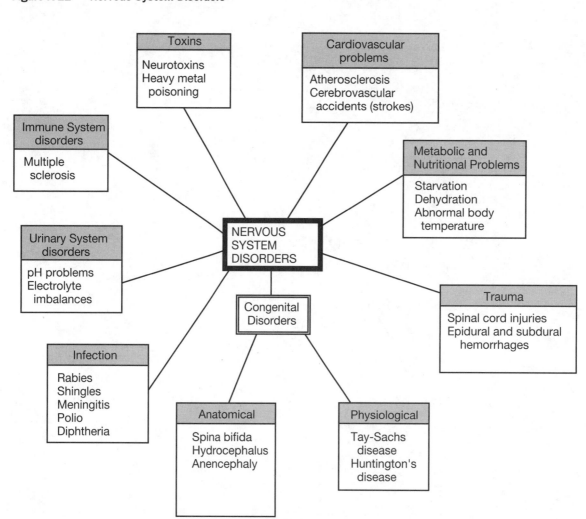

☐ THE NEUROLOGICAL EXAMINATION

During a physical examination, information about the nervous system is obtained indirectly, by assessing sensory, motor, and intellectual functions. Examples of factors noted in the physical examination include

- *State of consciousness:* There are many different levels of consciousness, ranging from unconscious and incapable of being aroused, to fully alert and attentive, to hyperexcitable.

- *Reflex activity:* The general state of the nervous system, and especially the state of peripheral sensory and motor innervation, can be checked by testing specific reflexes (p. 73). For example, the *knee-jerk reflex* will not be normal if there has been damage to associated segments of the lumbar spinal cord, their spinal nerve roots, or the peripheral nerves involved in the reflex.

- *Abnormal speech patterns:* Normal speech involves intellectual processing, motor coordination at the speech centers of the brain, precise respiratory control, regulation of tension in the vocal cords, and adjustment of the musculature of the palate and face. Problems with the selection, production, or use of words often follows damage to the cerebral hemispheres, as in a stroke (p. 108).

- *Abnormal motor patterns:* An individual's posture, balance, and mode of walking, or *gait*, are useful indicators of the level of motor coordination. Clinicians also ask about abnormal involuntary movements that may indicate a *seizure*, a temporary disorder of cerebral function (p. 70).

☤ Demyelination Disorders   EAP *p. 208*

Demyelination disorders are linked by a common symptom: the destruction of myelinated axons in the CNS and PNS. The mechanism responsible for this symptom differs in each of these disorders. We will consider only the major categories of demyelination disorders in this section.

- *Heavy metal poisoning:* Chronic exposure to heavy metal ions, such as arsenic, lead, or mercury, can lead to glial cell damage and demyelination. As demyelination occurs, the affected axons deteriorate, and the condition becomes irreversible. Historians note several interesting examples of heavy metal poisoning with widespread impact. For example, lead contamination of drinking water has been cited as one factor in the decline of the Roman Empire. In the seventeenth century, the great physicist Sir Isaac Newton is thought to have suffered several episodes of physical illness and mental instability brought on by his use of mercury in chemical experiments. Well into the nineteenth century, mercury used in the preparation of felt presented a serious occupational hazard for those employed in the manufacture of stylish hats. Over time, mercury absorbed through the skin and across the lungs accumulated in the CNS, producing neurological damage that affected both physical and mental function. (This effect is the source of the expression "mad as a hatter.") More recently, Japanese fishermen working in Minamata Bay, Japan, collected and consumed seafood contaminated with mercury discharged from a nearby chemical plant. Levels of mercury in their systems gradually rose to the point that clinical symptoms appeared in hundreds of people. Making matters worse, mercury contamination of developing embryos caused severe, crippling birth defects.

- *Diphtheria:* **Diphtheria** (dif-THĒ-rē-a; *diphtheria*, membrane + *-ia*, disease) is a disease that results from a bacterial infection of the respiratory tract. In addition to restricting airflow and sometimes damaging the respiratory surfaces, the bacteria produce a powerful toxin that injures the kidneys and adrenal glands, among other tissues. In the nervous system, diphtheria toxin damages Schwann cells and destroys myelin sheaths in the PNS. This demyelination leads to sensory and motor problems that may ultimately produce a fatal paralysis. The toxin also affects cardiac muscle cells, and heart enlargement and failure may occur. The fatality rate for untreated cases ranges from 35 to 90 percent, depending on the site of infection and the subspecies of bacterium. Because an effective vaccine exists, cases are relatively rare in countries with adequate health care.

- *Multiple sclerosis:* **Multiple sclerosis** (skler-Ō-sis; *sklerosis*, hardness), or **MS,** is a disease characterized by recurrent incidents of demyelination affecting axons in the optic nerve, brain, and/or spinal cord. Common symptoms include partial loss of vision and problems with speech, balance, and general motor coordination. The time between incidents and the degree of recovery varies from case to case. In about one-third of all cases, the disorder is progressive, and each incident leaves a greater degree of functional impairment. The average age at the first attack is 30-40; the incidence in women is 1.5 times that among men. There is no effective treatment at present, although corticosteroid injections and interferon may slow the progression of the disease. MS is discussed in more detail in a later section (p. 72).

- *Guillain-Barré syndrome:* **Guillain-Barré syndrome** is characterized by a progressive but reversible demyelination. Symptoms initially involve weakness of the legs, which spreads rapidly to muscles of the trunk and arms. These symptoms usually increase in intensity for 1-2 weeks before subsiding. The mortality rate is low (under 5 percent), but there may be some permanent loss of motor function. The cause is unknown, but because roughly two-thirds of Guillain-Barré patients develop symptoms within two months after a viral infection, it is suspected that the condition may result from a malfunction of the immune system. (The mechanism involved is considered in Chapter 15 of the text; see *p. 400.)*

☤ Drugs and Synaptic Function

Many drugs interfere with key steps in the process of synaptic transmission. These drugs may (1) interfere with transmitter synthesis, (2) alter the rate of transmitter release, (3) prevent transmitter inactivation, or (4) prevent transmitter binding to receptors. The discussion that follows is limited to clinically important compounds that exert their effects at cholinergic synapses. Their sites of activity are indicated in Figure A-23.

Botulinus toxin is responsible for the primary symptom of *botulism,* a widespread paralysis of skeletal muscles. Botulinus toxin blocks the release of ACh at the presynaptic membrane of cholinergic neurons. The venom of the black widow spider has the opposite effect. It causes a massive release of ACh that produces intense muscular cramps and spasms.

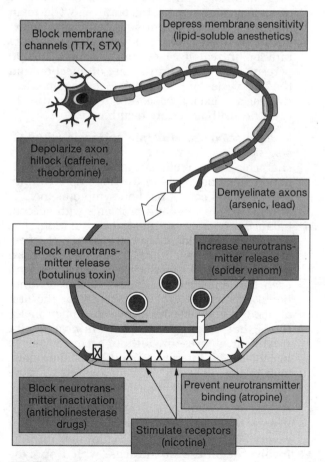

FIGURE A-23
Mechanism of Drug Action at a Cholinergic Synapse. Factors that facilitate neural function and make neurons more excitable are shown in violet. Factors that inhibit or depress neural function are shown in blue.

Anticholinesterase drugs, sometimes called *cholinesterase inhibitors,* block the breakdown of ACh by acetylcholinesterase. The result is an exaggerated and prolonged stimulation of the postsynaptic membrane. At the neuromuscular junctions, this abnormal stimulation produces an extended and extreme state of contraction. Military nerve gases block cholinesterase activity for weeks, although few persons exposed are likely to live long enough to regain normal synaptic function. Most animals utilize ACh as a neurotransmitter, and anticho-linesterase drugs, such as *malathion,* are in widespread use in pest-control projects.

Drugs such as **atropine** or **d-tubocurarine** prevent ACh from binding to the postsynaptic receptors. The latter compound is a derivative of *curare,* a plant extract used by certain South American tribes to paralyze their prey. Curare and related compounds induce paralysis by preventing stimulation of the neuromuscular junction by ACh. Atropine can also be administered intentionally to counteract the effects of anticholinesterase poisoning. Other compounds, including **nicotine,** an active ingredient in cigarette smoke, bind to the receptor sites and stimulate the postsynaptic membrane. There are no enzymes to remove these compounds, and the effects are relatively prolonged.

Table A-14 provides additional information on specific chemical compounds, their uses, and their sites of action.

℞ Spinal Anesthesia   EAP *p. 210*

Injecting a local anesthetic around a nerve produces a temporary blockage of nerve function. This can be done peripherally, as when sewing up skin lacerations, or at sites around the spinal cord to obtain more widespread anesthetic effects. Although an *epidural block,* the injection of an anesthetic into the epidural space, has the advantage of affecting only the spinal nerves in the immediate area of the injection, epidural anesthesia may be difficult to achieve in the upper cervical, midthoracic, and lumbar regions, where the epidural space is extremely narrow. **Caudal anesthesia** involves the introduction of anesthetics into the epidural space of the sacrum. Injection at this site paralyzes lower abdominal and perineal structures. Caudal anesthesia may be used instead of epidural blocks in the lower lumbar or sacral regions to control pain during childbirth.

Local anesthetics may also be introduced into the subarachnoid space of the spinal cord. However, the effects spread as CSF circulation and diffusion distributes the anesthetic along the spinal cord. As a result, precise control of the regional effects can be difficult to achieve. Problems with overdosing are seldom serious, because the diaphragmatic breathing muscles are controlled by upper cervical spinal nerves. Thus respiration continues even when the thoracic and abdominal segments have been paralyzed.

⚕ Epidural and Subdural Hemorrhages
EAP *p. 210*

The most common cases of epidural bleeding, or **epidural hemorrhage,** involve an arterial break. The arterial blood pressure usually forces considerable quantities of blood into the epidural space, distorting the underlying soft tissues of the brain. The individual loses consciousness from minutes to hours after the injury, and death follows in untreated cases.

An epidural hemorrhage involving a damaged vein does not produce massive symptoms immediately, and the individual may not develop symptoms until several hours to several days or even weeks after the original incident. Consequently, the problem may not be noticed until the nervous tissue has been severely damaged by distortion, compression, and secondary hemorrhaging. Epidural hemorrhages are rare, occurring in fewer than 1 percent of head injuries. This is rather fortunate, for the mortality rate is 100 percent in untreated cases and over 50 percent even after removal of the blood pool and closure of the damaged vessels.

In a **subdural hemorrhage** the blood accumulates between the dura and the arachnoid. Subdural hemorrhages are roughly twice as common as epidural hemorrhages. The most common source of blood is a small meningeal vein or one of the dural sinuses. Because the blood pressure is somewhat lower in the venous system, the extent and effects of this condition are more variable than those of the epidural hemorrhages.

⚕ Meningitis
EAP *p. 212*

The warm, dark, nutrient-rich environment of the meninges provides ideal conditions for a variety of bacteria and viruses. Microorganisms that cause meningitis include bacteria associated with middle

Table A-14 Drugs Affecting Acetylcholine Activity at Synapses

Drug	Mechanism	Effects	Remarks
Hemicholinium	Blocks ACh synthesis	Produces symptoms of synaptic fatigue	
Botulinus toxin	Blocks ACh release directly	Paralyzes voluntary muscles	Produced by bacteria; responsible for a deadly type of food poisoning
Barbiturates	Decrease rate of ACh release	Muscular weakness, depression of CNS activity	Administered as sedatives and anesthetics
Procaine (Novocain)	Reduces membrane permeability to sodium	Prevents stimulation of sensory neurons	Used as a local anesthetic
Tetrodotoxin (TTX) Saxitoxin (STX) Ciguatoxin (CTX)	Blocks sodium ion channels	Eliminates production of action potentials	Produced by some marine organisms during normal metabolic activity
Neostigmine	Prevents ACh inactivation by cholinesterase	Sustained contraction of skeletal muscles; other effects on cardiac muscle, smooth muscle, and glands	Used clinically to treat myasthenia gravis and to counteract overdoses of tubocurarine; related compound produced by Calabar bean
Insecticides (malathion, parathion, etc.), and nerve gases	As above	As above	Related compounds used in military nerve gases
d-tubocurarine	Prevents ACh binding to postsynaptic receptor sites	Paralysis of voluntary muscles	Curare produced by South American plant
Nicotine	Binds to ACh receptor sites	Low doses facilitate voluntary muscles; high doses cause paralysis	An active ingredient in cigarette smoke
Succinylcholine	Reduces sensitivity to ACh	Paralysis of voluntary muscles	Used to produce muscular relaxation during surgery
Atropine	Competes with ACh for binding sites on postsynaptic membrane	Reduced heart rate, smooth muscle activity; skeletal muscle weakness develops at high doses	Produced by deadly nightshade plant

ear infections; pneumonia, streptococcal ("strep"), staphylococcal ("staph"), or meningococcal infections; and tuberculosis. These pathogens may gain access to the meninges by traveling within blood vessels or by entering at sites of vertebral or cranial injury. Headache, chills, high fever, disorientation, and rapid heart and respiratory rates appear as higher centers are affected. Without treatment, delirium, coma, convulsions, and death may follow within hours.

The most common clinical assessment involves checking for a "stiff neck" by asking the patient to touch chin to chest. Meningitis affecting the cervical portion of the spinal cord results in a marked increase in the muscle tone of the extensor muscles of the neck. So many motor units become activated that voluntary or involuntary flexion of the neck becomes painfully difficult if not impossible.

The mortality rate for viral and bacterial meningitis ranges from 1 to 50 percent or higher, depending on the type of virus or bacteria, the age and health of the patient, and other factors. There is no effective treatment for viral meningitis, but bacterial meningitis can be combatted with antibiotics and the maintenance of proper fluid and electrolyte balance.

(R) Spinal Cord Injuries and Experimental Treatments EAP *p. 214*

At the outset, any severe injury to the spinal cord produces a period of sensory and motor paralysis termed **spinal shock.** The skeletal muscles become flaccid; neither somatic nor visceral reflexes function; and the brain no longer receives sensations of touch, pain, heat, or cold. The location and severity of the injury determine how long these symptoms persist and how completely the individual recovers.

Violent jolts, such as those associated with blows or gunshot wounds near the spinal cord, may cause **spinal concussion** without visibly damaging the spinal cord. Spinal concussion produces a period of spinal shock, but the symptoms are only temporary and recovery may be complete in a matter of hours. More serious injuries, such as vertebral fractures, usually involve physical damage to the spinal cord. In a **spinal contusion** hemorrhages occur in the meninges, pressure rises in the cerebrospinal fluid, and the white matter of the spinal cord may degenerate at the site of injury. Gradual recovery over a period of weeks may leave some functional losses. Recovery from a **spinal laceration** by vertebral fragments or other foreign bodies will usually be far slower and less complete. **Spinal compression** occurs when the spinal cord becomes physically squeezed or distorted within the vertebral canal. In a **spinal transection** the spinal cord is completely severed. At present surgical procedures cannot repair a severed spinal cord, but experimental techniques may restore partial function.

Spinal cord injuries often involve some combination of compression, laceration, contusion, and partial transection. Relieving pressure and stabilizing the affected area through surgery may prevent further damage and allow the injured spinal cord to recover as much as possible.

Two avenues of research are being pursued, one biological and the other electronic. A major biological line of investigation involves the biochemical control of nerve growth and regeneration. Neurons are influenced by a combination of growth promoters and growth inhibitors. Damaged myelin sheaths apparently release an inhibitory factor that slows the repair process. Researchers have made an antibody, *IN-1*, that will inactivate the inhibitory factor released in the damaged spinal cords of rats. The treatment stimulates repairs, even in severed spinal cords.

Nerve growth and regeneration is stimulated by a variety of recently identified chemicals. A partial listing includes **nerve growth factor** (NGF), **brain-derived neutrophic factor** (BDNF), **neurotrophin-3** (NT-3), **neurotrophin-4** (NT-4), **glial growth factor, glial maturation factor, ciliary neurotrophic factor,** and **growth-associated protein 43** (GAP-43). Many of these factors have now been synthesized using gene-splicing techniques, and sufficient quantities are available to permit their use in experiments on humans and other mammals. Initial results are promising, and these factors in various combination are being evaluated for treatment of CNS injuries and the chronic degeneration seen in Alzheimer's disease and Parkinson's disease.

In the meantime, other research teams are experimenting with the use of computers to stimulate specific muscles and muscle groups electrically. The technique is called *functional electrical stimulation,* or FES. This approach often involves implanting a network of wires beneath the skin with their tips in skeletal muscle tissue. The wires are connected to a computer small enough to be worn at the waist. The wires deliver minute electrical stimuli to the muscles, depolarizing their membranes and causing contractions. With this equipment and lightweight braces, quadriplegics have walked several hundred yards and paraplegics several thousand. The Parastep™ system, which uses a microcomputer controller, is now undergoing clinical trials.

Equally impressive results have been obtained using a network of wires woven into the fabric of close-fitting garments. This provides the necessary stimulation without the complications and maintenance problems that accompany implanted wires. A paraplegic woman in a set of electronic "hot pants" completed several miles of the 1985 Honolulu Marathon, and more recently a paraplegic woman walked down the aisle at her wedding.

Such technological solutions can provide only a degree of motor control without accompanying sensation. Everyone would prefer a biological procedure that would restore the functional integrity of the nervous system. For now, however, computer-assisted programs such as FES can improve the quality of life for thousands of paralyzed individuals.

☤ Hydrocephalus

EAP *p. 217*

The adult brain is surrounded by the inflexible bones of the cranium. The cranial cavity contains two fluids, blood and cerebrospinal fluid (CSF), and the relatively firm tissues of the brain. Because the total volume cannot change, when the volume of blood or CSF increases, the volume of the brain must decrease. In a subdural or epidural hemorrhage the fluid volume increases as blood collects within the cranial cavity. The rising **intracranial pressure** compresses the brain, leading to neural dysfunction that often ends in unconsciousness and death.

Any alteration in the rate of cerebrospinal fluid production is normally matched by an increase in the rate of removal at the arachnoid villi. If this equilibrium is disturbed, clinical problems appear as the intracranial pressure changes. The volume of cerebrospinal fluid will increase if the rate of formation accelerates or the rate of removal decreases. In either event the increased fluid volume leads to compression and distortion of the brain. Increased rates of formation may accompany head injuries, but the most common problems arise from masses, such as tumors or abscesses, or from developmental abnormalities. These conditions have the same effect: they restrict the normal circulation and reabsorption of CSF. Because CSF production continues, the ventricles gradually expand, distorting the surrounding neural tissues and causing the deterioration of brain function.

Infants are especially sensitive to alterations in CSF volume that increase intracranial pressure, because the arachnoid villi which reabsorb CSF, do not appear until roughly 3 years of age. As in an adult, if intracranial pressure becomes abnormally high, the ventricles will expand. But in an infant the cranial sutures have yet to fuse, and the skull can enlarge to accommodate the extra fluid volume. This can produce an enormously expanded skull, a condition called **hydrocephalus,** or "water on the brain." Infant hydrocephalus (Figure A-24) often results from some interference with normal CSF circulation, such as blockage of the mesencephalic aqueduct or constriction of the connection between the subarachnoid spaces of the cranial and spinal meninges. Untreated infants often suffer some degree of mental retardation. Successful treatment usually involves the installation of a **shunt,** a bypass that either bypasses the blockage site or drains the excess cerebrospinal fluid. In either case, the goal is reduction of the intracranial pressure. The shunt may be removed if (1) further growth of the brain eliminates the blockage or (2) the intracranial

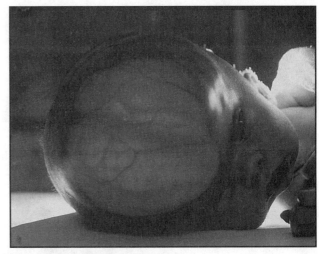

Figure A-24   Hydrocephalus.
This infant suffers from hydrocephalus, a condition usually caused by impaired circulation and removal of cerebrospinal fluid. CSF buildup in infancy leads to distortion of the brain and enlargement of the cranium.

pressure decreases following the development of the arachnoid villi at 3 years of age.

📋 Lumbar Puncture and Myelography

EAP *p. 217*

Tissue samples, or biopsies, are taken from many organs to assist in diagnosis. For example, when a liver or skin disorder is suspected, small plugs of tissue are removed and examined for signs of infection or cell damage, or used to identify the bacteria causing an infection. Unlike many other tissues, however, neural tissue consists largely of cells rather than extracellular fluids or fibers. Tissue samples are seldom removed for analysis because any extracted or damaged neurons will not be replaced. Instead, small volumes of cerebrospinal fluid (CSF) are extracted via a spinal tap and analyzed. CSF is intimately associated with the neural tissue of the CNS, and pathogens, cell debris, or metabolic wastes in the CNS will therefore be detectable in the CSF.

With the vertebral column flexed, a needle can be inserted between the lower lumbar vertebrae and into the subarachnoid spaces with minimal risk to the cauda equina. This procedure, known as a **lumbar puncture,** can be used to remove 3-9 ml of CSF from the subarachnoid space or to introduce anesthetic drugs.

Myelography involves the introduction of radiopaque dyes into the CSF of the subarachnoid space. Because the dyes are opaque to X-rays, the CSF appears white on an X-ray photograph, as in Figure A-25. Any tumors, inflammations, or adhesions that distort or divert CSF circulation will be shown in silhouette.

In the event of severe infection, inflammation, or leukemia (cancer of the white blood cells), antibiotics, steroids, or anticancer drugs can be injected into the subarachnoid space.

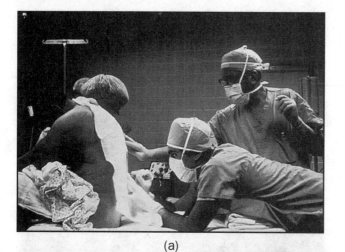

(a)

(b)

Figure A-25   Myelography.
(a) Position and procedure used to introduce spinal anesthetic or radiopaque dye. (b) A myelogram—an X-ray photograph of the spinal cord after introduction of a radiopaque dye into the CSF—showing the cauda equina in the lower lumbar region.

✝ Amnesia   EAP *p. 220*

Amnesia may occur suddenly or progressively, and recovery may be complete, partial, or nonexistent, depending on the nature of the problem. In **retrograde amnesia** *(retro-,* behind), the individual loses memories of past events. Some degree of retrograde amnesia often follows a head injury, and accident victims are frequently unable to remember the moments preceding a car wreck. In **anterograde amnesia** *(antero-,* ahead), an individual may be unable to store additional memories, but earlier memories are intact and accessible. The problem appears to involve an inability to generate long-term memories. At least two drugs—*diazepam (Valium)* and *Halcion*—have been known to cause brief periods of anterograde amnesia. A person with permanent anterograde amnesia lives in surroundings that are always new. Magazines can be read, chuckled over, and then reread a few minutes

later with equal pleasure, as if they had never been seen before. Physicians and nurses must introduce themselves at every meeting, even if they have been visiting the patient for years.

Posttraumatic amnesia (PTA) often develops after a head injury. The duration of the amnesia varies depending on the severity of the injury. PTA combines the characteristics of retrograde and anterograde amnesia; the individual can neither remember the past nor consolidate memories of the present.

✝ Seizures and Epilepsy   EAP *p. 222*

A *seizure* is a temporary disorder of cerebral function, accompanied by abnormal, involuntary movements, unusual sensations, and/or inappropriate behavior. The individual may or may not lose consciousness for the duration of the attack. There are many different types of seizures. Clinical conditions characterized by seizures are known as *seizure disorders,* or *epilepsies.* The term **epilepsy** refers to more than 40 different conditions characterized by a recurring pattern of seizures over extended periods. In roughly 75 percent of patients, no obvious cause can be determined.

Seizures of all kinds are accompanied by a marked change in the pattern of electrical activity monitored in an electroencephalogram. The alteration begins in one portion of the cerebral cortex but may subsequently spread to adjacent regions, potentially involving the entire cortical surface. The neurons at the site of origin are abnormally sensitive. When they become active, they may facilitate and subsequently stimulate adjacent neurons. As a result, the abnormal electrical activity can spread across the entire cerebral cortex.

The extent of the cortical involvement determines the nature of the observed symptoms. A **focal seizure** affects a relatively restricted cortical area, producing sensory and/or motor symptoms. The individual usually remains conscious throughout the attack. If the seizure occurs within a portion of the primary motor cortex, the activation of pyramidal cells will produce uncontrollable movements. The muscles affected or the specific sensations experienced provide an indication of the precise region involved. In a **temporal lobe seizure** the disturbance spreads to the sensory cortex and association areas, so the individual also experiences unusual memories, sights, smells, or sounds. Involvement of the limbic system may also produce sudden emotional changes. Often the individual will lose consciousness at some point during the incident.

Convulsive seizures are associated with uncontrolled muscle contractions. In a **generalized seizure** the entire cortical surface is involved. Generalized seizures may range from prolonged, major events to brief, almost unnoticed incidents. Only two examples will be considered here, grand mal and petit mal seizures.

Most readers will think of an epileptic attack as involving powerful, uncoordinated muscular contractions affecting the face, eyes, and limbs. These are symptoms of a **grand mal seizure.** During a grand mal attack the cortical activation begins at a single focus and then spreads across the entire surface. There may be no warning, but some individuals experience a vague apprehension or awareness that a seizure is about to begin. There follows a sudden loss of consciousness, and the individual drops to the floor as major muscle groups go into tonic contraction. The body remains rigid for several seconds before a rhythmic series of contractions occurs in the limb muscles. Incontinence may occur. After the attack subsides, the individual may appear disoriented or sleep for several hours. Muscles or bones subjected to extreme stresses may be damaged, and the person will probably be rather sore for days after the incident.

Petit mal epileptic attacks are very brief (under 10 seconds in duration) and involve few motor abnormalities. Typically the individual simply loses consciousness suddenly, with no warning. It is as if an internal switch were thrown and the conscious mind turned off. Because the individual is "not there" for brief periods during petit mal attacks, the incidents are known as *absence seizures.* During the seizure there may be small motor activities, such as fluttering of the eyelids or trembling of the hands.

Petit mal attacks usually begin between ages 6 and 14. They can occur hundreds of times per day, so that the child lives each day in small segments separated by blank periods. The victim is aware of brief losses of consciousness that occur without warning, but seldom seeks help because of embarrassment. Often he or she becomes extremely anxious about the timing of future attacks. However, the motor signs are so minor as to go completely unnoticed by other family members, and the psychological stress caused by this condition is often overlooked. The initial diagnosis is frequently made during counseling for learning problems. (You have probably taken an exam after missing 1 or 2 lectures of 20. Imagine taking an exam after missing every third minute of every lecture.)

Both petit mal and grand mal epilepsy can be treated with barbiturates or other anticonvulsive drugs, such as phenytoin sodium *(dilantin).*

✢ The Cerebral Nuclei and Parkinson's Disease   EAP *p. 223*

The cerebral nuclei contain two discrete populations of neurons. One group stimulates motor neurons by releasing ACh, and the other inhibits motor neurons by the release of GABA. Under normal conditions the excitatory neurons remain inactive, and the descending tracts are primarily responsible for inhibiting motor neuron activity. If the descending tracts are severed in an accident, the loss of inhibitory control leads to a generalized state of muscular contraction known as **decerebrate rigidity.**

The excitatory neurons are quiet because they are continually exposed to the inhibitory effects of the neurotransmitter dopamine. This compound is manufactured by neurons in the substantia nigra and carried by axoplasmic flow to synapses in the cerebral nuclei. If the ascending tract or the dopamine-producing neurons are damaged, this inhibition is lost, and the excitatory neurons become increasingly active. This increased activity produces the motor symptoms of **Parkinson's disease,** or *paralysis agitans.*

Parkinson's disease is characterized by a pronounced increase in muscle tone. Voluntary movements become hesitant and jerky, a condition called **spasticity,** for a movement cannot occur until one muscle group manages to overpower its antagonists. Individuals with Parkinson's disease show spasticity during voluntary movement and a continual **tremor** when at rest. A tremor represents a tug of war between antagonistic muscle groups that produces a background shaking of the limbs, in this case at a frequency of 4-6 cycles per second. Individuals with Parkinson's disease also have difficulty starting voluntary movements. Even changing one's facial expression requires intense concentration, and the individual acquires a blank, static expression. Finally, the positioning and preparatory adjustments normally performed automatically no longer occur. Every aspect of each movement must be voluntarily controlled, and the extra effort requires intense concentration that may prove tiring and extremely frustrating. In the late stages of this condition, other CNS effects, such as depression, hallucinations, and dementia often appear.

Providing the cerebral nuclei with dopamine can significantly reduce the symptoms for two-thirds of Parkinson's patients, but intravenous dopamine injection is not effective because the molecule cannot cross the blood-brain barrier. The most common procedure involves the oral administration of the drug L-DOPA (levodopa), a related compound that crosses the capillaries and is then converted to dopamine. Unfortunately, it appears that with repeated treatment, the capillaries become less permeable to L-DOPA, and so the required dosage increases. The effectiveness of L-DOPA can be increased by giving it in combination with other drugs, such as *Amantadine* or *Bromocriptine.* Amantadine accelerates dopamine release at synaptic terminals, and Bromocriptine, a dopamine agonist, stimulates dopamine receptors on postsynaptic membranes.

Surgery to control Parkinson's symptoms focuses on the destruction of large areas within the cerebral nuclei or thalamus to control the motor symptoms of tremor and rigidity. The high rate of success for drug therapy has greatly reduced the number of surgical procedures. Recent attempts to transplant tissues producing dopamine or related compounds into the cerebral nuclei have met with limited success. Variable results have been obtained with the transplantation of tissue from the adrenal gland. The transplantation of fetal tissue into adult brains has been more successful, and in 1993 three different research groups reported relatively long-term (up to three years after surgery) improvement in motor skills.

Individuals with Parkinson's disease are usually elderly. However, since 1983 an increasing number of young people have developed this condition. In that year a drug appeared on the streets rumored to be "synthetic heroin." In addition to the compound that produced the "high" sought by users, the drug contained several contaminants, including a complex molecule with the abbreviated name **MPTP.** This accidental byproduct of the synthetic process destroys neurons of the substantia nigra, eliminating the manufacture and transport of dopamine to the cerebral nuclei. As a result of exposure to this drug, approximately 200 young, healthy adults have developed symptoms of severe Parkinson's disease. Why MPTP targets these particular neurons, and not all of the CNS neurons that produce dopamine, remains a mystery.

☤ Cerebellar Dysfunction   EAP *p. 000*

Cerebellar function may be permanently altered by trauma or a stroke, or temporarily by drugs such as alcohol. Such alterations can produce disturbances in motor control. In severe ataxia balance problems are so great that the individual cannot sit or stand upright. Less severe conditions cause an obvious unsteadiness and irregular patterns of movement. The individual often watches his or her feet to see where they are going and controls ongoing movements by intense concentration and voluntary effort. Reaching for something becomes a major exertion, for the only information available must be gathered by sight or touch while the movement is taking place. Without the cerebellar ability to adjust movements while they are occurring, the individual becomes unable to anticipate the time course of a movement. Most often, a reaching movement ends with the hand overshooting the target. This inability to anticipate and stop a movement precisely is called **dysmetria** (dis-MET-rē-a; *dys-,* bad + *metron,* measure). In attempting to correct the situation, the hand usually overshoots again in the opposite direction, and then again. This leaves the hand oscillating back and forth until either the object can be grasped or the attempt is abandoned. This oscillatory movement is known as an **intention tremor.**

Clinicians check for ataxia by watching an individual walk in a straight line; the usual test for dysmetria involves touching the tip of the index finger to the tip of the nose. Because many drugs impair cerebellar performance, the same tests are used by police officers to check drivers suspected of alcohol or other drug abuse.

☤ Shingles and Hansen's Disease

EAP *p. 233*

In **shingles,** or *Herpes zoster,* the *Herpes varicella-zoster* virus attacks neurons within the dorsal roots of spinal nerves and sensory ganglia of cranial nerves. This disorder produces a painful rash whose distribution corresponds to that of the affected sensory nerves. Shingles develops in adults who were first exposed to the virus as children. The initial infection produces symptoms known as chicken pox. After this encounter the virus remains dormant within neurons of the anterior gray horns of the spinal cord. It is not known what triggers reactivation of this pathogen. Fortunately for those affected, attacks of shingles usually heal and leave behind only unpleasant memories.

Most people suffer only a single episode of shingles in their adult lives. However, the problem may recur in people with weakened immune systems, including those with AIDS or some forms of cancer. Treatment typically involves large doses of the antiviral drug *acyclovir (Zovirax).*

The condition traditionally called **leprosy,** now more commonly known as **Hansen's disease,** is an infectious disease caused by a bacterium, Mycobacterium leprae. It is a disease that progresses slowly, and symptoms may not appear for up to 30 years after infection. The bacterium invades peripheral nerves, especially those in the skin, producing initial sensory losses. Over time motor paralysis develops, and the combination of sensory and motor loss can lead to recurring injuries and infections. The eyes, nose, hands, and feet may develop deformities as a result of neglected injuries. There are several forms of this disease; peripheral nerves are always affected, but some forms also involve extensive skin and mucous membrane lesions.

Only about 5 percent of those exposed develop symptoms; people living in the tropics are at greatest risk. There are about 2,000 cases in the United States, and an estimated 12-20 million cases worldwide. If detected before deformities occur, the disease can usually be treated successfully with drugs such as rifampin and dapsone. Treated individuals are not infectious, and the practice of confining "lepers" in isolated compounds has been discontinued.

☤ Multiple Sclerosis   EAP *p. 233*

Multiple sclerosis (MS), introduced in the discussion of demyelination disorders, is a disease that produces muscular paralysis and sensory losses through demyelination. The initial symptoms appear as the result of myelin degeneration within the white matter of the lateral and posterior columns of the spinal cord or along tracts within the brain. For example, spinal cord involvement may produce weakness, tingling sensations, and a loss of "position sense" for the limbs. During subsequent attacks the effects become more widespread, and the cumulative sensory and motor losses may eventually lead to a generalized muscular paralysis.

Recent evidence suggests that this condition may be linked to a defect in the immune system that causes it to attack myelin sheaths. MS patients have lymphocytes that do not respond normally to foreign proteins, and because several viral proteins have amino acid sequences similar to those of normal myelin, it has been proposed that MS results from a case of mistaken identity. For unknown reasons MS appears to be associated with cold and temperate climates. It has been suggested that individuals developing MS may have an inherited susceptibility to the virus that is exaggerated by environmental conditions. The yearly incidence within the United States averages around 50 cases for every 100,000 in the population. Improvement has been noted in some patients treated with *interferon*, a peptide secreted by cells of the immune system, and recently corticosteroid treatment has been linked to a slowdown in the progression of MS.

⚕ Palsies
EAP *p. 233*

Peripheral nerve palsies, or peripheral neuropathies, are characterized by regional losses of sensory and motor function as the result of nerve trauma or compression. **Brachial palsies** result from injuries to the brachial plexus or its branches. **Crural palsies** involve the nerves of the lumbosacral plexus.

Although palsies may appear for several reasons, the pressure palsies are especially interesting. A familiar but mild example is the experience of having an arm or leg "fall asleep." The limb becomes numb, and afterwards an uncomfortable "pins-and-needles" sensation, or **paresthesia,** accompanies the return to normal function.

These incidents are seldom of clinical significance, but they provide graphic examples of the effects of more serious palsies that can last for days to months. In **radial nerve palsy,** pressure on the back of the arm interrupts the function of the radial nerve, so that the extensors of the wrist and fingers are paralyzed. This condition is also known as "Saturday night palsy," for falling asleep on a couch with your arm over the seat back (or beneath someone's head) can produce the right combination of pressures. Students may also be familiar with **ulnar palsy,** which can result from prolonged contact between elbow and desk. The ring and little fingers lose sensation, and the fingers cannot be adducted.

Men with large wallets in their hip pockets may develop symptoms of **sciatic compression** after driving or sitting in one position for extended periods. As nerve function declines, the individuals notice some lumbar or glutial pain, a numbness along the back of the leg, and a weakness in the leg muscles. Similar symptoms result from compression of the sciatic nerve by a distorted lumbar intervertebral disc. This condition is termed **sciatica,** and one or both legs may be affected, depending on the site of compression. Finally, sitting with your legs crossed may produce symptoms of a **peroneal palsy.** Sensory losses from the top of the foot and side of the leg are accompanied by a decreased ability to dorsiflex or evert the foot.

📋 Reflexes and Diagnostic Testing
EAP *p. 236*

Many reflexes can be assessed through careful observation and the use of simple tools. The procedures are easy to perform, and the results can provide valuable information about damage to the spinal cord or spinal nerves. By testing a series of spinal and cranial reflexes, a physician can assess the function of sensory pathways and motor centers throughout the spinal cord and brain.

Neurologists test many different reflexes; only a few are so generally useful that physicians make them part of a standard physical examination. These reflexes are shown in Figure A-26.

The *jerk* (Figure A-26a), *biceps reflex* (Figure A-26b), and *triceps reflex* (Figure A-26c) are stretch reflexes controlled by specific segments of the spinal cord. Testing these reflexes provides information about the corresponding spinal segments. For example, a normal patellar reflex, or knee jerk, indicates that spinal nerves and spinal segments L_2-L_4 are undamaged. The **abdominal reflex** (Figure A-26d), present in the normal adult, results from descending spinal facilitation. In this reflex, a light stroking of the skin produces a reflexive twitch in the abdominal muscles that moves the navel toward the stimulus. This reflex disappears following damage to descending tracts.

📋 ABNORMAL REFLEX ACTIVITY
EAP *p. 236*

In **hyporeflexia** normal reflexes are weak, but apparent, especially with reinforcement. In **areflexia** (ā-rē-FLEK-sē-a; *a-*, without) normal reflexes fail to appear, even with reinforcement. Hyporeflexia or areflexia may indicate temporary or permanent damage to skeletal muscles, dorsal or ventral nerve roots, spinal nerves, the spinal cord, or the brain.

Hyperreflexia occurs when higher centers maintain a high degree of facilitation along the spinal cord. Under these conditions reflexes are easily triggered, and the responses may be gross-

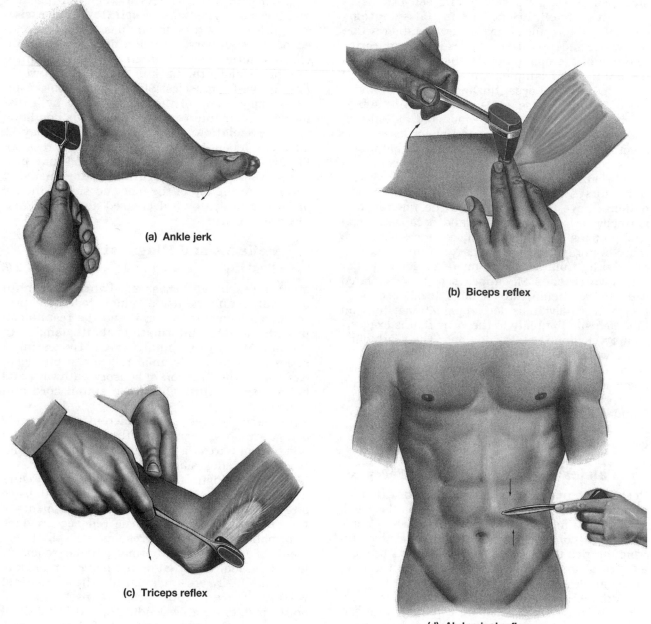

(a) Ankle jerk

(b) Biceps reflex

(c) Triceps reflex

(d) Abdominal reflex

Figure A-24   Reflexes and Diagnostic Testing.

ly exaggerated. This effect can also result from spinal cord compression or diseases that target higher centers or descending tracts. One potential result of hyperreflexia is the appearance of alternating contractions in opposing muscles. When one muscle contracts, it stimulates the stretch receptors in the other. The stretch reflex then triggers a contraction in that muscle, and this stretches receptors in the original muscle. This self-perpetuating sequence, which can be repeated indefinitely, is called **clonus** (KLŌ-nus). In a hyperreflexive person, a tap on the patellar tendon will set up a cycle of kicks, rather than just one or two.

A more extreme hyperreflexia develops if the motor neurons of the spinal cord lose contact with higher centers. Often, following a severe spinal injury, the individual first experiences a temporary period of areflexia known as spinal shock, discussed on p. 68. When the reflexes return, they respond in an exaggerated fashion, even to mild stimuli. For example, the lightest touch on the skin surface may produce a massive withdrawal reflex. The reflex contractions may occur in a series of intense muscle spasms potentially strong enough to break bones. In the **mass reflex** the entire spinal cord becomes hyperactive for several minutes, issuing exaggerated skeletal muscle and visceral motor commands.

☤ Huntington's Disease   EAP *p. 238*

Huntington's disease is an inherited disease marked by a progressive deterioration of mental abilities. There are approximately 25,000 Americans with this condition. In Huntington's disease the cerebral nuclei show degenerative changes, as do the frontal lobes of the cerebral cortex. The basic problem is the destruction of ACh-secreting and GABA-secreting neurons in the cerebral nuclei. The cause of this deterioration is not known. The first signs of the disease usually appear in early adulthood. As you would expect in view of the areas affected, the symptoms involve difficulties in performing voluntary and involuntary patterns of movement and a gradual decline in intellectual abilities leading eventually to dementia and death.

Screening tests can now detect the presence of the gene for Huntington's disease, which is an autosomal dominant gene located on chromosome 4. In people with Huntington's disease, a gene of uncertain function contains a variable number of repetitions of the nucleotide sequence CAG. This DNA segment appears to be unstable, and the number of repetitions can change from generation to generation. The duplication or deletion is thought to occur during gamete formation. The larger the number of repetitions, the earlier in life the symptoms appear, and more severe the symptoms. The link between the multiple copies of the CAG nucleotide and the disorder has yet to be understood. There is no effective treatment. A victim's children have a 50% risk of receiving the gene and developing Huntington's disease.

☤ Alzheimer's Disease   EAP *p. 244*

In its characteristic form, Alzheimer's disease produces a gradual deterioration of mental organization. The afflicted individual loses memories, verbal and reading skills, and emotional control. Initial symptoms are subtle—moodiness, irritability, depression, and a general lack of energy. These symptoms are often ignored, overlooked, or dismissed. Elderly relations are viewed as "eccentric" or "irrascible," and humored whenever possible.

As the condition progresses, however, it becomes more difficult to ignore or accommodate. The victim has difficulty making decisions, even minor ones. Mistakes—sometimes dangerous ones—are made, either through bad judgment or simple forgetfulness. For example, the person might decide to make dinner, light the gas burner, place a pot on the stove top, and go into the living room. Two hours later, the pot, still on the stove, melts into a shapeless blob and starts a fire that destroys the house.

As memory losses continue, the problems become more severe. The affected person may for-get relatives, her home address, or how to use the telephone. The memory loss often starts with an inability to store long-term memories, followed by the loss of recently stored memories, and eventually the loss of basic long-term memories, such as the sound of the victim's own name. The loss of memory affects both intellectual and motor abilities, and a patient with severe Alzheimer's disease has difficulty in performing even the simplest motor tasks. Although by this time victims are relatively unconcerned about their mental state or motor abilities, the condition can have devastating emotional effects on the immediate family.

Individuals with Alzheimer's disease show a pronounced decrease in the number of cortical neurons, especially in the frontal and temporal lobes. This loss is correlated with inadequate ACh production in the *nucleus basalis* of the cerebrum. Axons leaving this region project throughout the cerebral cortex, and when ACh production declines, cortical function deteriorates.

Most cases of Alzheimer's disease are associated with unusually large concentrations of plaques and neurofibrillary tangles in the nucleus basalis, hippocampus, and parahippocampal gyrus. In addition, an abnormal protein, called **Alzheimer's disease associated protein** (ADAP) appears in brain regions, such as the hippocampus, specifically associated with memory processing. Because this protein also appears in small quantities in the cerebrospinal fluid of many Alzheimer's patients, a blood screening test is now being developed to detect the condition before mental deterioration becomes pronounced.

📋 PROBLEMS WITH SENSORY SYSTEMS   EAP *p. 250*

A recurring theme of the text is that an understanding of how a system works enables you to predict how things might go wrong. You are already familiar with the organization and physiology of sensory systems, and some of the most important clinical problems were discussed in clinical comments on the preceding pages. Placing the entire array into categories provides an excellent example of a strategy that can be used to analyze any system in the body.

Every sensory system contains peripheral receptors, afferent fibers, ascending tracts, nuclei, and areas of the cerebral cortex. Any malfunction affecting the system must involve one of those components. Any clinical diagnosis requires seeking answers to a series of yes or no questions, eliminating one possibility at a time until the nature of the problem becomes apparent. Figure A-27 organizes the disorders considered in this chapter into a "trouble-shooting" format similar to that used to diagnose problems with automobiles or other mechanical devices.

9

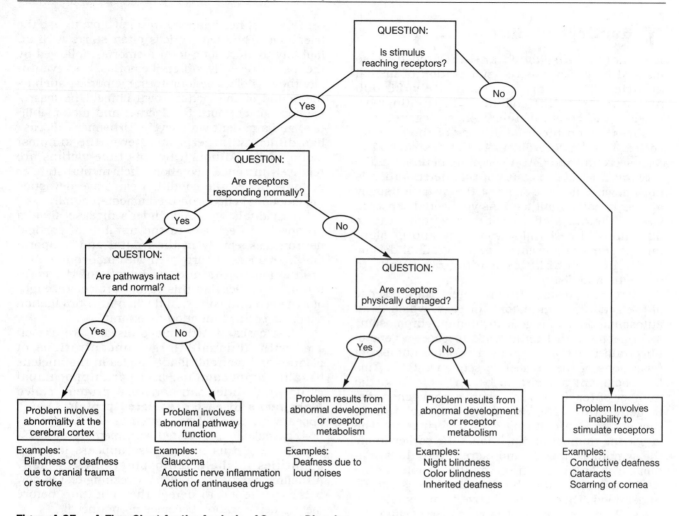

Figure A-27   A Flow Chart for the Analysis of Sensory Disorders

℞ The Control of Pain

EAP p. 251

Pain management poses a number of problems for clinicians. Painful sensations can result from tissue damage or sensory nerve irritation; it may originate where it is perceived, be referred from another location, or represent a false signal generated along the sensory pathway (see Figure A-2, p. 11). The treatment differs in each case, and an accurate diagnosis is an essential first step.

When pain results from tissue damage, the most effective solution is to stop the damage and end the stimulation. This is not always possible. Alternatively, the painful sensations can be suppressed at the injury site. Topical or locally injected anesthetics inactivate nociceptors in the immediate area. Aspirin and related analgesics reduce inflammation and suppress the release of irritating chemicals, such as enzymes or prostaglandins, in damaged tissues.

Pain can also be suppressed by inhibition of the pain pathway. Analgesics related to morphine reduce pain by mimicking the action of endorphins. Surgical steps can be taken to control severe pain, including (1) the sensory innervation of an area can be destroyed by an electric current, (2) the dorsal roots carrying the painful sensations can be cut (a *rhizotomy*), (3) the ascending tracts in the spinal cord can be severed (a *tractotomy*), or (4) thalamic or limbic centers can be stimulated or destroyed. These options, listed in order of increasing degree of effect, surgical complexity, and associated risk, are used only when other methods of pain control have failed to provide relief.

The Chinese technique of acupuncture to control pain has recently received considerable attention. Fine needles are inserted at specific locations and are either heated or twirled by the therapist. Several theories have been proposed to account for the positive effects, but none is widely accepted. It has been suggested that the pain relief may follow endorphin release, but it is not known how acupuncture stimulates endorphin release; the acupuncture points do not correspond to the distribution of any of the major peripheral nerves.

Many other aspects of pain generation and control remain a mystery. Up to 30 percent of patients experience a significant reduction in pain after receiving a nonfunctional medication. It has been suggested that this "placebo effect" results from endorphin release triggered by the expectation of pain relief. Although the medication has no direct effect, the indirect effect is quite significant.

Assessment of Tactile Sensitivities

EAP p. 252

Regional sensitivity to light touch can be checked by gentle contact with a fingertip or a slender wisp of cotton. The **two-point discrimination test** provides a more detailed sensory map for tactile receptors. Two fine points of a drawing compass, bent paper clip, or other object are applied to the skin surface simultaneously. The subject then describes the contact. When the points fall within a single receptive field, the individual will report only one point of contact. A normal individual loses two-point discrimination at 1 mm (0.04 in.) on the surface of the tongue, at 2-3 mm (0.08-0.12 in.) on the lips, at 3-5 mm (0.12-0.20 in.) on the backs of the hands and feet, and at 4-7 cm (1.6-2.75 in.) over the general body surface.

Vibration receptors are tested by applying the base of a tuning fork to the skin. Damage to an individual spinal nerve produces insensitivity to vibration along the paths of the related sensory nerves. If the sensory loss results from spinal cord damage, the injury site can often be located by walking the tuning fork down the spinal column, resting its base on the vertebral spines.

Descriptive terms are used to indicate the degree of sensitivity in the area considered. **Anesthesia** implies a total loss of sensation; the individual cannot perceive touch, pressure, pain, or temperature sensations from that area. **Hypesthesia** is a reduction in sensitivity, and **paresthesia** is the presence of abnormal sensations, such as the pins-and-needles sensation when an arm or leg "falls asleep" due to pressure on a peripheral nerve. (Several types of *pressure palsies* producing temporary paresthesia were discussed on p. 73.)

Conjunctivitis

EAP p. 256

Problems with the accessory structures of the eye and ear are more common causes of abnormal sensory function than damage to the receptors or the innervation. The term *conjunctivitis* is more useful as the description of a symptom than as a name for a specific disease. A great variety of pathogens, including bacteria, viruses, and fungi, can cause conjunctivitis, and a temporary form of the condition may be produced by chemical or physical irritation (including even such mundane experiences as prolonged crying or peeling an onion).

Chronic conjunctivitis, or **trachoma,** results from bacterial or viral invasion of the conjunctiva. Trachoma is often highly contagious. Severe cases may disrupt the corneal surface and affect vision. The pathogen most often involved is *Chlamydia trachomatis.* Trachoma is a relatively common problem in southwestern North America, North Africa, and the Middle East. The condition must be treated with topical and systemic antibiotics to prevent scleral damage, eventual corneal damage, and vision loss.

Corneal Transplants

EAP p. 258

The cornea has a very restricted ability to repair itself, so corneal injuries must be treated immediately to prevent serious visual losses. To restore vision after corneal scarring it is usually necessary to replace the cornea through a **corneal transplant**. Corneal replacement is probably the most common form of transplant surgery. Corneal transplants can be performed between unrelated individuals because there are no corneal blood vessels, and white blood cells that would otherwise reject the graft are unlikely to enter the area. Corneal grafts are obtained by posthumous donation; for best results the tissues must be removed within 5 hours after the donor's death.

Glaucoma

EAP p. 261

If aqueous humor cannot enter the canal of Schlemm, the condition of **glaucoma** develops. Although drainage is impaired, production of aqueous humor continues, and the intraocular pressure begins to rise. The fibrous scleral coat cannot expand significantly, so the increasing pressure begins to distort soft tissues within the eye.

The optic nerve is not wrapped in connective tissue, for it penetrates all three tunics. When intraocular pressures have risen to roughly twice normal levels, distortion of the nerve fibers begins to affect visual perception. If this condition is not corrected, blindness eventually results.

Glaucoma affects roughly 2 percent of the population over 35, and in most cases the primary factors responsible cannot be determined. Because it is a relatively common condition—over 2 million cases in the United States alone—most eye exams include a test of intraocular pressure. Glaucoma may be treated by the application of drugs that constrict the pupil and tense the edge of the iris, making the surface more permeable to aqueous humor. Surgical correction involves perforating the wall of the anterior chamber to encourage drainage. This procedure is now performed by laser surgery on an outpatient basis.

Otitis Media and Mastoiditis

EAP p. 268

Otitis media is an infection of the middle ear, most often of bacterial origin. *Acute otitis media* typically affects infants and children, and is occasionally seen in adults. The pathogens usually gain access via the pharyngotympanic tube, usually during an upper respiratory infection. As the pathogen population rises in the tympanic cavity, white blood cells rush to the site, and the middle ear becomes filled with pus. Eventually the tympanum may rupture, producing a characteristic oozing from the external auditory canal. The bacteria can usually be controlled by antibiotics, the pain reduced by analgesics, and the swelling reduced by decongestants. In the United States it is rare for otitis media to progress to the stage at which tympanic rupture occurs.

10

Otitis media is extremely common in underdeveloped countries where medical care and antibiotics are not readily available. Both children and adults in these countries often suffer from *chronic otitis media,* a condition characterized by chronic or recurring bouts of infection. This condition produces scarring or perforation of the tympanic membrane, which leads to some degree of hearing loss. Resulting damage to the inner ear or the auditory ossicles may further reduce auditory sensitivity.

If the pathogens leave the middle ear and invade the air cells within the mastoid process, **mastoiditis** develops. The connecting passageways are very narrow, and as the infection progresses, the subject experiences severe earaches, fever, and swelling behind the ear in addition to symptoms of otitis media. The same antibiotic treatment is used to deal with both conditions, the particular antibiotic selected depending on the identity of the bacterium involved. The major risk of mastoiditis is the spread of the infection to the brain via the connective tissue sheath of the facial nerve (N VII). Should this occur, prompt antibiotic therapy is needed, and if the problem remains, the person may have to undergo *mastoidectomy* (opening and drainage of the mastoid sinuses) or *myringotomy* (drainage of the middle ear through a surgical opening in the tympanic membrane).

⚕ Vertigo EAP *p. 271*

The term **vertigo** describes an inappropriate sense of motion. This meaning distinguishes it from "dizziness," a sensation of light-headedness and disorientation that often precedes a fainting spell. Vertigo can result from abnormal conditions in the inner ear or from problems elsewhere along the sensory pathway. It often accompanies CNS infection, and many people experience vertigo when they have high fevers.

Any event that sets endolymph into motion can stimulate the equilibrium receptors and produce vertigo. Placing an ice pack in contact with the temporal bone or flushing the external auditory canal with cold water may chill the endolymph in the outermost portions of the labyrinth and establish a temperature-related circulation of fluid. A mild and temporary vertigo is the result. Consumption of excessive quantities of alchohol and exposure to certain drugs can also produce vertigo by changing the composition of the endolymph or disturbing the hair cells.

Acute vertigo can also result from damage caused by abnormal endolymph production, as in Mèniére's disease. Probably the most common cause of vertigo is *motion sickness.* Motion sickness appears to develop when central processing stations receive conflicting sensory information. When you read in a boat or plane, for example, your eyes (which are tracking lines on a page) report that the book isn't moving, but your inner ear reports that your body is lurching and turning. Why and how these conflicting reports result in nausea, vomiting,

📋 Testing and Treating Hearing Deficits EAP *p. 276*

In the most common hearing test, a subject listens to sounds of varying frequency and intensity generated at irregular intervals. A record is kept of the responses, and the graphed record, or **audiogram,** is compared with that of an individual with normal hearing (see Figure A-28). **Bone conduction tests** are used to discriminate between conductive and nerve deafness. If you put your fingers in your ears and talk quietly, you can still hear yourself because the bones of the skull conduct the sound waves to the cochlea, bypassing the middle ear. In a bone conduction test the physician places a vibrating tuning fork against the skull. If the subject hears the sound of the tuning fork in contact with the skull, but not when held next to the auditory meatus, the problem must lie within the external or middle ear. If the subject remains unresponsive to either stimulus, the problem must be at the receptors or along the auditory pathway.

Several effective treatments exist for conductive deafness. A hearing aid overcomes the loss in sensitivity by simply increasing the intensity of stimulation. Surgery may repair the tympanic

Figure A-28 An Audiogram.

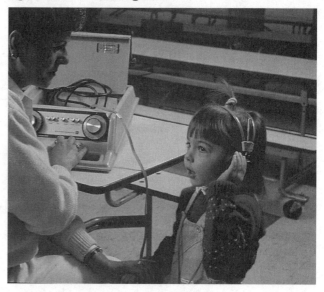

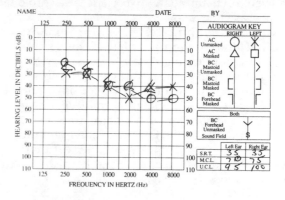

membrane or free damaged or immobilized ossicles. Artificial ossicles may also be implanted if the originals are damaged beyond repair.

There are few possible treatments for nerve deafness. Mild conditions may be overcome by the use of a hearing aid if some functional hair cells remain. In a **cochlear implant** a small battery-powered device is inserted beneath the skin behind the mastoid process. Small wires run through the round window to reach the cochlear nerve, and when the implant "hears" a sound it stimulates the nerve directly. Increasing the number of wires and varying their implantation sites make it possible to create a number of different frequency sensations. Those sensations do not approximate normal hearing because there is as yet no way to target the specific afferent fibers responsible for the perception of a particular sound. Instead, a random assortment of afferent fibers are stimulated, and the individual must learn to recognize the meaning and probable origin of the perceived sound. Although obviously not perfect, improvements in electrode placement and computer processing are being made. There are at present approximately 2,000 people in the United States using cochlear implants.

A new approach involves inducing the regeneration of hair cells of the organ of Corti. Researchers working with other mammals have been able to induce hair cell regeneration both in cultured hair cells and in living animals. This is a very exciting area of research, and there is hope that it may ultimately lead to an effective treatment for human nerve deafness.

CRITICAL THINKING QUESTIONS

4-1. Ten-year-old Christina falls while climbing a tree and lands on her back. Her frightened parents take her to the local emergency room where she is examined. Her knee-jerk reflex is normal, and she exhibits a plantar reflex (negative Babinski reflex). These results suggest that:

    a. Christina has injured one of her descending nerve tracts

    b. Christina has injured one of her ascending nerve tracts

    c. Christina has a spinal injury in the lumbar region

    d. Christina has a spinal injury in the cervical region

    e. Christina has suffered no damage to her spinal cord

4-2. Susan brings her husband Jim, age 32, to the emergency room. Jim has been complaining of a severe headache for the last 12 hours. He has a temperature of 102°F, a "stiff" neck, and complains of pain when moving his chin to chest. Reflexes are normal. A lumbar puncture is performed, and 48 hours later the results of CSF analysis are reported as follows:

Analysis of CSF

Culture: presence of *Streptococcus pneumoniae* (bacteria)

- Pressure of CSF: 201 cm H_2O
- Color of CSF: cloudy
- Glucose: 50 mg/dl
- Protein: 47 mg/dl
- Cell count (lymphocytes): 550 mm^3

What is the likely diagnosis?

    a. brain tumor

    b. meningitis

    c. cerebrovascular accident

    d. multiple sclerosis

4-3. Tapping the calcaneal tendon of a normal individual with a rubber hammer will produce a reflex response. What response would you expect? What type of reflex is this? Describe the steps involved in the reflex. Discuss other reflexes of this type, and what can be learned by reflex testing.

4-4. Mrs. Glenn, 73 years old, has recently had trouble controlling her movements. Even when resting, she has continual, slight tremors and increased muscle tone to the point of rigidity. She visits her physician, who conducts a series of tests, including an assessment of her reflexes. What change or changes would you expect to see in her spinal reflexes?

4-5. Chelsea is mountain climbing with a group of friends when she slips, falls, and bumps the left side of her head on a rock. She gets up slowly, and is dazed but otherwise appears unhurt. She feels able to proceed, and the climb continues. An hour later, Chelsea gets a severe headache and experiences a ringing in her ears. She starts having trouble speaking, and soon loses consciousness. Before medical personnel can reach the scene, Chelsea dies. What was the likely cause of death?

4-6. Dave has a hypothalamic tumor that compresses the right medial surface of the optic chiasm posterior to the decussation. How would this condition affect his vision?

4-7. Mr. Romero, 62 years old, has trouble hearing people during conversations, and his family persuades him to have his hearing tested. How can the physician determine whether Mr. Romero's problem results from nerve deafness or conductive deafness?

The Endocrine System

The endocrine system provides long-term regulation and adjustment of homeostatic mechanisms and a variety of body functions. For example, the endocrine system is responsible for the regulation of fluid and electrolyte balance, cell and tissue metabolism, growth and development, and reproductive functions. The endocrine system also assists the nervous system in responding to stressful stimuli.

The endocrine system is composed of nine major endocrine glands and several other organs, such as the heart and kidneys, that have other important functions. The hormones secreted by these endocrine organs are distributed by the circulatory system to target tissues throughout the body. Each hormone affects a specific set of target tissues that may differ from that of other hormones. The selectivity is based on the presence or absence of hormone-specific receptors in the cell membrane, cytoplasm, or nucleus of the target cells.

Homeostatic regulation of circulating hormone levels primarily involves negative feedback control mechanisms. The feedback loop involves an interplay between the endocrine organ and its target tissues. Release of a particular hormone by an endocrine gland may occur in response to one of three different types of stimuli:

1. Some hormones are released in response to variations in the concentrations of specific substances in the body fluids. Parathyroid hormone, for example, is released when calcium levels decline.

2. Some hormones are released only when the gland cells receive hormonal instructions from other endocrine organs. For example, the rate of production and release of T_3 and T_4 by the thyroid gland is controlled by thyroid stimulating hormone (TSH) from the anterior pituitary gland. The secretion of TSH is in turn regulated by the release of thyrotropic releasing hormone (TRH) from the hypothalamus.

3. Some hormones are released in response to neural stimulation. The release of epinephrine and norepinephrine from the adrenal medulla during sympathetic activation is an example.

Endocrine disorders can therefore develop due to abnormalities in (a) the endocrine gland, (b) the endocrine or neural regulatory mechanisms, or (c) the target tissues. Figure A-29 provides an overview of the major classes of endocrine disorders. In the discussion that follows, we will use the thyroid gland as an example because the text introduces major types of thyroid gland disorders. These *primary disorders* may result in overproduction (hypersecretion) or underproduction (hyposecretion of hormones. For example, clinicians may categorize a thyroid disorder as *primary hyperthyroidism* or *primary hypothyroidism* if the problem originates within the thyroid gland.

1. Most endocrine disorders are the result of problems within the endocrine gland itself. Causes of hyposecretion include the following:

   • *Metabolic factors:* Hyposecretion may result from a deficiency in some key substrate needed to synthesize that hormone. For example, hypothyroidism can be caused by inadequate dietary iodine levels, or exposure to drugs that inhibit iodine transport or utilization at the thyroid gland.

   • *Physical damage:* Any condition that interrupts the normal circulatory supply or that physically damages the endocrine cells in some other way will suppress hormone production temporarily. If the damage is severe, the condition may be permanent. Examples of problems that may cause temporary or permanent hypothyroidism include infection or inflammation of the gland *(thyroiditis),* interruption of normal circulation, and exposure to radiation as part of treatment for cancer of the thyroid or adjacent tissues. The thyroid may also be damaged in an *autoimmune disorder* that results in the production of antibodies that attack and destroy normal follicle cells.

   • *Congenital disorders:* The individual may be unable to produce normal amounts of a particular hormone because (a) the gland itself is too small, (b) the required enzymes are abnormal in some way, or (c) the gland cells lack the receptors normally involved in stimulating secretory activity.

2. Endocrine disorders can also result from problems with other endocrine organs involved in the negative feedback control mechanism. For example:

   • *Secondary hypothyroidism* can be caused by inadequate TSH production at the pituitary gland or by inadequate TRH secretion at the hypothalamus.

   • *Secondary hyperthyroidism* can be caused by excessive TRH or TSH production; secondary hyperthyroidism may develop in individuals with tumors of the pituitary gland.

3. Endocrine abnormalities can also be caused by the presence of abnormal hormonal receptors in target tissues. In this case the gland and the regulatory mechanisms may be normal, but the peripheral cells are unable to respond to the circulating hormone. The best example of this type of abnormality is *Type II diabetes* (maturity-onset, NIDDM), where peripheral cells do not respond normally to insulin. (Maturity-onset diabetes is discussed further on p. 84.)

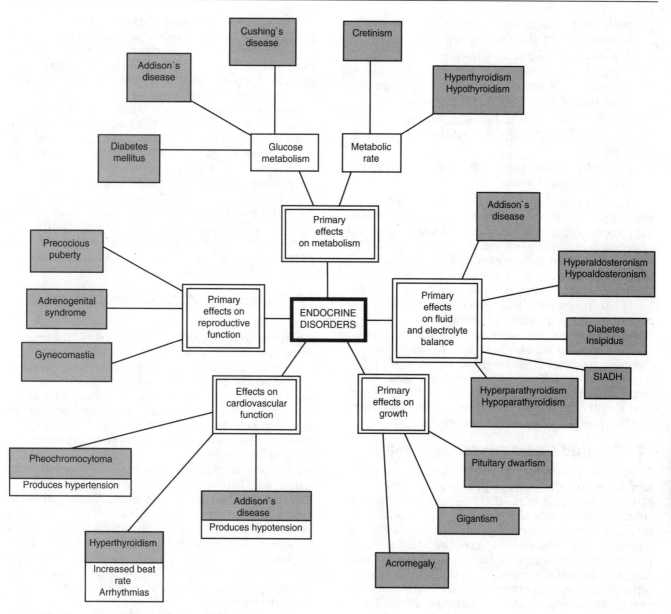

Figure A-29   Disorders of the Endocrine System

THE SYMPTOMS OF ENDOCRINE DISORDERS

Knowledge of the individual endocrine organs and their functions make predictions possible about the symptoms of specific endocrine disorders. For example, thyroid hormones increase basal metabolic rate, body heat production, perspiration, restlessness, and heart rate. An elevated metabolic rate, increased body temperature, weight loss, nervousness, excessive perspiration, and an increased or irregular heartbeat are common symptoms of hyperthyroidism. On the other hand, a low metabolic rate, decreased body temperature, weight gain, lethargy, dry skin, and a reduced heart rate often accompany hypothyroidism. The symptoms associated with over- and underproduction of major hormones are summarized in Table A-15, p. 86.

The Diagnosis of Endocrine Disorders

The first step in the diagnosis of an endocrine disorder is the physical examination. Several disorders produce characteristic physical signs that reflect abnormal hormone activities. Several examples were introduced in the text:

- *Cushing's disease* results from an oversecretion of glucocorticoids by the adrenal cortex. As the condition progresses there is a shift away from the normal pattern of fat distribution in the body. Adipose tissue accumulates in the abdominal area, the lower cervical area (causing a "humpback"), and in the face, but the extremities become relatively thin.

- *Acromegaly* results from oversecretion of growth hormone in the adult. In this condition the facial features become distorted due to excessive cartilage growth, and the lower jaw protrudes, a sign known as *prognathism.* The hands and feet also become enlarged.

- *Adrenogenital syndrome* results from the over-secretion of androgens by the adrenal glands of a female. Hair growth patterns change, and the condition of *hirsutism* (p. 41) develops.

- Hypothyroidism due to iodine deficiency produces a distinctively enlarged thyroid gland, or *goiter.*

- Hyperthyroidism can produce protrusion of the eyes, or *exophthalmos.*

These signs are very useful, but many other signs and symptoms related to endocrine disorders are less definitive. For example, the condition of *polyuria,* or increased urine production, may be the result of hyposecretion of ADH *(diabetes insipidus),* or a form of *diabetes mellitus,* and a symptom such as hypertension (high blood pressure) can be caused by a variety of cardiovascular or endocrine problems. In these instances diagnostic decisions are often based on blood tests, which can confirm the presence of an endocrine disorder by detecting abnormal levels of circulating hormones, followed by procedures that determine whether the primary cause of the problem lies within the endocrine gland, the regulatory mechanism(s), or the target tissues.

✝ Thyroid Gland Disorders   EAP p. 292

Hypothyroidism typically results from some problem involving the thyroid gland rather than with pituitary production of TSH. In primary hypothyroidism TSH levels are elevated, but levels of T_3 (triiodothyronine) and T_4 (tetraiodothyronine or *thyroxine)* are depressed. Treatment of chronic hypothyroidism, such as the hypothyroidism that follows radiation exposure, usually involves the administration of synthetic thyroid hormones (thyroxine) to maintain normal blood concentrations.

A **goiter** is an enlargement of the thyroid gland. The enlargement usually indicates increased follicular size, despite a decrease in the rate of thyroid hormone production. A goiter usually develops when the thyroid gland is unable to synthesize and release adequate amounts of thyroid hormones. Under continuing TSH stimulation, thyroglobulin production accelerates and the thyroid follicles enlarge. One type of goiter occurs if the thyroid fails to obtain enough iodine to meet its synthetic requirements. (This condition is now rare in the U.S. due to the use of iodized table salt.) Administering iodine may not solve the problem entirely, for the sudden availability of iodine may produce symptoms of *hyperthyroidism* as the stored thyroglobulin becomes activated. The usual therapy involves taking thyroxine, which has a negative feedback effect on the hypothalamus and pituitary, inhibiting the production of TSH. Over time the resting thyroid may return to its normal size.

Thyrotoxicosis, or *hyperthyroidism,* occurs when thyroid hormones are produced in excessive quantities. In **Graves' disease** excessive thyroid activity leads to goiter and the symptoms of hyperthyroidism. Protrusion of the eyes, or **exophthalmos** (eks-ahf-THAL-mōs) may also appear, for unknown reasons. Graves' disease has a genetic autoimmune basis and affects women much more often than men. Treatment may involve the use of antithyroid drugs, surgical removal of portions of the glandular mass, or destruction of part of the gland by exposure to radioactive iodine.

Hyperthyroidism may also result from inflammation or, rarely, thyroid tumors. In extreme cases the individual's metabolic processes accelerate out of control. During a *thyrotoxic crisis,* or "thyroid storm," the subject experiences an extremely high fever, rapid heart rate, and the malfunctioning of a variety of physiological systems.

✝ Disorders of Parathyroid Function
EAP p. 293

When the parathyroid gland secretes inadequate or excessive amounts of parathyroid hormone, calcium concentrations move outside of normal homeostatic limits. Hypoparathyroidism may develop after neck surgery, especially a thyroidectomy, if the blood supply to the parathyroid glands is restricted. In many other cases the primary cause of the condition is uncertain. Treatment is difficult because at present PTH can be obtained only by extraction from the blood of normal individuals. Thus PTH is extremely costly, and because supplies are very limited, PTH administration is not used to treat this condition, despite its probable effectiveness. As an alternative, a dietary combination of vitamin D and calcium can be used to elevate body fluid calcium concentrations. (As noted in Chapter 6, vitamin D stimulates the absorption of calcium ions across the lining of the digestive tract.)

In *hyperparathyroidism* calcium concentrations become abnormally high. Calcium salts in the skeleton are mobilized, and bones are weakened. On X-rays the bones have a light, airy appearance because the dense calcium salts no longer dominate the tissue. CNS function is depressed, thinking slows, memory is impaired, and the individual often experiences emotional swings and depression. Nausea and vomiting occur, and in severe cases the patient may become comatose. Muscle function deteriorates, and skeletal muscles become weak. Other tissues are often affected as calcium salts crystallize in joints, tendons, and the dermis, and calcium deposits may produce masses, called *kidney stones,* that block filtration and conduction passages in the kidney.

Hyperparathyroidism most often results from a tumor of the parathyroid gland. Treatment involves the surgical removal of the overactive tissue. Fortunately there are four parathyroids, and the secretion of even a portion of one gland can maintain normal calcium concentrations.

☤ Disorders of the Adrenal Cortex

EAP *p. 294*

Clinical problems related to the adrenal gland vary depending on which groups of adrenal cells becomes involved. The conditions may result from changes in the functional capabilities of the adrenal cells (primary conditions) or disorders affecting the regulatory mechanisms (secondary conditions). In **hypoaldosteronism** adrenal cells fail to produce enough aldosterone, usually either as an early sign of adrenal insufficiency or because the kidneys are not releasing adequate amounts of renin. Low aldosterone levels lead to excessive losses of water and sodium ions at the kidneys, and the water loss in turn leads to low blood volume and a fall in blood pressure. The resulting changes in electrolyte concentrations, including *hyperkalemia* (high extracellular K^+ levels) affect transmembrane potentials, eventually causing dysfunctions in neural and muscular tissues.

Hypersecretion of aldosterone results in the condition of **aldosteronism,** or *hyperaldosteronism.* Under continued aldosterone stimulation, the kidneys retain sodium ions in exchange for potassium ions that are lost in the urine. Hypertension and hypokalemia occur as extracellular potassium levels decline, it increases the concentration gradient for potassium ions across cell membranes. This leads to an acceleration in the rate of potassium diffusion out of the cells and into the interstitial fluids. The reduction in intracellular and extracellular potassium levels eventually interferes with the function of excitable membranes, especially cardiac muscle cells and neurons, and kidney cells.

Addison's disease may result from inadequate stimulation of the adrenal cells by ACTH or from their inability to synthesize the necessary hormones usually due to autoimmune problems or infection. Affected individuals become weak and lose weight, due to a combination of appetite loss, hypotension, and hypovolemia. They cannot adequately mobilize energy reserves, and their blood glucose concentrations fall sharply within hours after a meal. Stresses cannot be tolerated, and a minor infection or injury may lead to a sharp and fatal decline in blood pressure. A particularly interesting symptom is the increased melanin pigmentation in the skin. The ACTH molecule and the MSH molecule are similar in structure, and at high concentrations ACTH stimulates the MSH receptors on melanocytes. President John F. Kennedy suffered from this disorder.

Cushing's disease results from overproduction of glucocorticoids. The symptoms resemble those of a protracted and exaggerated response to stress. (The stress response is discussed in the text on EAP *p. 301*.) Glucose metabolism is suppressed, lipid reserves are mobilized, and peripheral proteins are broken down. Lipids and amino acids are mobilized in excess of the existing demand. The energy reserves are shuffled around, and the distribution of body fat changes. Adipose tissues in the cheeks and around the base of the neck become enlarged at the expense of other areas, producing a "moon-faced" appearance. The demand for amino acids falls most heavily on the skeletal muscles, which respond by breaking down their contractile proteins. This response reduces muscular power and endurance.

The chronic administration of large doses of steroids can produce symptoms similar to those of Cushing's disease, but such treatment is usually avoided. Roughly 75 percent of cases result from an overproduction of ACTH, and afflicted individuals may also show changes in skin pigmentation.

☤ Disorders of the Adrenal Medulla

EAP *p. 295*

The overproduction of epinephrine by the adrenal medulla may reflect chronic sympathetic activation. A **pheochromocytoma** (fē-ō-krō-mō-sī-TŌ-mah) is a tumor that produces catecholamines in massive quantities. The tumor usually develops within the adrenal medulla, but it may also involve other sympathetic ganglia. The most dangerous symptoms are rapid and irregular heartbeat and high blood pressure; other symptoms include uneasiness, sweating, blurred vision, and headaches. This condition is rare, and surgical removal of the tumor is the most effective treatment.

☤ Diabetes mellitus

EAP *p. 298*

There are two major types of diabetes mellitus: *insulin-dependent (Type I) diabetes* and *non-insulin-dependent (Type II) diabetes.*

In **insulin-dependent diabetes mellitus (IDDM)**, or **Type I diabetes**, the primary cause is inadequate insulin production by the beta cells of the pancreatic islets. In this condition insulin production declines due to a drop in the number of beta cells. As noted in Chapter 3, glucose normally enters peripheral cells through facilitated diffusion. (p. 55) This transport mechanism does not function in the absence of insulin, and when insulin concentrations decline, cells can no longer absorb glucose from their surroundings. Under these conditions peripheral tissues remain glucose-starved despite the presence of adequate or even excessive amounts of glucose in the circulation. After a meal rich in glucose, blood concentrations may become so elevated that the kidney cells cannot reclaim all of the glucose molecules entering the urine. The high urinary concentration of glucose limits the ability of the kidney to conserve water, so the individual urinates frequently and may become dehydrated. The chronic dehydration leads to disturbances of neural function (blurred vision, tingling sensations, disorientation, fatigue) and muscle weakness.

Long-term treatment involves a combination of dietary control and administration of insulin, either by periodic injection or through more continual subcutaneous delivery with an **insulin pump**. The treatment is complicated by the fact that tissue glucose demands cycle up and down, depending on physical activity, emotional state, stress, and other factors that are hard to assess or predict. It is therefore diffi-

11

cult to maintain stable and normal blood glucose levels over long periods of time.

Type I diabetes most often appears in individuals under 40 years of age. Because it frequently appears in childhood, it has been called **juvenile-onset diabetes**. Most people with this type of diabetes (roughly 80 percent) have circulating antibodies that target the surfaces of beta cells. The disease may therefore be an example of an autoimmune disorder, a condition that results when the immune system attacks normal body cells. (Possible mechanisms responsible for *autoimmune disorders* will be discussed in Chapter 15.) Consequently, attempts have been made to prevent the appearance of Type I diabetes with *azathioprine* (Imuran), a drug that suppresses the immune system. This procedure is somewhat dangerous, however, because compromising immune function increases the risk of acquiring serious infections or developing cancer.

Non-insulin-dependent diabetes mellitus (NIDDM), or Type II diabetes, typically affects obese individuals over 40 years of age. Because of the age factor this condition is also called **maturity-onset diabetes**. Insulin levels are normal or elevated, but peripheral tissues no longer respond normally, often due to a reduction in the number of insulin receptors. Treatment consists of weight loss and dietary restrictions that may elevate insulin production and tissue response.

Both types of diabetes mellitus may affect 1-2 percent of the U.S. population. Maturity-onset diabetes is roughly three times as common as insulin-dependent diabetes. The standard testing procedures check the primary diabetic signs: high fasting blood glucose concentrations, the appearance of glucose in the urine, and an inability to reduce elevated glucose levels. The latter capability is examined by a *glucose tolerance test*. Blood concentrations are monitored after a fasting subject consumes roughly 75 g (2.6 oz) of glucose. In a normal individual the glucose will enter the circulation, insulin production will rise, and peripheral tissues will absorb the glucose so rapidly that blood concentrations will remain relatively normal. Without adequate insulin production, glucose concentrations skyrocket to more than twice normal levels.

Probably because glucose levels cannot be stabilized adequately, even with treatment, patients with diabetes mellitus often develop chronic medical problems. In general, these problems are related to circulatory system abnormalities. The most common examples include the following:

1. Vascular changes at the retina, including proliferation of capillaries and hemorrhaging, often cause disturbances of vision. This condition is called **diabetic retinopathy**.

2. Changes occur in the clarity of the lens, producing cataracts.

3. Small hemorrhages and inflammation at the kidneys cause degenerative changes that can lead to kidney failure. This condition is called **diabetic nephropathy**.

4. A variety of neural problems appear, including nerve palsies, paresthesias, and autonomic dysfunction. These disorders, collectively termed **diabetic neuropathy**, are probably related to disturbances in the blood supply to neural tissues, since neurons don't require insulin to absorb and utilize glucose.

5. Degenerative changes in cardiac circulation can lead to early heart attacks. For a given age group, heart attacks are 3-5 times more likely in diabetic individuals.

6. Other peripheral changes in the vascular system can disrupt normal circulation to the extremities. For example, a reduction in blood flow to the feet can lead to tissue death, ulceration, infection, and loss of toes or a major portion of one or both feet.

(R) Endocrinology and Athletic Performance EAP *p. 298*

One of the first endocrinological "mass experiments" on humans occurred early in World War II, when the German government administered testosterone to Nazi SS officers in an attempt to make them more aggressive. (There is no evidence that the experiment succeeded.) Nowadays, such practices are universally deplored in all civilized societies, and medical research involving humans is generally subject to tight ethical constraints and meticulous scientific scrutiny. Yet a clandestine, unscientific, and potentially quite dangerous program of "experimentation" with hormones is today being pursued by athletes in many countries. Despite being banned by the International Olympic Committee, the United States Olympic Committee, the NCAA, and the NFL, and condemned by the American Medical Association and the American College of Sports Medicine, a significant number of amateur and professional athletes continue to use hormones to improve their performance. Although synthetic forms of testosterone are used most often, young athletes may use any combination of testosterone, growth hormone, and a variety of synthetic hormones.

Androgen Abuse

The use of androgens, or "anabolic steroids," has become popular with many athletes, both amateur and professional. The goal of steroid use is to increase muscle mass, endurance, and "competitive spirit." It has been suggested that as many as 30 percent of college and professional athletes and 10-20 percent of male high school athletes may be using anabolic steroids (with or without growth hormone) to improve their performance. Among body builders, the proportion using steroids in this country may be as high as 80 percent.

Black market sales of anabolic steroids probably exceed $100 million annually. The compounds are administered orally or by injection, typically in doses 10-1000 times higher than those normally prescribed in medical treatment.

One supposed justification for this practice has been the unfounded opinion that compounds manufactured in the body are not only safe, but "good for you." In reality the administration of natural or synthetic androgens in abnormal amounts carries unacceptable health risks. Androgens affect many tissues in a variety of ways. Known complications include (1) premature epiphyseal closure; (2) various liver problems (including jaundice and hepatic tumors); (3) prostate enlargement and urinary tract obstruction; and (4) testicular atrophy and infertility. A link to heart attacks, impaired cardiac function, and strokes has also been suggested. Moreover, the normal regulation of androgen production involves a feedback mechanism comparable to that described for adrenal steroids earlier in this chapter. A releasing hormone stimulates the production of LH, and LH in turn stimulates the secretion of testosterone and other androgens by the interstitial cells of the testes. The circulating androgens then inhibit the production of both the releasing hormone and LH, as indicated in Figure 11-7a, p. 288. Thus when synthetic androgens are administered in high doses, they can (1) suppress the normal production of testosterone, and (2) depress the manufacture of the associated releasing hormone by the hypothalamus. *This suppression may be permanent.*

The use of androgenic "bulking agents" by female body builders may not only add muscle mass, but alter muscular proportions and secondary sexual characteristics. For example, women taking steroids can develop irregular menstrual periods and changes in body hair distribution (including baldness). Finally, androgen abuse may cause a generalized depression of the immune system.

EPO Abuse

Because it is now being synthesized using recombinant DNA techniques, erythropoietin, or EPO, is readily available. Athletes engaged in endurance sports, such as cycling or marathon running, may use it to boost the number of oxygen-carrying red blood cells in circulation. Although this improves the oxygen content of the blood, it also makes the blood more dense, and the heart must work harder to push it around the circulatory system. Over the last 4 years the deaths of 18 young, otherwise healthy European cyclists have been attributed to heart failure related to EPO abuse.

GHB and Clenbuterol

Androgens and EPO are known hormones with reasonably well understood effects. Because drug testing is now widespread in amateur and professional sports, people interested in "getting an edge" are experimenting with other drugs whose long-term and short-term effects are difficult to predict. Two examples are the recent use of GHB and clenbuterol by amateur athletes.

Gamma-hydroxybutyrate, or GHB, was tested for use as an anaesthetic 30 years ago, and rejected in part because it was linked to epileptic seizures. In 1990 this drug appeared in health-food stores, where it was sold as an anabolic agent and diet aid. During a 5 month period in 1990, 16 cases of severe reaction to GHB were treated in San Francisco alone. Symptoms experienced included confusion, hallucination, seizures, and coma at doses from 0.25 teaspoons to 4 tablespoons.

Clenbuterol abuse is reportedly widespread, although exact numbers are difficult to obtain. Clenbuterol, sometimes used to treat asthma, mimics epinephrine by increasing the diameter of the respiratory passageways and accelerating blood flow through active skeletal muscles. Although it is also rumored to have anabolic properties, there is no evidence to support this. Heavy usage can cause severe headaches, tremors, insomnia, and potentially dangerous abnormal heartbeats. During the 1992 Olympics in Barcelona, Spain, two American athletes were disqualified because they tested positive for this drug.

Light and Behavior   EAP p. 299

Exposure to sunlight can do more than stimulate a tan or promote the formation of vitamin D_3. There is evidence that daily light-dark cycles have widespread effects on the central nervous system, with melatonin playing a key role. Several studies have indicated that residents of temperate and higher latitudes in the northern hemisphere undergo seasonal changes in mood and activity patterns. These people feel most energetic from June through September, whereas the period of December through March finds them with relatively low spirits. (The situation in the southern hemisphere, where the winter and summer seasons are reversed, is just the opposite.) The degree of seasonal variation differs from individual to individual; some people are affected so severely that they seek medical attention. The observed symptoms have recently been termed **seasonal affective disorder, or SAD.** Individuals with SAD experience depression and lethargy, and find it difficult to concentrate. Often they sleep for long periods, perhaps 10 hours or more per day. They may also go on eating binges and have a craving for carbohydrates.

Melatonin secretion appears to be regulated by sunlight exposure, not simply by light exposure. Normal interior lights are apparently not strong enough or do not release the right mixture of light wavelengths to depress melatonin production. Because many people spend very little time outdoors, melatonin production increases in the winter, and the depression, lethargy, and concentration problems appear to be linked to elevated melatonin levels in the blood. Comparable symptoms can be produced in a normal experimental subject by an injection of melatonin.

11

Table A-15   Clinical Implications of Endocrine Malfunctions

Hormone	Under-production Syndrome	Principal Symptoms	Over-production Syndrome	Principal Symptoms
Growth hormone (GH)	Pituitary growth failure (pituitary dwarfism)	Retarded growth, abnormal fat distribution, low blood glucose hours after a meal	Gigantism, acromegaly	Excessive growth
Antidiuretic hormone (ADH)	Diabetes insipidus	Polyuria, dehydration, thirst	SIADH (syndrome of inappropriate ADH secretion)	Increased body weight and water content
Thyroxine (T4), triiodothyronine (T3)	Myxedema, cretinism	Low metabolic rate, body temperature; impaired physical and mental development	Hyperthyroidism, Graves' disease	High metabolic rate and body temperature
Parathyroid hormone (PTH)	Hypopara-thyroidism	Muscular weakness, neurological problems, formation of dense bones, tetany due to low blood calcium concentrations	Hyperparathyroidism	Neurological, mental, muscular problems due to high blood calcium concen-trations; weak and brittle bones
Insulin	Diabetes mellitus (Type I)	High blood glucose, impaired glucose utilization, dependence on lipids for energy,	Excess insulin production or administration glycosuria	Low blood glucose levels, possibly causing coma
Mineralocorticoids (MC)	Hypoaldo-steronism	Polyuria, low blood volume, high blood potassium concentrations	Aldosteronism	Increased body weight due to water retention; low blood potassium concentration
Glucocorticoids (GC)	Addison's disease	Inability to tolerate stress, mobilize	Cushing's disease energy reserves, or maintain normal blood glucose concentrations	Excessive breakdown of tissue proteins and lipid reserves; impaired glucose metabolism
Epinephrine (E), norepinephrine (NE)	None identified		Pheochromocytoma	High metabolic rate, body temperature, and heart rate; elevated blood glucose levels
Estrogens (female)	Hypogonadism	Sterility, lack of secondary sexual characteristics	Adrenogenital syndrome	Overproduction of androgens by z. reticularis of adrenal leads to masculinization
	Menopause	Cessation of ovulation	Precocious puberty	Premature sexual maturation and related behavioral changes
Androgens	Hypogonadism (male)	Sterility, lack of secondary sexual characteristics	Adrenogenital syndrome (gynecomastia)	Abnormal production of estrogen, some-times due to adrenal or interstitial cell tumors; leads to breast enlargement
			Precocious puberty	Premature sexual maturation and related behavioral changes

Many SAD patients may be successfully treated by exposure to sun lamps that produce full-spectrum light. Experiments are under way to define exactly how intense the light must be and determine the minimal effective time of exposure.

FIGURE A-30
Endocrine Abnormalities

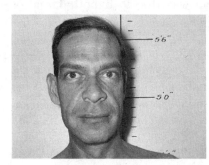

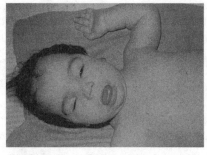

(b) *Cretinism*, which results from thyroid hormone insufficiency in infancy.

(c) An enlarged thyroid gland, or goiter, is usually associated with thyroid hyposecretion due to iodine insufficiency.

(a) *Acromegaly* results from the overproduction of growth hormone after the epiphyseal plates have fused. Bone shapes change and cartilaginous areas of the skeleton enlarge. Note the broad facial features and the enlarged lower jaw.

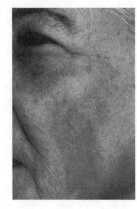

(d) *Addison's disease* is caused by hyposecretion of corticosteroids, especially glucocorticoids. Pigment changes result from stimulation of melanocytes by ACTH, which is structurally similar to MSH.

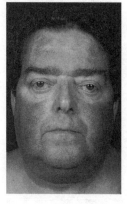

(e) *Cushing's disease* is caused by hypersecretion of glucocorticoids. Lipid reserves are mobilized, and adipose tissue accumulates in the cheeks and at the base of the neck.

Endocrine Disorders   EAP *p. 302*

Major endocrine disorders are presented in Table A-15, and the appearance of individuals with representative disorders is indicated in Figure A-30.

CRITICAL THINKING QUESTIONS

5-1. Pheochromocytomas are tumors of the adrenal medulla that cause hypersecretion of the catecholamines produced by this region of the adrenal gland. What symptoms would you associate with a person who suffers from this condition?

   a. hypertension
   b. sweating
   c. nervousness
   d. elevated metabolic rate
   e. all of the above

5-2. Fifty-year-old Barbara B. reports to the emergency room with heart palpitations (rapid, irregular heart beat). EKG recording shows atrial fibrillation. Barbara's medical history includes recent weight loss of 10 pounds and complaints of increased irritability and nervousness within the last month. Some of Barbara's abnormal diagnostic and lab test results are as follows:

Radioactive iodine uptake test (RAIU) at 2 hours: 20% absorption (normal: 1-13%)

Serum thyroxine (T_4) test: 13 ng/dl

Serum triiodothyronine (T_3) test: 210 ng/dl

Serum TSH: <0.1

What is the probable diagnosis?

   a. hypothyroidism
   b. Graves' disease
   c. myxedema
   d. goiter

5-3. Bill develops a benign tumor of the parathyroid glands that causes the level of parathyroid hormone in his blood to be higher than normal. Which of the following would you expect to occur as a result of this condition?

   a. decreases in the blood levels of calcium
   b. convulsions
   c. decreased muscle strength
   d. increased bone density
   e. all of the above

5-4. Sixteen-year-old John is a promising athlete who is below the average height for his age. He wants to play football in college but is convinced that he needs to be taller and stronger in order to accomplish his dream. He persuades his parents to visit a doctor and inquire about GH treatments. If you were his physician, what would you tell him about the potential risks and benefits of such treatments?

5-5. Angie is diagnosed with diabetic retinopathy. After performing several tests, her physician decides to remove her anterior pituitary gland. Shortly after the surgery, her eyesight returns to normal. What was the apparent cause of Angie's problem?

5-6 A patient is suffering from secondary hypothyroidism. Assuming you had available all the modern testing materials, how could you determine if the problem was due to hypothalamic or pituitary dysfunction?

NOTES

The Cardiovascular System

The components of the cardiovascular system include the blood, heart, and blood vessels. Blood flows through a network of thousands of miles of vessels within the body, transporting nutrients, gases, wastes, hormones, and electrolytes and redistributing the heat generated by active tissues. The exchange of materials between the blood and peripheral tissues occurs across the walls of tiny capillaries that are situated between the arterial and venous systems. The total capillary surface area for exchange is truly enormous, averaging around 6,300 square meters, 50 percent larger than the area of a football field.

Because it plays a key role in supporting all other systems, disorders of the cardiovascular system will affect virtually every cell in the body. One method of organizing the many potential disorders involving this system is by the nature of the primary problem, whether it affects the blood, the heart, or the vascular network. Figure A-31 provides an introductory overview of major cardiovascular disorders that are discussed in the text and in later sections of the *Applications Manual.*

PHYSICAL EXAMINATION AND THE CARDIOVASCULAR SYSTEM

Individuals with cardiovascular problems often seek medical attention with one or more of the following as chief complaints:

Figure A-31   Disorders of the Cardiovascular System

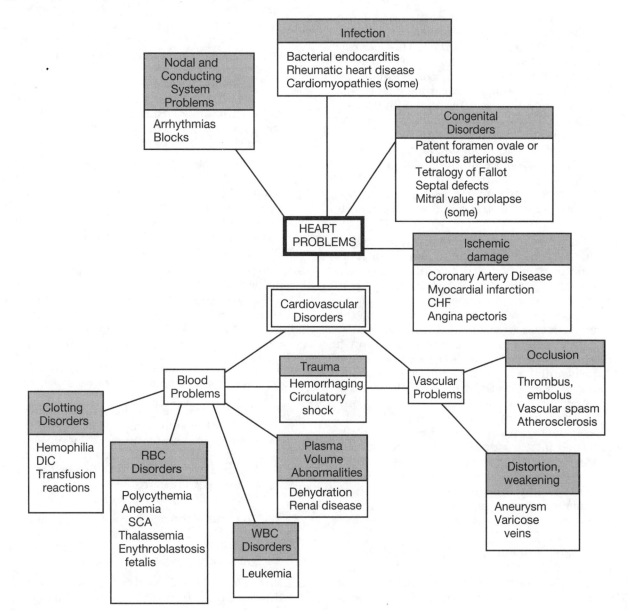

1. *Weakness and fatigue:* These symptoms develop when the cardiovascular system can no longer meet tissue demands for oxygen and nutrients. These symptoms may occur because cardiac function is impaired, as in *heart failure* (p. 109) or *cardiomyopathy* (p. 97), or because the blood is unable to carry normal amounts of oxygen, as in the various forms of *anemia* (p. 95). In the early stages of these conditions, the individual feels healthy at rest, but becomes weak and fatigued with any significant degree of exertion because the cardiovascular system cannot keep pace with the rising tissue oxygen demands. In more advanced stages of these disorders, weakness and fatigue are chronic problems that continue, even at rest.

2. *Cardiac pain:* This is a deep pain felt in the substernal region and often radiating down the left arm or up into the shoulder and neck. There are two major causes of cardiac pain:

   • Constant severe pain can result from inflammation of the pericardial sac, a condition known as *pericarditis.* This *pericardial pain* may superficially resembles the pain experienced in a *myocardial infarction* (MI), or heart attack. Pericardial pain differs from the pain of an MI in that (a) it may be relieved by leaning forward, (b) a fever may be present, and (c) the pain does not respond to the administration of drugs, such as *nitroglycerin,* that dilate coronary blood vessels. Nitroglycerin, which is effective in relieving angina pectoris, does not relieve the pain associated with pericarditis.

   • Cardiac pain can also result from inadequate blood flow to the myocardium. This type of pain is called *myocardial ischemic pain.* Ischemic pain occurs in *angina pectoris* and in a myocardial infarction. Angina pectoris *(p. 100)* most often results from the constriction of coronary blood vessels by atherosclerosis. The associated pain appears during physical exertion, when myocardial oxygen demands increase, and the pain is relieved by drugs such as nitroglycerin, which dilate coronary vessels and improve coronary blood flow. The pain associated with a myocardial infarction is felt as a heavy weight or a constriction of the chest. The pain of an MI is also distinctive because (a) it is not linked to exertion, (b) it is not relieved by nitroglycerin or other coronary vasodilators, and (c) nausea, vomiting and sweating may occur during the attack.

3. *Palpitations:* Palpitations are a person's perception of an altered heart rate. The individual may complain of the heart "skipping a beat" or "racing." The most likely cause of palpitations

is an abnormal pattern of cardiac activity known as an *arrhythmia.* The detection and analysis of arrhythmias are considered in a later section (p. 102).

4. *Pain on movement:* Individuals with advanced athrosclerosis often experience pain in the extremities during movement. The pain may become so severe that the person is unwilling or unable to walk or perform other common activities. The underlying problem is constriction or partial occlusion of major arteries, such as the external iliac arteries to the lower limbs, by plaque formation.

These are only a few of the many symptoms that can be caused by cardiovascular disorders. In addition, the individual may notice the appearance of characteristic signs of underlying cardiovascular problems. A partial listing of important cardiovascular signs includes the following:

1. *Edema* is an increase of fluid in the tissues which occurs when (a) the pumping efficiency of the heart is decreased, (b) the plasma protein content of the blood is reduced, or (c) venous pressures are abnormally high. The tissues of the extremities are most often affected, and individuals experience swollen feet, ankles, and legs. When edema is so severe that pressing on the affected area leaves an indention, the sign is called *pitting edema.* Edema is discussed in Chapter 14 of the text (p. *355).*

2. Breathlessness, or *dyspnea,* occurs when cardiac output is inadequate for tissue oxygen demands. Dyspnea may also occur with *pulmonary edema,* a buildup of fluid within the alveoli of the lungs. Pulmonary edema and dyspnea are often associated with *congestive heart failure* (p. 109).

3. *Varicose veins* are dilated superficial veins that are visible at the skin surface. This condition, which develops when venous valves malfunction, can be caused or exaggerated by increased systemic venous pressures. Varicose veins are considered further on p. 105.

4. There may be characteristic and distinctive changes in skin coloration. For example,

   • *Pallor* is the lack of normal red or pinkish color to the skin of a Caucasian or the conjunctiva and oral mucosa of darker-skinned individuals. Pallor accompanies many forms of anemia, but may also be the result of inadequate cardiac output, shock (p. 106), or circulatory collapse.

   • *Cyanosis* is the bluish color of the skin occurring with a deficiency of oxygen to the tissues. Cyanosis usually results from either cardiovascular or respiratory disorders.

5. *Vascular skin lesions* were introduced in the discussion of skin disorders on p. 37.

Characteristic vascular lesions may occur in clotting disorders (p. 96) and several forms of *leukemia* (p. 96). For example,

- Abnormal bruising may be the result of a disorder affecting the clotting system, platelet production, or vessel structure.

- *Petechiae*, which appear as purple spots on the skin surface, are often seen in certain types of leukemia or other diseases associated with low platelet counts.

CARDIOVASCULAR DISORDERS AND DIAGNOSTIC PROCEDURES

Often the initial detection of a cardiovascular disorder occurs during the assessment stage of a physical examination.

1. When the vital signs are taken, the pulse is checked for vigor, rate, and rhythm. Weak or irregular heart beats will often be noticed at this time.

2. The blood pressure is monitored with a stethoscope, blood pressure cuff, and sphygmomanometer. Unusually high or low readings can alert the examiner to potential problems with cardiac or vascular function. However, a diagnosis of hypotension or hypertension is not made on the basis of a single reading, but after several readings over a period of time. Hypertension and hypotension are discussed in detail on p. 105.

3. The heart sounds are monitored by auscultation with a stethoscope.

- Cardiac rate and rhythm can be checked and arrhythmias detected.

- Abnormal heart sound, or *murmurs*, may indicate problems with atrioventricular or semilunar valves. Murmurs are noted in relation to their location in the heart (as determined by the position of the stethoscope on the chest wall), the time of occurrence in the cardiac cycle, and whether the sound is low or high pitched.

- Nothing is usually heard during auscultation of normal vessels of the circulatory system. *Bruits* are the sounds resulting from turbulent blood flow around an obstruction within a vessel. Bruits are often heard where large atherosclerotic plaques have formed.

Functional abnormalities of the heart and blood vessels can often be detected through physical assessment and the recognition of characteristic signs and symptoms. The structural basis for these problems is usually determined through the use of scans, X rays, and the monitoring of electrical activity in the heart. For problems with a hema-

tological basis, laboratory tests performed on blood samples usually provide the information necessary to reach a tentative diagnosis.

⚕ Polycythemia
EAP *p. 311*

An elevated hematocrit with a normal blood volume constitutes **polycythemia** (po-lē-sī-THĒ-mē-ah). There are several different types of polycythemia. *Erythrocytosis* (e-rith-rō-sī-TŌ-sis), a polycythemia affecting only red blood cells, will be considered later in the chapter. **Polycythemia vera** ("true polycythemia") results from an increase in the numbers of all blood cells. The hematocrit may reach 80-90, at which point the tissues become oxygen-starved because red blood cells are blocking the smaller vessels. This condition seldom strikes young people; most cases involve patients age 60-80. There are several treatment options, but none cures the condition. The cause of polycythemia vera is unknown, although there is some evidence that the condition is linked to radiation exposure.

⚕ Thalassemia
EAP *p. 313*

The *thalassemias* are a diverse group of inherited blood disorders caused by an inability to produce adequate amounts of alpha or beta hemoglobin chains. A specific condition is categorized as an **alpha-thalassemia** or **beta-thalassemia** depending on whether the α or β hemoglobin subunits are affected. Normal individuals inherit two copies of alpha chain genes from each parent, and alpha-thalassemia develops when one or more of these genes are missing or inactive. The severity of the symptoms varies depending on how many normal alpha chain genes remain functional. For example, an individual with three normal alpha chain genes will not develop symptoms at all, but this person can be a carrier, passing the defect to the next generation. A child born of parents who are both carriers is likely to develop a more severe form of the disease. (The genetic mechanisms involved are considered in Chapter 21 of the text.)

- Individuals with two copies of the normal alpha chain gene have somewhat impaired hemoglobin synthesis. The red blood cells are small, and contain less than the normal quantity of hemoglobin.

- Individuals with one copy of the alpha chain gene have very small *(microcytic)* red blood cells that are relatively fragile.

- Individuals with no functional copies of the alpha chain gene usually die shortly after birth, because the hemoglobin synthesized cannot bind and transport oxygen normally.

There are regional and racial differences in the frequency of alpha-thalassemia. For example, roughly 2 percent of African Americans have only two copies of the normal gene, one received from each parent.

12

However, the incidence of severe alpha-thalassemia is highest among Southeast Asians.

Each person inherits only one gene for the beta hemoglobin chain from each parent. If an individual does not receive a copy of the normal gene from either parent, the condition of **beta-thalassemia major,** or *Cooley's disease,* develops. Symptoms of this condition include severe anemia; microcytosis; a low hematocrit (under 20); and enlargement of the spleen, liver, heart, and areas of red bone marrow. Potential treatments for those with severe symptoms include transfusions, splenectomy (to slow the rate of RBC recycling), and bone marrow transplantation. *Beta-thalassemia minor,* or *beta-thalassemia trait,* seldom produces clinical symptoms. The rates of hemoglobin synthesis are depressed by roughly 15 percent, but this decrease does not affect their functional abilities, and no treatment is necessary.

🩺 Sickle Cell Anemia   EAP *p. 313*

Sickle cell anemia results from the production of an abnormal form of hemoglobin. The β chains are involved, and the abnormal subunit is called *hemoglobin S.* Sickle cell anemia affects 60,000-80,000 African Americans today; this represents roughly 0.14 percent of the African-American population.

An individual with sickle cell anemia carries two copies of the abnormal gene. If only one copy is present, the individual has a *sickling trait.* One African American in 12 carries the sickling trait. Although it is now known that the genes are present in Americans of Mediterranean, Middle Eastern, and East Indian ancestry, statistics on the incidence of sickling trait and SCA in these groups are as yet unavailable.

To develop sickle cell anemia an individual must have two copies of the sickling gene, one from each parent. If only one sickling gene is present, the individual has the *sickling trait.* In such cases most of the hemoglobin is of the normal form, and the erythrocytes function normally. But the presence of the abnormal hemoglobin gives the individual the ability to resist the parasitic infections that cause **malaria,** a mosquito-borne illness. The malaria parasites enter the bloodstream when an individual is bitten by an infected mosquito. The microorganisms then invade and reproduce within the erythrocytes. But when they enter an erythrocyte from a person with the sickling trait, the cell responds by sickling. Either the sickling itself kills the parasite, or the sickling attracts the attention of a phagocyte that engulfs the RBC and kills the parasite. In either event the individual remains unaffected by the disease, while normal individuals sicken and often die.

Symptoms of sickle cell anemia include pain and damage to a variety of organs and systems, depending on the location of the obstructions. In addition, the trapped red blood cells eventually die and break down, producing a characteristic anemia. Transfusions of normal blood can temporarily prevent additional complications, and there are experimental drugs that can control or reduce sick-

ling. *Hydroxyurea* is an anticancer drug that stimulates production of fetal hemoglobin, a slightly different form of hemoglobin produced during development. It is effective, but has toxic side effects (not surprising in an anticancer drug). The food additive **butyrate,** found in butter and other foods, appears to be even more effective in promoting the synthesis of fetal hemoglobin. In clinical trials it has been effective in treating sickle cell anemia and other conditions caused by abnormal hemoglobin structure, such as β-*thalassemia.*

📋 Bilirubin Tests and Jaundice

EAP *p.313*

When hemoglobin is broken down, the heme units (minus the iron) are converted to bilirubin. Normal plasma bilirubin concentrations range from 0.5 to 1.2 mg/dl. Of that amount, roughly 85 percent is unconjugated bilirubin that will be absorbed by the liver. Several different clinical conditions are characterized by an increase in the total plasma bilirubin concentration. In such conditions, bilirubin diffuses into peripheral tissues, giving them a yellow coloration that is most apparent in the skin and over the sclera of the eyes. This combination of signs (yellow skin and eyes) is called **jaundice** (JAWN-dis).

Jaundice can have many different causes, but blood tests that determine the concentration of unconjugated and conjugated forms of bilirubin can provide useful diagnostic clues. For example, **hemolytic jaundice** results from the destruction of large numbers of red blood cells. When this occurs, phagocytes release massive quantities of unconjugated bilirubin into the blood. Because the liver cells then accelerate the secretion of bilirubin in the bile, the blood concentration of conjugated bilirubin does not increase proportionately. A blood test from a patient with hemolytic jaundice would reveal (1) elevated total bilirubin, (2) high concentrations of unconjugated bilirubin, and (3) conjugated bilirubin contributing much less than 15 percent to the total bilirubin concentration.

These results are quite different from those seen in **obstructive jaundice.** In this condition, the ducts that remove bile from the liver are constricted or blocked. Liver cells cannot get rid of conjugated bilirubin, and large quantities diffuse into the blood. In this case diagnostic tests would show (1) elevated total bilirubin, (2) unconjugated bilirubin contributing much less than 85 percent to the total bilirubin concentration, and (3) high concentrations of conjugated bilirubin.

⚕ Iron Deficiencies and Excesses

EAP *p. 314*

If dietary supplies of iron are inadequate, hemoglobin production slows down, and symptoms of *iron deficiency anemia* appear. This form of anemia can also be caused by any condition that produces a blood loss, since the iron in the lost blood cannot be recycled. As the red blood cells are replaced, iron reserves must be mobilized for use in the synthesis of new hemoglobin molecules. If those reserves are exhausted, or dietary sources are inadequate, symptoms of iron deficiency appear. In iron deficiency anemia, the red blood cells are unable to synthesize functional hemoglobin, and they are unusually small when they enter the circulation. The hematocrit declines, and the hemoglobin content and oxygen-carrying capacity of the blood are substantially reduced. Symptoms include weakness and a tendency to fatigue easily.

Women are especially dependent on a normal dietary supply of iron, because their iron reserves are smaller than those of men. The body of a normal man contains around 3.5 g of iron in the ionic form Fe^{2+}. Of that amount, 2.5 g are bound to the hemoglobin of circulating red blood cells, and the rest is stored in the liver and bone marrow. In women, the total body iron content averages 2.4 g, with roughly 1.9 g incorporated into red blood cells. Thus a woman's iron reserves consist of only 0.5 g, half that of a typical man.

Because their reserves are relatively small, women are dependent on a reliable dietary supply of iron. When the demand for iron increases out of proportion with dietary supplies, iron deficiency develops. An estimated 20 percent of menstruating women in the United States show signs of iron deficiency. Pregnancy also stresses iron reserves, for the woman must provide the iron needed to produce both maternal and fetal erythrocytes.

Good dietary sources of iron include liver, red meats, kidney beans, egg yolks, spinach, and carrots. Iron supplements can help prevent iron deficiency, but too much iron can be as dangerous as too little. Iron absorption across the digestive tract normally keeps pace with physiological demands. When the diet contains abnormally high concentrations of iron, or hereditary factors increase the rate of absorption, the excess iron gets stored in peripheral tissues. This is called *iron loading*. Eventually cells begin to malfunction as massive iron deposits accumulate in the cytoplasm. For example, iron deposits in pancreatic cells can lead to diabetes mellitus; deposits in cardiac muscle cells lead to abnormal heart contractions and heart failure. (There is evidence that iron deposits in the heart caused by the overconsumption of red meats may contribute to heart disease.) Liver cells become nonfunctional, and liver cancers may develop.

Comparable symptoms of iron loading may appear following repeated transfusions of whole blood, because each unit of whole blood contains roughly 250 mg of iron. For example, as noted above, the various forms of *thalassemia* result from a genetic inability to produce adequate amounts of one of the four globin chains in hemoglobin. Erythrocyte production and survival are reduced, and so is the oxygen-carrying capacity of the blood. Individuals with severe untreated thalassemia usually die in their twenties, but not because of the anemia. These patients are treated for severe anemia with frequent blood transfusions which prolong life, but the excessive iron loading eventually leads to fatal heart problems.

Hemolytic Disease of the Newborn

Hemolytic disease of the newborn results from the maternal production of anti-Rh antibodies that cross the placenta to attack fetal Rh-positive red blood cells. Within 6 months after delivery, roughly 20 percent of Rh-negative mothers who carried Rh-positive children have become sensitized, and produce anti-Rh antibodies. For the entire sequence of events, See Figure A-32. Without treatment, the fetus will probably die before delivery or shortly thereafter.

A newborn with severe HDN is anemic, and the high concentration of circulating bilirubin produces jaundice. Because the maternal antibodies will remain active for 1 to 2 months after delivery, the infant may need to have its entire blood volume replaced. This blood replacement removes most of the maternal antibodies as well as the affected erythrocytes, reducing the complications and the chance that the infant will die.

When there is a danger that the fetus may not survive to full term, premature delivery may be induced after 7 to 8 months of development. In a severe case affecting a fetus at an earlier stage, one or more transfusions can be given while the fetus continues to develop within the uterus.

To avoid the problem completely, the maternal production of Rh antibodies is prevented by administering Rh antibodies (available under the name RhoGam) during pregnancy and following delivery. These "foreign" antibodies quickly destroy any fetal red blood cells that cross the placental barrier. Thus there are no exposed antigens to stimulate the maternal immune system, sensitization does not occur, and Rh antibodies are not produced. This relatively simple procedure could almost entirely prevent HDN mortality caused by Rh incompatibilities.

Erythrocytosis and Blood Doping   EAP *p. 315*

In **erythrocytosis** (e-rith-rō-sī-TŌ-sis), the blood contains abnormally large numbers of red blood cells. Erythrocytosis usually results from the massive release of erythropoietin by tissues (especially the kidneys) deprived of oxygen. People moving to high altitudes usually experience erythrocytosis following their arrival, because the air contains less oxygen than it does at sea level. The increased number of red blood cells compensates for the fact that individually each RBC is carrying less oxygen than it would at sea level. Mountaineers and those living at altitudes of 10,000-12,000 feet may have hematocrits as high as 65.

Individuals whose hearts or lungs are functioning inadequately may also develop erythrocytosis. For example, this condition is often seen in

12

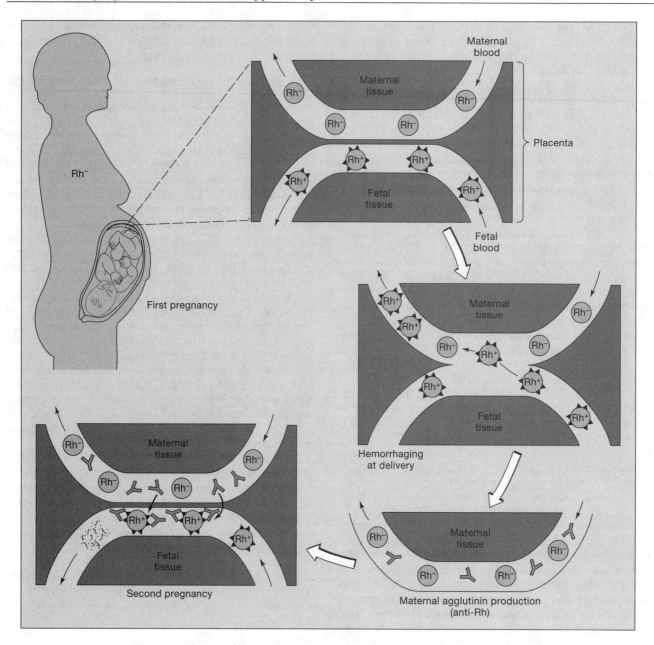

FIGURE A-32

Rh Factors and Pregnancy. When an Rh-negative woman has her first Rh-positive child, mixing of fetal and maternal blood occurs at delivery when the placental connection breaks down. The appearance of Rh-positive blood cells in the maternal circulation sensitizes the mother, stimulating the production of anti-Rh agglutinins. If another pregnancy occurs with an Rh-positive fetus, maternal agglutinins can cross the placental barrier and attack fetal blood cells, producing symptoms of HDN (hemolytic disease of the newborn).

heart failure and emphysema, two conditions discussed in later chapters. Whether the blood fails to circulate efficiently or the lungs do not deliver enough oxygen to the blood, peripheral tissues remain oxygen-poor despite the rising hematocrit. Having a higher concentration of red blood cells increases the oxygen-carrying capacity of the blood, but it also makes the blood thicker and harder to push around the circulatory system. This increases the work load on the heart, making a bad situation even worse.

The practice of **blood doping** has become widespread among competitive athletes involved with endurance sports such as cycling. The procedure entails removing whole blood from the athlete in the weeks before an event. The packed red cells are separated from the plasma and stored. By the time of the race, the competitor's bone marrow will have replaced the lost blood. Immediately before the event the packed red cells are reinfused, increasing the hematocrit. The objective is to elevate the oxygen-carrying capacity of the blood, and so increase endurance. The consequence is that the athlete's heart is placed under a tremendous strain. The long-term effects are unknown, but the practice obviously carries a significant risk; it has recently been banned in amateur sports. Attempts to circumvent this rule by the use of EPO in 1992-1993 resulted in the tragic deaths of 18 European cyclists.

Blood Tests and RBCs

EAP *p. 315*

This section describes several common blood tests that assess circulating RBCs.

RETICULOCYTE COUNT. *Reticulocytes* are immature red blood cells that are still synthesizing hemoglobin. Most reticulocytes remain in the bone marrow until they complete their maturation, but some enter the circulation. Reticulocytes normally account for around 0.8 percent of the erythrocyte population. Values above 1.5 percent or below 0.5 percent indicate that something is wrong with the rates of RBC survival or maturation.

HEMATOCRIT (Hct). The hematocrit value is the percentage of whole blood occupied by cells. Normal adult hematocrits average 46 for men and 42 for women, with ranges of 40-54 for men and 37-47 for women.

HEMOGLOBIN CONCENTRATION (Hb). This test determines the amount of hemoglobin in the blood, expressed in grams per deciliter (g/dl). Normal ranges are 14-18 g/dl in males and 12-16 g/dl in females. The differences in hemoglobin concentration reflect the differences in hematocrit. For both sexes, a single RBC contains 27-33 picograms (pg) of hemoglobin.

RBC COUNT. Calculations of the RBC count, the number of RBCs per microliter of blood, are based on the hematocrit and hemoglobin content, and can be used to develop a better picture of the condition of the RBCs. Values often reported in blood screens include

- *Mean corpuscular volume (MCV),* the average volume of an individual red blood cell, in cubic micrometers. It is calculated by dividing the volume of red cells per microliter by the RBC count, using the formula

$$MCV = Hct/RBC \text{ count (in millions)} \times 10$$

Normal values range from 82.2 to 100.6. Using a representative RBC count of 5.2 million, the mean corpuscular volume would be

$$MCV = 46/5.2 \times 10 = 88.5 \ \mu m^3$$

Cells of normal size are **normocytic,** while larger or smaller than normal RBCs are called **macrocytic** or **microcytic,** respectively.

- *Mean corpuscular hemoglobin concentration (MCHC),* the amount of hemoglobin within a sin-

gle RBC, expressed in picograms. Normal values range from 27 to 34 pg. MCHC is calculated as

$$MCHC = Hb/RBC \text{ count (in millions)} \times 10$$

RBCs containing normal amounts of hemoglobin are termed **normochromic** while **hyperchromic** and **hypochromic** indicate higher or lower than normal hemoglobin content, respectively.

Anemia (a-NĒ-mē-ah) exists when the oxygen-carrying capacity of the blood is reduced, diminishing the delivery of oxygen to peripheral tissues. Such a reduction causes a variety of symptoms, including premature muscle fatigue, weakness, lethargy, and a general lack of energy. Anemia may exist because the hematocrit is abnormally low or because the amount of hemoglobin in the RBCs is reduced. Standard laboratory tests can be used to differentiate between the various forms of anemia on the basis of the number, size, shape, and hemoglobin content of red blood cells. As an example, Table A-16 shows how this information can be used to distinguish among four major types of anemia.

1. **Hemorrhagic anemia** results from severe blood loss. Erythrocytes are of normal size, each contains a normal amount of hemoglobin, and reticulocytes are present in normal concentrations, at least initially. Blood tests would therefore show a low hematocrit and low hemoglobin, but the MCV, MCHC, and reticulocyte counts would be normal.

2. In **aplastic** (ā-PLAS-tik) **anemia,** the bone marrow fails to produce new red blood cells. The 1986 nuclear accident in Chernobyl (U.S.S.R.) caused a number of cases of aplastic anemia. The condition is usually fatal unless surviving stem cells repopulate the marrow or a bone marrow transplant is performed. In aplastic anemia the circulating red blood cells are normal in all respects, but because new RBCs are not being produced, the reticulocyte count is extremely low.

3. In **iron deficiency anemia,** normal hemoglobin synthesis cannot occur because iron reserves are inadequate. Developing red blood cells cannot synthesize functional hemoglobin, and as a result they are unusually small. A blood test therefore shows a low hematocrit, low hemoglobin content, low MCV, and low MCHC, but a normal reticulocyte count. An estimated 60 mil-

Table A-16   RBC Tests and Anemias

Anemia type	Hct	Hb	Reticulocyte count	MCV	MCHC
Hemorrhagic	low	low	normal	normal	normal
Aplastic	low	low	very low	normal	normal
Iron deficiency	low	low	normal	low	low
Pernicious	low	low	very low	high	high

lion women worldwide have symptoms of iron deficiency anemia. (See the discussion on iron deficiencies and excesses on p. 92.)

4. In **pernicious** (per-NISH-us) **anemia,** normal red blood cell maturation ceases because of an inadequate supply of vitamin B_{12}. Erythrocyte production declines, and the red blood cells are abnormally large and may develop a variety of bizarre shapes. Blood tests from a person with pernicious anemia indicate a low hematocrit with a very high MCV and a low reticulocyte count.

✠ The Leukemias

EAP *p. 319*

Leukemias characterized by the presence of abnormal granulocytes or other cells of the bone marrow are called **myeloid,** and those involving lymphocytes are termed **lymphoid.** The first symptoms appear as immature, and abnormal white blood cells appear in the circulation. As their numbers increase, they travel via the circulation, invading tissues and organs throughout the body.

These cells are extremely active, and they require abnormally large amounts of energy. As in other cancers, described in Chapter 4 and elsewhere in this *Manual* (p. 115), invading leukemic cells gradually replace the normal cells, especially in the bone marrow. Red blood cell, normal WBC, and platelet formation decline, with resulting anemia, infection, and impaired blood clotting, and untreated leukemias are invariably fatal.

Leukemias may be classified as *acute* (short and severe) or *chronic* (prolonged). Acute leukemias may be linked to radiation exposure, hereditary susceptibility, viral infections, or unknown causes. Chronic leukemias may be related to chromosomal abnormalities or immune system malfunctions. Survival in untreated acute leukemia averages about three months; individuals with chronic leukemia may survive for years.

Effective treatments exist for some forms of leukemia and not others. For example, when acute lymphoid leukemia is detected early, 85-90 percent of patients can be held in remission for 5 years or longer, but only 10-15 percent of patients with acute myeloid leukemia survive 5 years or more. The yearly mortality rate for leukemia (all types) in the United States has not declined appreciably in the past 30 years, remaining at around 6.8 per 100,000 population. However, new treatments are being developed that show promise when used against specific forms of leukemia. For example, administration of α-*interferon,* a hormone of the immune system, has been very effective in treating hairy cell leukemia and chronic myeloid leukemia.

One option for treating acute leukemias is to perform a bone marrow transplant. In this procedure massive chemotherapy or radiation treatment is given, enough to kill all the cancerous cells. Unfortunately, this also destroys the patient's blood cells and stem cells in the bone marrow and other blood-forming tissues. The individual then receives an infusion of healthy bone marrow cells that repopulate the blood and marrow tissues.

If the bone marrow is extracted from another person (a **heterologous marrow transplant),** care must be taken to ensure that the blood types and tissue types are compatible (see Chapters 12 and 15 of the text). If they are not, the new lymphocytes may attack the patient's tissues, with potentially fatal results. Best results are obtained when the donor is a close relative. In an **autologous marrow transplant** bone marrow is removed from the patient, cleansed of cancer cells, and reintroduced after radiation or chemotherapy treatment. Although there are fewer complications, the preparation and cleansing of the marrow are technically difficult and time consuming.

Bone marrow transplants are also performed to treat patients whose bone marrow has been destroyed by toxic chemicals or radiation. For example, heterologous transplants were used successfully in the U.S.S.R. to treat survivors of the Chernobyl nuclear reactor accident in 1986.

▯ Testing the Clotting System

EAP *p. 324*

Several clinical tests check the efficiency of the clotting system:

BLEEDING TIME. This test measures the time it takes for a small skin wound to seal itself. There are several variations on this procedure, with normal values ranging from 1 to 9 minutes. The nonprescription drug *aspirin* prolongs the bleeding time by affecting platelet function and suppressing the extrinsic pathway.

COAGULATION TIME. In this test, a sample of whole blood is allowed to stand under controlled conditions until a visible clot has formed. Normal values range from 3 to 15 minutes. The test has several potential sources of error, and so is not very accurate. It is nevertheless of value because it is the simplest test that can be performed on a blood sample. More sophisticated tests begin by adding citrate ions to the sample. Citrate ties up the calcium ions in the plasma and prevents premature clotting.

PARTIAL THROMBOPLASTIN TIME (PTT). In this test a plasma sample is mixed with chemicals that mimic the effects of activated platelets. Calcium ions are then introduced, and the clotting time is recorded. Clotting normally occurs in 35-50 seconds if the enzymes and clotting factors of the intrinsic pathway are present in normal concentrations.

PLASMA PROTHROMBIN TIME (PROTHROMBIN TIME, PT). This test checks the performance of the extrinsic pathway. The procedure is similar to that in the PTT test, but the clotting process is triggered by exposure to a combination of tissue thromboplastin and calcium ions. Clotting normally occurs in 12-14 seconds.

✝ The Cardiomyopathies   EAP *p. 333*

The **cardiomyopathies** (kar-dē-ō-mī-OP-a-thēz) include an assortment of diseases with a common symptom: the progressive, irreversible degeneration of the myocardium. Cardiac muscle fibers are damaged and replaced by fibrous tissue, and the muscular walls of the heart become thin and weak. As muscle tone declines, the ventricular chambers become greatly enlarged. When the remaining fibers cannot develop enough force to maintain cardiac output, symptoms of heart failure develop.

Chronic alcoholism and coronary artery disease are probably the most common causes of cardiomyopathy in the United States. Infectious agents, including viruses, bacteria, fungi, and protozoans, can also produce cardiomyopathies. Diseases affecting neuromuscular performance, such as muscular dystrophy (discussed elsewhere in this manual), can also damage cardiac muscle fibers, as can starvation or chronic variations in the extracellular concentrations of calcium or potassium ions.

There are also several inherited forms of cardiomyopathy. **Hypertrophic cardiomyopathy (HCM)** is an inherited disorder that makes the wall of the left ventricle thicken to the point where it has difficulty pumping blood. Most people with HCM do not become aware of it until relatively late in life. However, HCM can also cause a fatal arrhythmia; it has been implicated in the sudden deaths of several young athletes, and Defense Secretary Les Aspin was hospitalized for this condition in 1993. Implantation of an electronic cardiac pacemaker has proven to be beneficial in controlling these arrhythmias.

Finally, there are a significant number of cases of *idiopathic cardiomyopathy*, a term used when the primary cause cannot be determined.

Ⓡ Heart Transplants and Assist Devices   EAP *p. 333*

Individuals suffering from severe cardiomyopathies may be considered as candidates for heart transplants. This surgery involves the complete removal of the weakened heart and its replacement by a heart taken from a suitable donor. To survive the surgery, the recipient must be in otherwise satisfactory health; because the number of suitable donors is limited, the available hearts are usually assigned to individuals younger than age 50. Out of the 8,000-10,000 U.S. patients each year suffering from potentially fatal cardiomyopathies, only around 1,000 receive heart transplants. There is an 80-85 percent one-year survival rate and a 50-70 percent five-year survival rate after successful transplantation. This rate is quite good, considering that these patients would have died if the transplant had not been performed. However, the procedure remains controversial due to the high cost involved. As health care dollars become managed more closely, society will need to decide whether the probable life extension is worth the expense. Needless to say, this will not be an easy decision.

Many individuals with cardiomyopathy who are initially selected for transplant surgery succumb to the disease before a suitable donor becomes available. For this reason there continues to be considerable interest in the development of an artificial heart. One model, the *Jarvik-7*, had limited clinical use in the 1980s. Attempts to implant it on a permanent basis were unsuccessful, primarily because of formation of blood clots on the mechanical valves. When these clots broke free, they formed drifting emboli that plugged peripheral vessels, producing strokes, kidney failure, and other complications. In 1989 the federal government prohibited further experimental use of the Jarvik-7 as a permanent heart implant. Modified versions of this unit and others now under development may still be used to maintain transplant candidates while awaiting the arrival of a donor organ. These are called *left ventricular assist devices* (LVAD). As the name implies, these devices assist, rather than replace, the damaged heart. A mechanical left ventricular assist device has been used to support a patient awaiting a transplant.

Another interesting approach involves using skeletal muscle tissue to apply permanent patches to injured hearts or to build small accessory pumps. For example, one procedure creates a biological LVAD by freeing a portion of the latissimus dorsi muscle from the side and placing this flap, with its circulation intact, into the thoracic cavity. There it is folded to form a sling around the heart, and an electronic pacemaker is used to stimulate its contraction. Each time it contracts, the sling squeezes the heart and helps push blood into the major arteries. These methods are less stressful than heart transplants because (1) they leave the damaged heart in place, and (2) the transplanted tissue is taken from the same individual, so it will not be attacked by the immune system. Although preliminary results have been encouraging, few procedures have been performed, and the use of skeletal muscle patches remains an experimental concept rather than a recognized treatment for cardiomyopathy.

A third experimental approach, which has yet to be tried with human patients, involves the insertion of fetal heart muscle cells in a damaged adult heart. The fetal cells appear to adapt to their surroundings and differentiate into functional contractile cells.

✝ Infection and Inflammation of the Heart   EAP *p. 333*

Many different microorganisms may infect heart tissue, leading to serious cardiac abnormalities. **Carditis** (kar-DĪ-tis) is a general term indicating inflammation of the heart. Clinical conditions resulting from cardiac infection are usually identified by the primary site of infection. For example, those affecting the endocardium produce symptoms of **endocarditis.** Endocarditis primarily affects the chordae tendineae and heart valves, and the mortality rate may reach 21-35 percent. The most severe complications result from the formation of blood clots on the damaged surfaces. These clots subsequently break free, entering the circulation as drifting emboli (see *p. 324* of the text) that may cause strokes, heart attacks, or kidney failure. Destruction of heart valves by infection may lead to valve leakage, heart failure, and death.

Bacteria, viruses, protozoa, and fungal pathogens that attack the myocardium produce **myocarditis.** The microorganisms implicated include those responsible for many of the conditions discussed in earlier chapters, including diphtheria, syphilis, polio, and malaria. The membranes of infected heart muscle cells become facilitated, and the heart rate rises dramatically. Over time, abnormal contractions may appear, the heart muscle weakens, and these may eventually prove fatal.

✚ RHD and Valvular Stenosis

EAP *p. 333*

Rheumatic (roo-MA-tik) **fever** is an inflammatory condition that may develop following infection by streptococcal bacteria. Rheumatic fever most often affects children of ages 5-15; symptoms include high fever, joint pain and stiffness, and a distinctive full-body rash. Obvious symptoms usually persist for less than 6 weeks, although severe cases may linger for 6 months or more. The longer the duration of the inflammation, the more likely it is that carditis will develop. The carditis that does develop in 50-60 percent of patients often escapes detection, and scar tissue forms gradually in the myocardium and the heart valves. Valve condition deteriorates over time, and valve problems serious enough to affect cardiac function may not appear until 10-20 years after the initial infection.

Over the interim the affected valves become thickened and often calcified to some degree. This thickening narrows the opening guarded by the valves, producing a condition called **valvular stenosis** (ste-NO-sis; *stenos*, narrow). The resulting clinical disorder is known as **rheumatic heart disease, or RHD.** The thickened cusps stiffen in a partially closed position, but the valves do not completely block the circulation because the edges of the cusps are rough and irregular (Figure A-33). Regurgitation may occur, and much of the blood pumped out of the heart may flow right back in. The abnormal valves are also much more susceptible to bacterial infection, a type of *endocarditis.*

Mitral stenosis and **aortic stenosis** are the most common forms of valvular heart disease. About 40 percent of patients with RHD develop mitral stenosis, and two-thirds of them are women. The reason for the correlation between female gender and mitral stenosis is unknown. In mitral stenosis blood enters the left ventricle at a slower than normal rate, and when the ventricle contracts blood flows back into the left atrium as well as into the aortic trunk. As a result, the left

ventricle has to work much harder to maintain adequate systemic circulation. The right and left ventricles discharge identical amounts of blood with each beat, and as the output of the left ventricle declines, blood "backs up" in the pulmonary circuit. Venous pressures then rise in the pulmonary circuit, and the right ventricle must develop greater pressures to force blood into the pulmonary trunk. In severe cases of mitral stenosis, the ventricular musculature is not up to the task. The heart weakens, and peripheral tissues begin to suffer from oxygen and nutrient deprivation. (This condition, called heart failure, is discussed in more detail in a later section.)

Symptoms of aortic stenosis develop in roughly 25 percent of patients with RHD; 80 percent of these individuals are males. Symptoms of aortic stenosis are initially less severe than those of mitral stenosis. Although the left ventricle enlarges and works harder, normal circulatory function can often be maintained for years. Clinical problems develop only when the opening narrows enough to prevent adequate blood flow. Symptoms then resemble those of mitral stenosis.

One reasonably successful treatment for severe stenosis involves the replacement of the damaged valve with a prosthetic (artificial) valve. Figure A-33 shows a stenotic heart valve (left) and two possible replacements: a valve from a pig (center) and a synthetic valve (right), one of a number of designs that have been employed. Pig valves do not require anticoagulant therapy, but may wear out and begin leaking after roughly 10 years in service. The plastic and/or stainless steel components of the artificial valve are more durable but may activate the immune system of the recipient, leading to inflammation, clot formation, and other potential complications. Synthetic valve recipients must take anticoagulant drugs to prevent strokes and other disorders caused by embolus formation. Valve replacement operations are quite successful, with about 95 percent of the surgical patients surviving for three years or more and 70 percent surviving over five years.

✚ Coronary Artery Disease

EAP *p. 336*

The term **coronary artery disease (CAD)** refers to degenerative changes in the coronary circulation. Cardiac muscle fibers need a constant supply of oxygen and nutrients, and any reduction in coronary circulation produces a corresponding reduction in cardiac performance. Such reduced circulatory supply, known as **coronary ischemia** (is-KĒ-mē-a), usually results from partial or complete blockage of the coronary arteries. The usual cause is the formation of a fatty deposit, or *plaque*, in the wall of a coronary vessel. The plaque, or an associated thrombus, then narrows the passageway and reduces blood flow. Spasms in the smooth muscles of the vessel wall can further decrease blood flow or even stop it altogether. Plaque development and growth will be considered in Chapter 14.

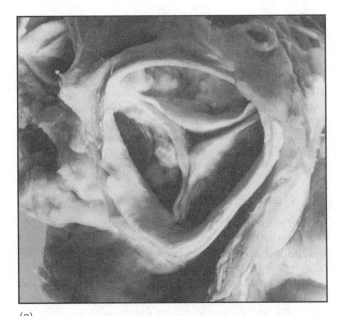

(a)

Figure A-33   Artificial Heart Valves.
(a) A stenotic semilunar valve; note the irregular, stiff cusps.
(b) Intact™ Bioprosthetic heart valve. (c) Medtronic Hall™
Prosthetic heart valve (d) Cloth-covered ball-in-cage valves.

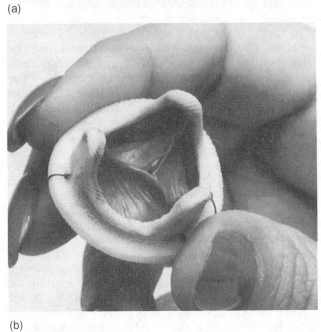

(b)

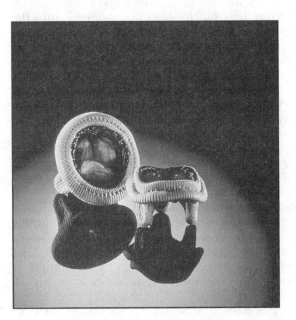

13

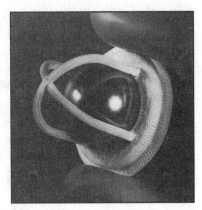

(c)   (d)

One of the first symptoms of CAD is often **angina pectoris** (an-JĪ-na PEK-tor-is; *angina*, pain spasm + *pectoris*, of the chest). In the most common form of angina, temporary insufficiency and ischemia develop when the workload of the heart increases. Although the individual may feel comfortable at rest, any unusual exertion or emotional stress can produce a sensation of pressure, chest constriction, and pain that may radiate from the sternal area to the arms, back, and neck.

Angina can often be controlled by a combination of drug treatment and changes in lifestyle. Lifestyle changes to combat angina include (1) limiting activities known to trigger angina attacks, such as strenuous exercise, and avoiding stressful situations; (2) stopping smoking; and (3) modifying the diet to lower fat consumption. Medications useful for controlling angina include drugs that block sympathetic stimulation (*propranolol* or *metoprolol*); vasodilators such as *nitroglycerin* (nī-trō-GLIS-er-in) and *atrial natriuretic peptide* (*ANP*); and drugs that block calcium movement into the cardiac muscle cells (*calcium channel blockers*).

Angina can also be treated surgically. A single, soft plaque may be reduced with the aid of a slender, elongate **catheter** (KATH-e-ter). The catheter, a small-diameter tube, is inserted into a large artery and guided into a coronary artery to the plaque. A variety of surgical tools can be slid into the catheter, and the plaque can then be removed with laser beams or chewed to pieces by a miniature version of the Roto-Rooter machine. Debris created during plaque destruction is sucked up by the catheter, preventing blockage of smaller vessels.

In **balloon angioplasty** (AN-jē-ō-plas-tē; *angeion*, vessel) the catheter tip contains an inflatable balloon. Once in position, the balloon is inflated, pressing the plaque against the vessel walls. This procedure works best in small (under 10 mm) soft plaques. Several factors make this a highly attractive treatment: (1) The mortality rate during surgery is only around 1 percent; (2) the success rate is over 90 percent; and (3) it can be performed on an outpatient basis. Although in about 20 percent of patients the plaque deposit returns to its original size within 6 months, the process can be repeated as needed. Unfortunately, only about 10 percent of severe angina patients have isolated problems suitable for balloon angioplasty.

A **coronary artery bypass graft (CABG)** involves taking a small section from either a small artery (often the *internal thoracic artery*) or a peripheral vein (such as the *great saphenous vein* of the leg) and using it to create a detour around the obstructed portion of a coronary artery. As many as four coronary arteries can be rerouted this way during a single operation. The procedures are named according to the number of vessels repaired, so one speaks of *single, double, triple,* or *quadruple coronary bypass operations.* The mortality rate during surgery for operations performed before significant heart damage has occurred is relatively low (1–2 percent). Under these conditions the procedure completely eliminates the angina symptoms in 70 percent of the cases and provides partial relief in another 20 percent.

Although it does offer certain advantages, recent studies have shown that for mild angina, coronary bypass surgery does not yield significantly better results than drug therapy. Current recommendations are that coronary bypass surgery be reserved for cases of severe angina that do not respond to other treatment.

⚕ Heart Attacks   EAP *p. 336*

In a **myocardial** (mī-ō-KAR-dē-al) **infarction (MI)**, or *heart attack*, the coronary circulation becomes blocked and the cardiac muscle cells die from lack of oxygen. The affected tissue then degenerates, creating a nonfunctional area known as an *infarct.* Heart attacks most often result from severe coronary artery disease. The consequences depend on the site and nature of the circulatory blockage. If it occurs near the base of one of the coronary arteries, the damage will be widespread and the heart will probably stop beating. If the blockage involves one of the smaller arterial branches, the individual may survive the immediate crisis, but there are many potential complications, all unpleasant. As scar tissue forms in the damaged area, the heartbeat may become irregular, and other vessels can become constricted, creating additional circulatory problems.

Myocardial infarctions are most often associated with fixed blockages, such as those seen in CAD. When the crisis develops because of thrombus (clot) formation at a plaque, the condition is called **coronary thrombosis.** A vessel already narrowed by plaque formation may also become blocked by a sudden spasm in the smooth muscles of the vascular wall. The individual then may experience intense pain, similar to that of an angina attack but persisting even at rest. However, pain does not always accompany a heart attack. These *silent heart attacks* may be even more dangerous, because the condition may not be diagnosed and treated before a fatal MI occurs. Roughly 25 percent of heart attacks are not recognized when they occur.

The cytoplasm of a damaged cardiac muscle cell differs from that of a normal muscle cell. As the supply of oxygen decreases, the cells become more dependent on anaerobic metabolism to meet their energy needs. Over time the cytoplasm accumulates large numbers of enzymes involved with anaerobic energy production.

As the cardiac muscle cell membranes deteriorate, these enzymes enter the surrounding intercellular fluids. The appearance of such enzymes in the circulation thus indicates that an infarct has occurred. The enzymes tested for in a diagnostic blood test include **lactate dehydrogenase (LDH), serum glutamic oxaloacetic transaminase (SGOT,** also called *aspartate aminotransferase*), **creatine** phosphokinase **(CPK,** or **CK),** and a special form of creatine phosphokinase found only in cardiac muscle **(CK-MB).**

Roughly 25 percent of MI patients die before obtaining medical assistance, and 65 percent of MI deaths among those under age 50 occur within an hour after the initial infarct. The goals of treatment are to limit the size of the infarct and prevent additional complications by preventing irregular contractions, improving circulation with vasodilators, providing additional oxygen, reducing the cardiac workload, and, if possible, eliminating the cause of the circulatory blockage. Anticoagulants may help prevent the formation of additional thrombi, and clot-dissolving enzymes may reduce the extent of the damage if they are administered within 6 hours after the MI has occurred. Controversy exists over the benefits of using t-PA, which is relatively expensive, versus other fibrinolytic agents, such as urokinase or streptokinase. A procedure that combines small amounts of t-PA with streptokinase has been proposed as a workable compromise that produces effects equivalent to high doses of t-PA alone. Follow-up treatment with heparin or aspirin or both is recommended; without further treatment the circulatory blockages will reappear in roughly 20 percent of patients.

There are roughly 1.3 million MIs in the United States each year, and half of the victims die within a year of the incident. A number of factors appear to increase the risk of a heart attack: smoking, high blood pressure, high blood cholesterol levels, high circulating levels of *low-density lipoproteins* (LDL), diabetes, male gender (below age 70), severe emotional stress, and obesity. The role of lipoproteins and cholesterol in plaque formation and heart disease will be considered in Chapter 14. Hereditary factors may also predispose an individual to coronary artery disease. Although the rate of heart attacks of women under age 70 is lower than that of men, their mortality rate is actually higher— perhaps because heart disease in women is neither diagnosed as early nor treated as aggressively as is heart disease in men.

The presence of two risk factors more than doubles the risk, so eliminating as many risk factors as possible will improve one's chances of preventing or surviving a heart attack. Changes in diet to limit cholesterol, exercise to lower weight, and seeking treatment for high blood pressure are steps in the right direction. It has been estimated that reduction of coronary risk factors could prevent 150,000 deaths each year in the United States alone.

📋 Interpreting Abnormal ECGs

EAP p. 339

Damage to the conduction pathways caused by mechanical distortion, ischemia, infection, or inflammation can affect the normal rhythm of the heart. The resulting condition is called a **conduction deficit,** or **heart block.** Heart blocks of vary-

ing severity are illustrated in Figure A-34. In a **first-degree heart block** (Figure A-34b), the AV node and proximal portion of the AV bundle slow the passage of impulses heading for the ventricular myocardium. As a result, a pause appears between the atrial and ventricular contractions. Although a delay exists, the regular rhythm of the heart continues, and each atrial beat is followed by a ventricular contraction.

If the delay lasts long enough, the nodal cells will still be repolarizing from the previous beat when the next impulse arrives from the pacemaker. The arriving impulse will then be ignored, the ventricles will not be stimulated, and the normal "atria-ventricles, atria-ventricles" pattern will disappear. This condition is a **second-degree heart block** (Figure A-32c). A mild second-degree block may produce only an occasional skipped beat, but with more substantial delays the ventricles will follow every second atrial beat. The resulting pattern of "atria, atria-ventricles, atria, atria-ventricles" is known as a **two-to-one (2:1) block.** Three-to-one or even four-to-one blocks are also encountered.

In a **third-degree heart block,** or **complete heart block,** the conducting pathway stops func-

Figure A-34   Heart Blocks

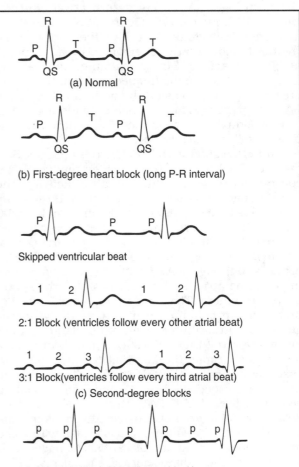

(a) Normal

(b) First-degree heart block (long P-R interval)

Skipped ventricular beat

2:1 Block (ventricles follow every other atrial beat)

3:1 Block (ventricles follow every third atrial beat)

(c) Second-degree blocks

(d) Complete block (third-degree block)
   (atrial beats occur regularly, ventricular beats occur at
   slower, unrelated pace)

13

tioning altogether (Figure A-34d). The atria and ventricles continue to beat, but their activities are no longer synchronized. The atria follow the pace set by the SA node, beating 70-80 times per minute, and the ventricles follow the commands of the AV node, beating at a rate of 40-60 per minute. A temporary third-degree block can be induced by stimulating the vagus nerve. In addition to slowing the rate of impulse generation by the SA node, such stimulation inhibits the AV nodal cells to the point that they cannot respond to normal stimulation. Comments such as "my heart stopped" or "my heart skipped a beat" usually refer to this phenomenon. The pause typically lasts for just a few seconds. Longer delays end when a conducting cell, usually one of the Purkinje fibers, depolarizes to threshold. This phenomenon is called **ventricular escape** because the ventricles are escaping from the control of the SA node. Ventricular escape can be a lifesaving event if the conduction system is damaged. Even without instructions from the SA or AV nodes, the ventricles will continue to pump blood at a slow but steady rate.

Tachycardia and Fibrillation

Additional important examples of arrhythmias are shown in Figure A-35. **Premature atrial contractions** (PACs), indicated in Figure A-35b, often occur in normal individuals. In a PAC the normal atrial rhythm is momentarily interrupted by a "surprise" atrial contraction. Stress, caffeine, and various drugs may increase the frequency of PAC incidence, presumably by increasing the permeabilities of the SA pacemakers. The impulse spreads along the conduction pathway, and a normal ventricular contraction follows the atrial beat.

In **paroxysmal atrial tachycardia** (par-ok-SIZ-mal), or **PAT** (Figure A-35c), a premature atrial contraction triggers a flurry of atrial activity. The ventricles are still able to keep pace, and the heart rate jumps to about 180 beats per minute. In **atrial flutter** the atria are contracting in a coordinated manner, but the contractions are occurring very frequently. During a bout of **atrial fibrillation** (fi-bri-LĀ-shun), Figure A-35d, the impulses are moving over the atrial surface at rates of perhaps 500 beats per minute. The atrial wall quivers instead of producing an organized contraction. The ventricular rate in atrial flutter or atrial fibrillation cannot follow the atrial rate, and may remain within normal limits. Despite the fact that the atria are now essentially nonfunctional, the condition may go unnoticed, especially in older individuals leading sedentary lives. PACs, PAT, atrial flutter, and even atrial fibrillation are not considered very dangerous unless they are prolonged, or associated with some more serious indications of cardiac damage, such as coronary artery disease or valve problems.

In contrast, ventricular arrhythmias may be serious and even fatal. Because the conduction system functions in one direction only, a ventricular arrhythmia is not linked to atrial activities. **Premature ventricular contractions** (PVCs; Figure A-35e) occur when a Purkinje cell or ventricular myocardial cell depolarizes to threshold and triggers a premature contraction. The cell responsible is called an *ectopic pacemaker.* The frequency of PVCs can be increased by exposure to epinephrine and other stimulatory drugs or to ionic changes that depolarize cardiac muscle fiber membranes. Similar factors may be responsible for periods of **ventricular tachycardia,** also known as **VT,** or *V-tach* (Figure A-35f).

PVCs and VT often precede the most serious arrhythmia, **ventricular fibrillation** (VF), shown in Figure A-35g. The resulting condition, also known as **cardiac arrest,** is rapidly fatal because the heart stops pumping blood. During ventricular fibrillation the cardiac muscle fibers are overly sensitive to stimulation, and the impulses are traveling from cell to cell around and around the ventricular walls. A normal rhythm cannot become established because the ventricular muscle fibers are stimulating one another at such a rapid rate. The problem

Figure A-35   Cardiac Arrhythmias.

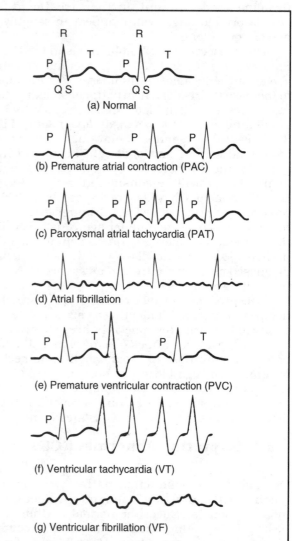

(a) Normal

(b) Premature atrial contraction (PAC)

(c) Paroxysmal atrial tachycardia (PAT)

(d) Atrial fibrillation

(e) Premature ventricular contraction (PVC)

(f) Ventricular tachycardia (VT)

(g) Ventricular fibrillation (VF)

is exaggerated by a sustained rise in free intracellular calcium ion concentrations, due to massive stimulation of alpha and beta receptors following sympathetic activation.

A **defibrillator** is a device that attempts to eliminate ventricular fibrillation and restore normal cardiac rhythm. Two electrodes are placed in contact with the chest and a powerful electrical shock is administered. The electrical stimulus depolarizes the entire myocardium simultaneously. With luck, after repolarization the SA node will be the first area of the heart to reach threshold. Thus the primary goal of defibrillation is not just to stop the fibrillation, but to give the ventricles a chance to respond to normal SA commands.

In treating arrhythmias there are several medications that can slow down rapid heart rates, or the abnormal portions of the conducting system can be destroyed. Pacemakers are used to accelerate slow heart rates. Implantable pacemakers able to sense ventricular fibrillation and deliver an immediate defibrillating shock have been successful in preventing sudden death in patients with previous episodes of ventricular tachycardia and ventricular fibrillation.

⚕ Aneurysms

EAP *p. 348*

An **aneurysm** (AN-ū-rizm) is a bulge in the weakened wall of a blood vessel, usually an artery. This bulge resembles a bubble in the wall of a tire, and like a bad tire, the affected artery may suffer a catastrophic blowout. The most dangerous aneurysms are those involving arteries of the brain, where they cause strokes, and of the aorta, where a blowout will cause fatal bleeding in a matter of seconds.

Aneurysms are most often caused by chronic high blood pressure, although any trauma or infection that weakens vessel walls can lead to an aneurysm. In addition, at least some aortic aneurysms have been linked to inherited disorders, such as *Marfan's syndrome,* that have weakened connective tissues in vessel walls. It is not known whether other genetic factors are involved in the development of other aneurysms.

An aneurysm usually forms gradually, as vessel walls become less elastic. When a weak point develops, the arterial pressures distort the wall, creating an aneurysm. Unfortunately, because they are often painless, they are likely to go undetected.

When aneurysms are detected by ultrasound or other scanning procedures, the risk of rupture can sometimes be estimated on the basis of their size. For example, an aortic aneurysm larger than 6 cm has a 50:50 chance of rupturing in the next 10 years. Treatment often begins with the reduction of blood pressure by means of vasodilators or beta-blockers (drugs which decrease heart rate and force of concentration). An aneurysm in an accessible area, such as the abdomen, may be surgically removed and the vessel repaired. Figure A-36 shows a large aortic aneurysm before and after surgical repair with a synthetic patch.

⚕ Arteriosclerosis

EAP *p. 350*

Arteriosclerosis (ar-tē-rē-ō-skle-RŌ-sis) is a thickening and toughening of arterial walls. Although this

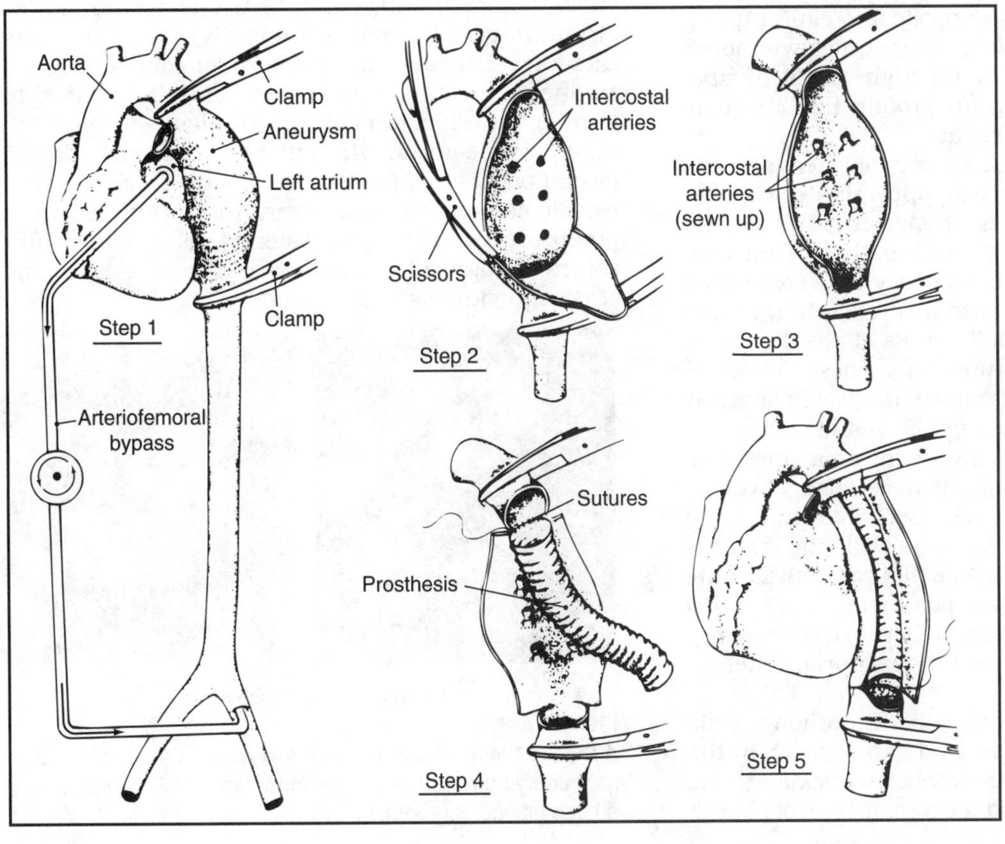

Figure A-36   Repair of an Aneurysm.

13

condition may not sound life-threatening, complications related to arteriosclerosis account for roughly one-half of all deaths in the United States. There are many different forms of arteriosclerosis; for example, arteriosclerosis of coronary vessels is respons-ible for *coronary artery disease* (CAD), and arteriosclerosis of arteries supplying the brain can lead to strokes.

There are two major forms of arteriosclerosis:

- **Focal calcification** is the gradual degeneration of smooth muscle in the tunica media and the subsequent deposition of calcium salts. This process typically involves arteries of the limbs and genital organs. Some focal calcification occurs as part of the aging process, and it may develop in association with atherosclerosis. Rapid and severe calcification may occur as a complication of diabetes mellitus, an endocrine disorder considered in Chapter 11.

- **Atherosclerosis** (ath-er-ō-skle-RŌ-sis) is associated with damage to the endothelial lining and the formation of lipid deposits in the tunica media. This is the most common form of arteriosclerosis.

Many factors may be involved in the development of atherosclerosis. One major factor is lipid levels in the blood. Atherosclerosis tends to develop in persons whose blood contains elevated levels of plasma lipids, specifically cholesterol. Circulating cholesterol is transported to peripheral tissues in lipoproteins, protein-lipid complexes. (The various types of lipoproteins and their interrelationships are discussed in Chapter 18.) Recent evidence indicates that many forms of atherosclerosis are associated with either (1) low levels of *apolipoprotein-E* (ApoE), a transport protein whose lipids are quickly removed by peripheral tissues, or (2) high levels of *lipoprotein(a)*, a *low-density lipoprotein* (LDL) that is removed at a much slower rate.

When ApoE levels are low, or lipoprotein(a) levels are high, cholesterol-rich lipoproteins remain in circulation for an extended period. Circulating monocytes then begin removing them from the bloodstream. Eventually the monocytes become filled with lipid droplets. Now called *foam cells,* they attach themselves to the endothelial walls of blood vessels, where they release growth factors. These cytokines stimulate the divisions of smooth muscle fibers near the tunica interna, thickening the vessel wall.

Other monocytes then invade the area, migrating between the endothelial cells. As these changes occur, the monocytes, smooth muscle fibers, and endothelial cells begin phagocytizing lipids as well. The result is a **plaque,** a fatty mass of tissue that projects into the lumen of the vessel. At this point the plaque has a relatively simple structure, and there is evidence that the process can be reversed if appropriate dietary adjustments are made.

If the conditions persist, the endothelial cells become swollen with lipids, and gaps appear in the endothelial lining. Platelets now begin sticking to the exposed collagen fibers, and the combination of platelet adhesion and aggregation leads to the formation of a localized blood clot that will further restrict blood flow through the artery. The structure of the plaque is now relatively complex. Plaque growth may be halted, but the structural changes are usually permanent.

Typical plaques can be seen in Figure A-37. Elderly individuals, especially elderly men, are most likely to develop atherosclerotic plaques. There is evidence that estrogens may slow plaque formation; this may account for the lower incidence of coronary artery disease, myocardial infarctions (MIs), and strokes in women. After menopause, when estrogen production declines, the risk of CAD, MIs, and strokes in women increases markedly.

In addition to advanced age and male sex, other important risk factors include high blood cholesterol levels, high blood pressure, and cigarette smoking. Roughly 20 percent of middle-aged men have all three of these risk factors; these individuals are four times more likely to experience an MI or cardiac arrest than are other men in their age group. Although fewer women develop this condition, elderly women smokers with high blood cholesterol and high blood pressure are at much greater risk than other women. Other factors that may promote development of atherosclerosis in both men and women include diabetes mellitus, obesity, and stress. There is also evidence that at least some forms of atherosclerosis may be linked to chronic infection with *Chlamydia pneumoniae,* a bacterium responsible for several types of respiratory infections, including some forms of pneumonia.

Potential treatments for atherosclerotic plaques, such as catheterization, balloon angioplasty, and bypass surgery, were discussed in on page 100. In cases where dietary modifications do not lower circulating LDL levels sufficiently, there are drug therapies that can bring them under control. Genetic engineering techniques have recently been used to treat an inherited form of *hypercholesterolemia* (high blood cholesterol) linked to extensive plaque formation. (The patients were unable to absorb and recycle cholesterol in the liver.) In this experimental procedure, circulating cholesterol levels declined after copies of appropriate genes were inserted into some of the individual's liver cells.

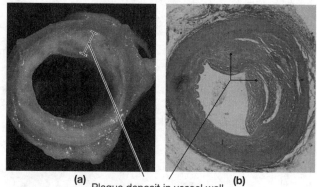

(a)　Plaque deposit in vessel wall　(b)

FIGURE A-37

A Plaque Blocking a Peripheral Artery. (a) A section of a coronary artery narrowed by plaque formation. (b) Sectional view of a large plaque. (LM × 18)

The Cardiovascular System   105

Without question, the best approach to atherosclerosis is to try and avoid it by eliminating or reducing associated risk factors. Suggestions include: (1) reducing the amount of dietary cholesterol and saturated fats by restricting consumption of fatty meats (such as beef, lamb, and pork), egg yolks, and cream; (2) giving up smoking (or never starting to begin with); (3) checking your blood pressure and taking steps to lower it if necessary; (4) having your blood cholesterol levels checked at annual physical examinations; (5) controlling your weight; and (6) exercising regularly.

✝ Problems with Venous Valve Function
EAP *p. 350*

In Chapter 4 of the text it was noted that one of the consequences of aging is a loss of elasticity and resilience in connective tissues throughout the body. Blood vessels are no exception, and with age the walls of veins begin to sag. This change usually affects the superficial veins of the legs first, because at these locations gravity opposes blood flow. The situation is aggravated by a lack of exercise or an occupation requiring long hours standing or sitting. Because there is no muscular activity to help keep the blood moving, venous blood pools on the proximal (heart) side of each valve. As the venous walls are distorted, the valves become less effective, and gravity can then pull blood back toward the capillaries. This further impedes normal blood flow, and the veins become grossly distended. These sagging, swollen vessels are called **varicose** (VAR-i-kōs) **veins.** Varicose veins are relatively harmless but unsightly; surgical procedures are sometimes used to remove or constrict the offending vessels.

Varicose veins are not limited to the extremities, and another common site involves a network of veins in the walls of the anus. Pressures within the abdominopelvic cavity rise dramatically when the abdominal muscles are tensed. Straining to force defecation can force blood into these veins, and repeated incidents leave them permanently distended. These distended veins, known as **hemorrhoids** (HEM-ō-roydz), can be uncomfortable and in severe cases extremely painful.

Hemorrhoids are often associated with pregnancy, due to changes in circulation and abdominal pressures. Minor cases can be treated by the topical application of drugs that promote contraction of smooth muscles within the venous walls. More severe cases may require the surgical removal or destruction of the distended veins.

✝ Hypertension and Hypotension
EAP *p. 353*

Elevated blood pressure is considered **primary hypertension,** or essential hypertension, if no obvious cause can be determined. Known risk factors include a hereditary history of hypertension, sex (males are at higher risk), high plasma cholesterol, obesity, chronic stresses, and cigarette smoking. **Secondary hypertension** appears as the result of abnormal hormonal production outside the cardiovascular system. For example, a condition resulting in excessive production of antidiuretic hormone (ADH), renin, aldosterone, or epinephrine will probably produce hypertension, and many forms of kidney disease will lead to hypertension caused by fluid retention or excessive renin production.

Hypertension significantly increases the work load on the heart, and the left ventricle gradually enlarges. More muscle mass requires a greater oxygen demand, and when the coronary circulation cannot keep pace symptoms of coronary ischemia appear.

Increased arterial pressures also place a physical stress on the walls of blood vessels throughout the body. This stress promotes or accelerates the development of arteriosclerosis and increases the risks of aneurysms, heart attacks, and strokes. Vessels supplying the retinas of the eyes are often affected, and hemorrhages and associated circulatory changes can produce disturbances in vision. Because these vessels are examined in a normal physical exam, retinal changes may provide the first evidence that hypertension is affecting peripheral circulation.

One of the most difficult aspects of hypertension is that there are usually no obvious symptoms. As a result, clinical problems do not appear until the condition has reached the crisis stage. There is, therefore, considerable interest in early detection and prompt treatment of hypertension.

Treatment consists of a combination of lifestyle changes and physiological therapies. Quitting smoking, getting regular exercise, and restricting dietary intake of salt, fats, and calories will improve peripheral circulation, prevent increases in blood volume and total body weight, and reduce plasma cholesterol levels. These strategies may be sufficient to control hypertension if it has been detected before significant cardiovascular damage has occurred. Therapies usually involve antihypertensive drugs, such as calcium channel blockers, beta-blockers, diuretics, and vasodilators, singly or in combination. Beta-blockers eliminate the effects of sympathetic stimulation on the heart, and the unopposed parasympathetic system lowers the resting heart rate and blood pressure. Diuretics promote the loss of water and sodium ions at the kidneys, lowering blood volume, and vasodilators further reduce blood pressure. A new class of antihypertensive drugs lowers blood pressure by preventing the conversion of angiotensin I to angiotensin II. These **angiotensin-converting enzyme (ACE) inhibitors,** such as *captopril,* are being used to treat chronic hypertension and congestive heart failure.

14

In hypotension blood pressure declines, and peripheral systems begin to suffer from oxygen and nutrient deprivation. One clinically important form of hypotension can develop following the administration of antihypertensive drugs. Problems may appear when the individual changes position, going from lying down to sitting, or sitting to standing. Normally each time you sit or stand, blood pressure in the carotid sinus drops, for the heart must suddenly counteract gravity to push blood up to the brain. The fall in pressure triggers the carotid reflex, and blood pressure returns to normal. But if the carotid response is prevented by beta-blockers or other drugs, blood pressure at the brain may fall so low that the individual becomes weak, dizzy, disoriented, or unconscious. This condition is known as **orthostatic hypotension** (ōr-tho-STAT-ik; *orthos*, straight + *statikos*, causing to stand), or simply **orthostasis** (ōr-thō-STĀ-sis). Most readers will have experienced brief episodes of orthostasis when standing up suddenly after reclining for an extended period. The carotid reflex frequently slows with age, and older people learn to sit and stand more carefully to avoid the effects of orthostatic hypotension.

☤ Shock

EAP *p. 362*

Shock is an acute circulatory crisis marked by low blood pressure (hypotension) and inadequate peripheral blood flow. Severe and potentially fatal symptoms develop as vital tissues become starved for oxygen and nutrients. Common causes of shock are (1) a fall in cardiac output after hemorrhaging or other fluid losses, (2) damage to the heart, (3) external pressure on the heart, and (4) extensive peripheral vasodilation.

CIRCULATORY SHOCK A severe reduction in blood volume produces symptoms of **circulatory shock.** Symptoms of circulatory shock appear after fluid losses of about 30 percent of the total blood volume. The cause can be hemorrhaging or fluid

FIGURE A-38

Shock. (a) The progressive stage is characterized by a gradual decline in systemic blood pressure, tissue blood flow, and cardiac output. (b) The irreversible stage involves a series of integrated chain reactions leading to a rapid decline in cardiac output, a dramatic and irreversible fall in blood pressure, circulatory collapse, and eventual death.

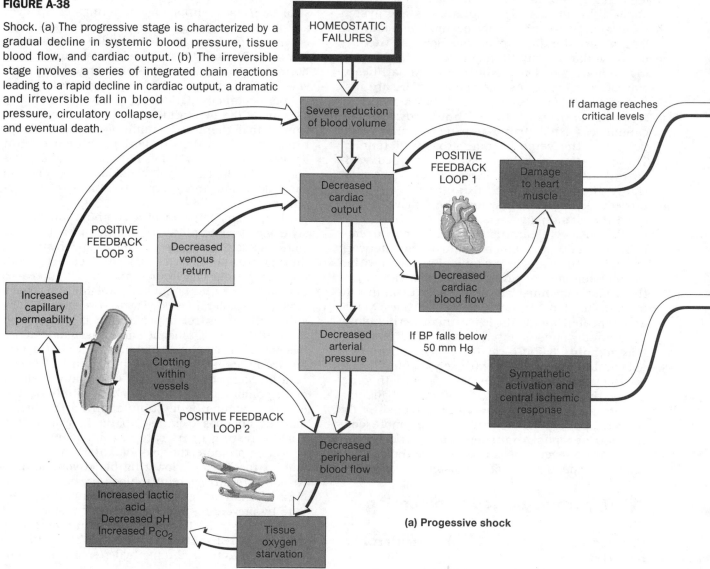

(a) Progressive shock

losses to the environment, as in dehydration or after severe burns. All cases of circulatory shock share the same basic symptoms:

1. Hypotension, with systolic pressures below 90 mm Hg.

2. Pale, cool, and moist ("clammy") skin. The skin is pale and cool because of peripheral vasoconstriction; the moisture reflects sympathetic activation of the sweat glands.

3. Frequent confusion and disorientation, due to a fall in blood pressure at the brain.

4. Rapid, weak pulse.

5. Cessation of urination, because the reduced blood to the kidneys slows or stops urine production.

6. A drop in blood pH (*acidosis*), due to lactic acid generation in oxygen-deprived tissues.

Circulatory shock is often divided into *compensated, progressive,* and *irreversible* stages.

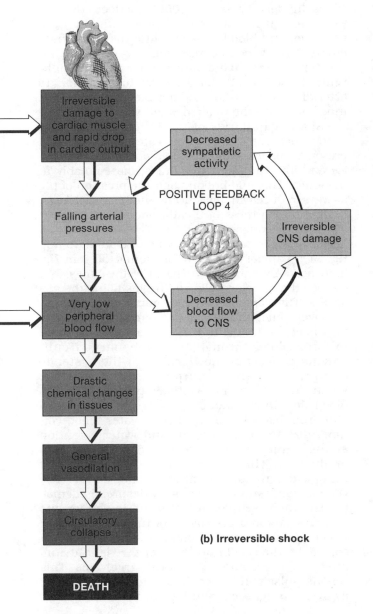

(b) Irreversible shock

THE COMPENSATED STAGE (STAGE I) During the **compensated stage** homeostatic adjustments can cope with the situation; the short-term and long-term responses detailed in Figure A-38 are part of the compensation process. During the period of compensation, peripheral blood flow is reduced but remains within tolerable limits.

THE PROGRESSIVE STAGE (STAGE II) When blood volume declines by more than 35 percent, the individual enters the **progressive stage** of circulatory shock. Homeostatic mechanisms are now unable to cope with the situation. Despite sustained vasoconstriction and the mobilization of the venous reserve, blood pressure remains abnormally low, venous return is reduced, and cardiac output is inadequate. A vicious cycle begins when the low cardiac output causes myocardial damage. This damage leads to a further reduction in cardiac output and subsequent reductions in blood pressure and venous return. The sequence is diagrammed in Figure A-38a.

When the mean arterial blood pressure falls to about 50 mm Hg, carotid sinus baroreceptors trigger a massive activation of the vasomotor centers. In essence, the goal now is to preserve the circulation to the brain at any cost. Blood flow to cerebral vessels is not affected, and the blood pressure in the carotid arteries remains relatively high (70 mm Hg). But in other organs sympathetic output causes a sustained and maximal vasoconstriction. This reflex, called the **central ischemic response,** reduces peripheral circulation to an absolute minimum.

The central ischemic response is a last-ditch effort that maintains adequate blood flow to the brain at the expense of other tissues. Unless prompt treatment is provided, the condition will soon prove fatal. Treatment must concentrate on (1) preventing further fluid losses and (2) giving a transfusion of whole blood, plasma expanders, or blood substitutes.

THE IRREVERSIBLE STAGE (STAGE III) In the absence of treatment, progressive shock will soon turn into **irreversible shock** (Figure A-38b). At this point, conditions in the heart, liver, kidneys, and CNS are rapidly deteriorating to the point that death will occur, even *with* medical treatment.

Irreversible shock begins when conditions in the tissues become so abnormal that the arteriolar smooth muscles and precapillary sphincters become unable to contract, despite the commands of the vasomotor center. The result is a widespread peripheral vasodilation and an immediate and fatal decline in blood pressure. This event is called **circulatory collapse.** The blood pressure in many tissues then falls so low that the capillaries collapse like deflating balloons. Blood flow through these capillary beds then stops completely, and the cells in the affected tissues die. The dying cells release more abnormal chemicals, and the effect quickly spreads throughout the body.

14

✝ Other Types of Shock EAP *p. 362*

Although the text focuses on circulatory shock caused by low blood volume, shock can develop when the blood volume is normal. **Cardiogenic** (kar-dē-ō-JEN-ik) **shock** occurs when the heart becomes unable to maintain a normal cardiac output. The most common cause is failure of the left ventricle as a result of a myocardial infarction. Between 5 and 10 percent of patients surviving a heart attack must be treated for cardiogenic shock. The use of thrombolytic drugs, such as t-PA and streptokinase, can be very effective in restoring coronary circulation and ventricular function, thereby relieving the peripheral symptoms. Cardiogenic shock may also be the result of arrhythmias, valvular heart disease, advanced coronary artery disease, cardiomyopathy, or ventricular arrhythmias (see p. 102).

In **obstructive shock,** ventricular output is reduced because tissues or fluids are restricting the expansion and contraction of the heart. For example, fluid buildup in the pericardial cavity (cardiac tamponade, discussed elsewhere in this *Manual* (see Infection and Inflammation of the Heart) can compress the heart and limit ventricular filling.

Distributive shock results from a widespread, uncontrolled vasodilation. This produces a dramatic fall in blood pressure that leads to a reduction in blood flow and the onset of shock. Three important examples are *neurogenic shock, septic shock*, and *anaphylactic shock.*

Neurogenic (noo-rō-JEN-ik) **shock** can be caused by general or spinal anesthesia and by trauma or inflammation of the brain stem. The underlying problem is damage to the vasomotor center or to the sympathetic tracts or nerves, leading to a loss of vasomotor tone.

Septic shock results from the massive release of endotoxins, poisons derived from the cell walls of bacteria during a systemic infection. These compounds cause a vasodilation of precapillary sphincters throughout the body, resulting in a drop in peripheral resistance and a decline in blood pressure. Symptoms of septic shock generally resemble those of other types of shock, but the skin is flushed, and the individual has a high fever. For this reason septic shock is also known as "warm shock." One interesting example of septic shock, called **toxic shock syndrome (TSS),** results from an infection by the bacterium *Staphylococcus aureus*. This disease was unrecognized before 1978, when it appeared in a group of children. Since that time there have been roughly 2,500 cases in the United States, most (95 percent) affecting women. Although other sources of infection are possible, infection most often appears to occur during menstruation, and the chances of infection are increased with the use of superabsorbent tampons. (The brands involved have been taken off the market, and the incidence has declined steadily since 1980.)

Extensive peripheral vasodilation also occurs in **anaphylactic** (an-a-fi-LAK-tik) **shock,** a dangerous allergic reaction. This type of shock is discussed on p. 123.

✝ Heart Failure EAP *p. 362*

A condition of heart failure exists when the cardiac output is insufficient to meet the circulatory demand. The initial symptoms of heart failure vary depending on whether the problem is restricted to the left ventricle or the right ventricle, or involves both. However, over time these differences are eliminated; for example, the major cause of right ventricular failure is left ventricular failure. Figure A-39 provides a simplified flow chart for heart failure and indicates potential therapies.

Suppose that the left ventricle cannot maintain normal cardiac output, due to damage to the ventricular muscle (see the discussion of myocardial infarctions on p. 100) or high arterial pressures (hypertension, p. 105). In effect, the left ventricle can no longer keep pace with the right ventricle, and blood backs up into the pulmonary circuit. This venous congestion is responsible for the term **congestive heart failure.** The right ventricle now works harder, elevating pulmonary arterial pressures and forcing blood through the lungs and into the weakened left ventricle.

At the capillaries of the lungs, arterial and venous pressures are now elevated. This elevated pressure pushes additional fluid out of the capillaries and into the interstitial fluids, most notably at the lungs. The fluid buildup and compression of the airways reduces the effectiveness of gas exchange, leading to shortness of breath, often the first obvious sign of congestive failure. This fluid buildup begins at a postcapillary pressure of around 25 mm Hg. At a capillary pressure of around 30 mm Hg, fluid not only enters the tissues of the lungs, but crosses the alveolar walls and begins filling the airspaces. This condition is called **pulmonary edema.**

Over time, the less muscular right ventricle may become unable to generate enough pressure to force blood through the pulmonary circuit. Venous congestion now occurs in the systemic circuit, and cardiac output declines further. When the reduction in systemic pressures lowers blood flow to the kidneys, renin and erythropoietin are released. This in turn elevates blood volume, due to increased salt and water retention at the kidneys, and accelerated red blood cell production. This rise in blood volume actually complicates the situation, as it tends to increase venous congestion and cause widespread edema.

The increased volume of blood in the venous system leads to a distension of the veins, making superficial veins more prominent. When the heart contracts, the rise in pressure at the right atrium produces a pressure pulse in the large veins. This venous pulse can be seen and palpated most easily over the right external jugular vein.

Treatment of congestive heart failure often includes

1. Restriction of salt intake. The expression "water follows salt" applies here, because when sodium and chloride ions are absorbed across the lining of the digestive tract, water is also absorbed by osmosis.

2. Administration of drugs to promote fluid loss. These drugs, called **diuretics** (dī-ū-RET-iks; *diouretikos*, promoting urine), increase salt and water losses at the kidneys.

3. Extended bed rest, to enhance venous return to the heart, coupled with physical therapy to maintain good venous circulation.

4. Administration of drugs that enhance cardiac output. These drugs may target the heart, the peripheral circulation, or some combination of the two. When the heart has been weak-ened, drugs related to digitalis, an extract from the leaves of the foxglove plant, are often selected. *Digitoxin, digoxin,* and *ouabain* are examples. These compounds increase the force of cardiac muscle cell contractions. When high blood pressure is a factor, some type of vasodilator is given to lower peripheral resistance.

5. Administration of drugs that reduce peripheral vascular resistance, such as *hydralazine* or *ACE* (angiotensin-converting enzyme) *inhibitors*. The drop in peripheral resistance reduces the work load of the left ventricle.

℞ Treatment of Cerebrovascular Disease   EAP *p. 367*

Symptoms of cerebrovascular disease usually appear when atherosclerosis reduces the circula-

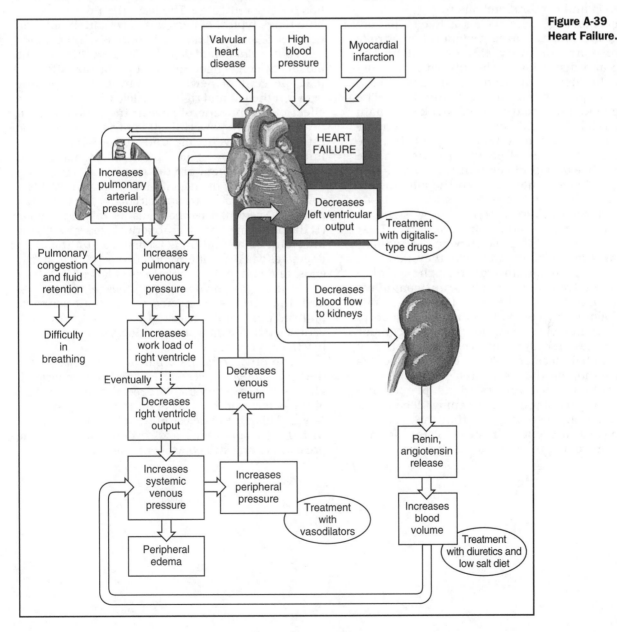

**Figure A-39
Heart Failure.**

tory supply to the brain. If the circulation to a portion of the brain is completely shut off, a *cerebrovascular accident (CVA)*, or *stroke*, occurs. The most common causes of strokes include **cerebral thrombosis** (clot formation at a plaque), **cerebral embolism** (drifting blood clots, fatty masses, or air bubbles), and **cerebral hemorrhages** (rupture of a blood vessel, often following formation of an aneurysm). The observed symptoms and their severity vary depending on the vessel involved and the location of the blockage.

If the circulatory blockage disappears in a matter of minutes, the effects are temporary, and the condition is called a **transient ischemic attack,** or **TIA**. TIAs often indicate that cerebrovascular disease exists, and preventive measures can be taken to forestall more serious incidents. For example, taking aspirin each day slows blood clot formation in patients experiencing TIAs, and this reduces the risks of cerebral thrombosis and cerebral embolism.

If the blockage persists for a longer period, neurons die and the area degenerates. Stroke symptoms are initially exaggerated by the swelling and distortion of the injured neural tissues; if the individual survives, there is often a gradual improvement in brain function. The management and treatment of strokes remain controversial. Surgical removal of the offending clot or blood mass may be attempted, but the results are variable. Recent progress in the emergency treatment of cerebral thromboses and cerebral embolisms has involved the administration of clot-dissolving enzymes such as *tissue plasminogen activator* (t-PA; now sold as *Alteplase*), *streptokinase,* or *urokinase.* Best results are obtained if these enzymes are administered within an hour, although they may still be of use up to 24 hours after the stroke. Subsequent treatment involves anticoagulant therapy, usually with heparin (for one to two weeks) followed by coumadin (for up to a year) to prevent further clot formation. A more complicated surgical procedure involves the insertion of a transplanted piece of a blood vessel that routes blood around the damaged area. None of these treatments is as successful as preventive surgery, where plaques are removed before a stroke. It should also be noted that the very best solution is to prevent or restrict plaque formation by controlling the risk factors involved.

⚕ Congenital Circulatory Problems
EAP *p. 376*

Minor individual variations in the circulatory network are quite common. For example, very few individuals have identical patterns of venous distribution. Congenital circulatory problems serious enough to represent a threat to homeostasis are relatively rare. They usually reflect abnormal formation of the heart or problems with the interconnections between the heart and the great vessels. Several examples of congenital circulatory defects are illustrated in Figure A-40. All of these conditions can be surgically corrected, although multiple surgeries may be required.

The incomplete closure of the foramen ovale or ductus arteriosus (Figure A-40a) results in the bypassing of the lungs and the recirculation of blood into the pulmonary circuit. Because normal blood oxygenation does not occur, the circulating blood has a deep red color. The skin then develops the blue tones typical of *cyanosis,* a condition noted in Chapter 5, and the infant is known as a "blue baby."

Ventricular septal defects (Figure A-40b) are the most common congenital heart problems, affecting 0.12 percent of newborn infants. The opening between the left and right ventricles has the reverse effect of a connection between the atria: When it beats, the more powerful left ventricle ejects blood into the right ventricle and pulmonary circuit. Pulmonary hypertension, pulmonary edema, and cardiac enlargement are the results.

The *tetralogy of Fallot* (fa-LŌ) (Figure A-40c) is a complex group of heart and circulatory defects that affect 0.10 percent of newborn infants. In this condition (1) the pulmonary trunk is abnormally narrow, (2) the interventricular septum is incomplete, (3) the aorta originates where the interventricular septum normally ends, and (4) the right ventricle is enlarged.

In the *transposition of great vessels* (Figure A-40d) the aorta is connected to the right ventricle and the pulmonary artery is connected to the left ventricle. This malformation affects 0.05 percent of newborn infants.

In an *atrioventricular septal defect* (Figure A-40e) the atria and ventricles are incompletely separated. The results are quite variable, depending on the extent of the defect and the effects on the atrioventricular valves. This type of defect most often affects infants with *Down syndrome,* a disorder caused by the presence of an extra copy of chromosome 21.

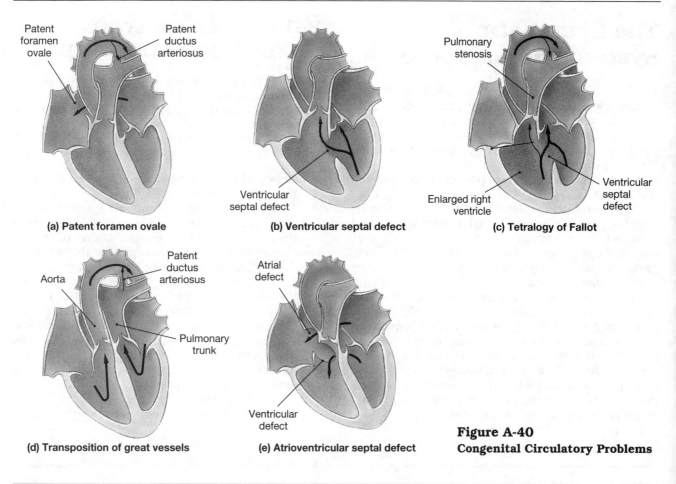

(a) Patent foramen ovale

Patent foramen ovale
Patent ductus arteriosus

(b) Ventricular septal defect

Ventricular septal defect

(c) Tetralogy of Fallot

Pulmonary stenosis
Enlarged right ventricle
Ventricular septal defect

(d) Transposition of great vessels

Aorta
Patent ductus arteriosus
Pulmonary trunk

(e) Atrioventricular septal defect

Atrial defect
Ventricular defect

Figure A-40
Congenital Circulatory Problems

CRITICAL THINKING QUESTIONS

6-1. Vickie has a tumor that causes her to release excess amounts of ADH. Possible symptoms arising from this elevated hormone would be:
 a. decreased blood volume
 b. increased blood pressure
 c. peripheral vasoconstriction
 d. polycythemia
 e. all of the above

6-2. Tammy visits her physician for a routine physical examination. The doctor notes she is pale and complains of being tired and weak. The physician orders blood tests and the results of the blood work follow:

RBC count: 3.5 million/mm^3

Hematocrit: 32

Hemoglobin: 10 g/dl

MCV: 70

MCH: 20 pg

MCHC: 28

WBC count: 8,000/mm^3

Platelet count: 200,000/mm^3

Reticulocyte count: 1% of total erythrocytes

These results would indicate
 a. hemorrhagic anemia
 b. aplastic anemia
 c. iron deficiency anemia
 d. pernicious anemia
 e. macrocytic anemia

6-3. Cathy has just given birth to a little girl. When the nurses take the infant back to the nursery and try to feed her, she becomes cyanotic. The episode passes, but when the infant is bathed, she becomes cyanotic again. Blood gas levels are taken, and they show the arterial blood is only 60% saturated. Physical examination indicates that there are no structural deformities involving the respiratory or digestive system. Echocardiography shows a heart defect. What is the specific cause of the problem?

The Lymphatic System and Immunity

The lymphatic system consists of a fluid, *lymph*, a network of *lymphatic vessels*, specialized cells called *lymphocytes*, and an array of *lymphoid tissues* and *lymphoid organs* scattered throughout the body. This system has two major functions: (1) the protection of the body via the *immune response*, and (2) the transport of fluid from the interstitial fluid to the bloodstream.

The immune response produced by activated lymphocytes is responsible for the detection and destruction of foreign or toxic substances that may disrupt homeostasis. For example, viruses, bacteria, and tumor cells are usully recognized and eliminated by cells of the lymphatic system. Immunity is the specific resistance to disease, and all of the cells and tissues involved with the production of immunity are sometimes considered to be part of an *immune system*. Whereas the lymphatic system is an anatomically distinct system, the immune system is a physiological system that includes not only the lymphatic system, but components of the integumentary, cardiovascular, respiratory, digestive, and other systems.

PHYSICAL EXAMINATION OF THE LYMPHATIC SYSTEM

Individuals with lymphatic system disorders may experience a variety of symptoms. The pattern observed varies depending on whether the problem affects the immune or circulatory functions of the lymphatic system (Figure A-41a). Important symptoms and signs include:

- *Enlarged lymph nodes* are often found during infection. They also develop in cancers of the lymphatic system, such as *lymphoma*, or when metastasis is underway from primary tumors in other tissues. The status of regional lymph nodes can therefore be important in the diagnosis and treatment of many different cancers. The onset and duration of swelling, the size, texture, and mobility of the nodes, the number of affected nodes, and the degree of tenderness are all important in diagnosis. For example, nodes containing cancer cells are often large, locked in place, and nontender. On palpation these nodes feel like dense masses, rather than individual lymph nodes. In contrast, infected lymph nodes are usually large, freely mobile, and very tender.

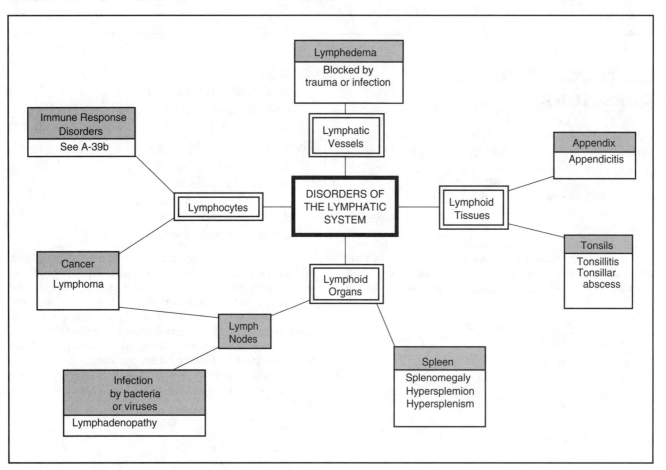

Figure A-41(a)   Disorders of the Lymphatic System

- *Lymphangitis* consists of *erythematous* (red) streaks on the skin that may develop with an inflammation of superficial lymph vessels. Lymphangitis often occurs in the lower limbs, and the reddened streaks originate at a site of pathogenic infection. Before the linkage to the lymphatic system was appreciated, this sign was called "blood poisoning."

- *Splenomegaly* is an enlargement of the spleen that may result from acute infections, such as *endocarditis* (p. 97) and *mononucleosis* (p. 116), or chronic infections such as *malaria* (p. 92) or *leukemia* (p. 96). The spleen can be examined through palpation or percussion (p. 12), and the border can be palpated to detect splenic enlargement. In percussion, an enlarged spleen produces a dull sound rather than the more normal tympanic sound. The patient history may also reveal important clues. For example, an individual with an enlarged spleen often reports a feeling of fullness after eating a small meal, probably because the enlarged spleen limits gastric expansion.

- *Weakness and fatigue* often accompany immunodeficiency disorders (p. 113), *Hodgkin's disease* and other *lymphomas* (p. 115), and *infectious mononucleosis* (p. 116).

- *Skin lesions* such as hives or urticaria (p. 121) can develop during allergic reactions. Immune responses to a variety of allergens, including animal hair, pollen, dust, medications, and some foods may cause such lesions. Skin lesions associated with *acquired immune deficiency syndrome* (AIDS) include pink or red elevated papules, and the skin tumors known as *Kaposi's sarcoma* (p. 118).

- *Respiratory problems*, including rhinitis and wheezing, may accompany the allergic response to allergens such as pollen, hay, dust, and mildew. *Bronchospasms* are smooth muscle contractions that constrict the airways and make breathing difficult. Bronchospasms, which often accompany severe allergic or asthmatic attacks, are a response to the appearance of antigens within the respiratory passageways.

- *Recurrent infections* may occur for a variety of reasons. *Tonsillitis and adenoiditis* (p. 114) are common recurrent infections in children. More serious chronic infections are common among persons with immunodeficiency disorders such as AIDS (p. 118) or *severe combined immunodeficiency* (SCID) (p. 30). When the immune response is inadequate, the individual cannot overcome even a minor infection. Infections of the respiratory system are very common, and recurring gastrointestinal infections may produce chronic diarrhea. The pathogens involved may not infect persons

with normal immune responses. Infections are also a problem for individuals taking medications that suppress the immune response. Examples of immunosuppressive drugs include anti-inflammatories such as the *corticosteroids (prednisone)*, as well as more specialized drugs such as *FK-506* or *cyclosporin*.

When circulatory functions are impaired, the most common sign is *lymphedema*, a tissue swelling caused by the buildup of interstitial fluid. Lymphedema can result from trauma to a lymphatic vessel or from a lymphatic blockage due to a tumor or infections, including parasitic infection such as *filariasis* (p. 26).

DISORDERS OF THE LYMPHATIC SYSTEM

Disorders of the lymphatic system can be sorted into three general categories that are diagrammed in Figure A-41b:

1. *Disorders resulting from an insufficient immune response.* This category includes conditions such as AIDS (p. 118) or SCID (p. 30). Individuals with depressed immune defenses may develop life-threatening diseases caused by microorganisms that are harmless to other individuals.

2. *Disorders resulting from an excessive immune response.* Conditions such as *allergies* (p. 113) and *immune complex disorders* (p. 123) can result from an immune response that is out of proportion with the size of the stimulus.

3. *Disorders resulting from an inappropriate immune response. Autoimmune disorders* result when normal tissues are mistakenly attacked by T cells or the antibodies produced by activated B cells.

⚕ Lymphedema   EAP *p. 385*

Blockage of the lymphatic drainage from a limb produces **lymphedema** (lim-fe-DĒ-ma). In this painless condition, interstitial fluids accumulate, and the limb gradually becomes swollen and grossly distended. If the condition persists, the connective tissues lose their elasticity and the swelling becomes permanent. Lymphedema by itself does not pose a major threat to life. The danger comes from the constant risk that an uncontrolled infection will develop in the affected area. Because the interstitial fluids are essentially stagnant, toxins and pathogens can accumulate and overwhelm the local defenses without fully activating the immune system.

Temporary lymphedema may result from tight clothing. Chronic lymphedema can result from scar tissue formation or from parasitic infections. In **filariasis** (fil-a-RĪ-a-sis), larvae of a parasitic nematode (roundworm), usually *Wucheria bancrofti*, is transmitted by mosquitoes or blackflies. The adult

15

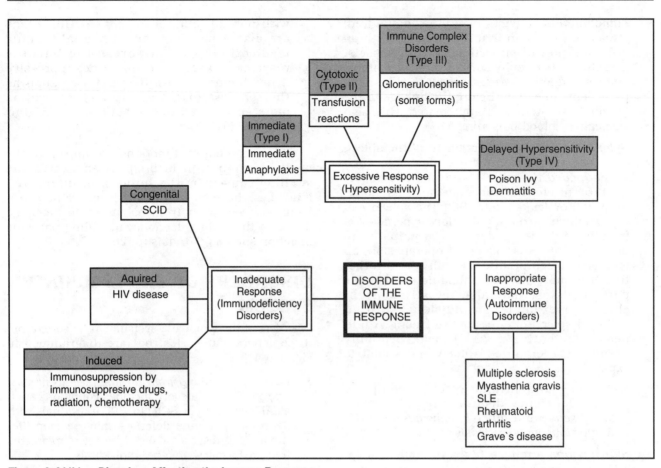

Figure A-41(b)   Disorders Affecting the Immune Response

worms form massive colonies within lymphatic vessels and lymph nodes. Repeated scarring of the passageways eventually blocks lymphatic drainage and produces extreme lymphedema with permanent distension of tissues. The limbs or external genitalia often become grossly distended, a condition known as **elephantiasis** (el-e-fan-TĪ-a-sis).

Therapy for chronic lymphedema consists of treating infections by the administration of antibiotics and (when possible) reducing the swelling. One possible treatment involves the application of elastic wrappings that squeeze the tissue. This external compression elevates the hydrostatic pressure of the interstitial fluids and opposes the entry of additional fluid from the capillaries.

⚕ Infected Lymphoid Nodules
EAP *p. 387*

Lymphoid nodules may be overwhelmed by a pathogenic invasion. The result is a localized infection accompanied by regional swelling and discomfort. An individual with **tonsillitis** has infected tonsils. Symptoms include a sore throat, high fever, and leukocytosis (an abnormally high white blood cell count). The affected tonsil (most often the pharyngeal) becomes swollen and inflamed, sometimes enlarging enough to partially block the entrance to the trachea. Breathing then becomes difficult, and in severe cases, impossible. As the infection pro-

ceeds, abscesses develop within the tonsilar tissues, and the bacteria may enter the bloodstream by passing through the lymphatic capillaries and vessels to the venous system.

In the early stages, antibiotics may control the infection, but once abscesses have formed, the best treatment involves surgical drainage of the abscesses. **Tonsillectomy**, the removal of the tonsil, was once highly recommended and frequently performed to prevent recurring tonsilar infections. The procedure does reduce the incidence and severity of subsequent infections, but questions have been raised concerning the overall cost to the individual. The tonsils represent a "first line" of defense against bacterial invasion of the pharyngeal walls. If they are removed, bacteria may not be detected until a truly severe infection is well under way.

Appendicitis usually follows an erosion of the epithelial lining of the appendix. Several factors may be responsible for the initial ulceration, notably bacterial or viral pathogens. Bacteria that normally inhabit the lumen of the large intestine then cross the epithelium and enter the underlying tissues. Inflammation occurs, and the opening between the appendix and the rest of the intestinal tract may become constricted. Mucus secretion accelerates, and the organ becomes increasingly distended. Eventually the swollen and inflamed appendix may rupture, or *perforate*. If this occurs, bacteria will be released into the warm, dark, moist confines of the abdominopelvic cavity, where they can cause a life-threatening infection. The most effective treatment for appendicitis is the surgical removal of the organ, a procedure known as an **appendectomy**.

☤ Lymphomas

EAP *p. 387*

Lymphomas are malignant tumors consisting of cancerous lymphocytes or lymphocytic stem cells. Over 30,000 cases of lymphoma are diagnosed in the United States each year, and that number has been steadily increasing. There are many different types of lymphoma. One form, called **Hodgkin's disease (HD)**, accounts for roughly 40 percent of all lymphoma cases. Hodgkin's disease most often strikes individuals at ages 15-35 or those over age 50. The reason for this pattern of incidence is unknown; although the cause of the disease is uncertain, an infectious agent (probably a virus) is suspected. Other types are usually grouped together under the heading of **non-Hodgkin's lymphoma (NHL)**. They are extremely diverse, and in most cases the primary cause remains a mystery. At least some forms reflect a combination of inherited and environmental factors. For example, one form, called **Burkitt's lymphoma**, most often affects male children in Africa and New Guinea. The affected children have been infected with the *Epstein-Barr virus (EBV)*. This highly variable virus is also responsible for infectious mononucleosis, (discussed further below), and it has been suggested as a possible cause of chronic fatigue syndrome.

The EBV infects B cells, but under normal circumstances the infected cells are destroyed by the immune system. EBV is widespread in the environment, and childhood exposure usually produces lasting immunity. Children developing Burkitt's lymphoma may have a genetic susceptibility to EBV infection; in addition, presence of another illness, such as malaria, may weaken their immune systems to the point that a lymphoma can develop.

The first symptom associated with any lymphoma is usually a painless enlargement of lymph nodes. The involved nodes have a firm, rubbery texture. Because the nodes are painless, the condition is often overlooked until it has progressed to the point that secondary symptoms appear. For example, patients seeking help for recurrent fever, night sweats, gastrointestinal or respiratory problems, or weight loss may be unaware of any underlying lymph node changes. In the late stages of the disease, symptoms can include liver or spleen enlargement, central nervous system (CNS) dysfunction, pneumonia, a variety of skin conditions, and anemia.

In planning treatment, clinicians consider the histological structure of the nodes and the stage of the disease. When examining a biopsy, the structure of the node is described as *nodular* or *diffuse*. A nodular node retains a semblance of normal structure, with follicles and germinal centers. In a diffuse node the interior of the node has changed, and follicular structure has broken down. In general, the nodular lymphomas progress more slowly than the diffuse forms, which tend to be more aggressive. On the other hand, the nodular lymphomas are more resistant to treatment and are more likely to recur even after remission has been achieved.

The most important factor influencing treatment selection is the stage of the disease. Table A-17 includes a simplified staging classification for lymphomas. When diagnosed early (stage I or II), localized therapies may be effective. For example, the cancerous node(s) may be surgically removed and the region(s) irradiated to kill residual cancer cells. Success rates are very high when a lymphoma is detected in these early stages. For Hodgkin's disease, localized radiation can produce remission lasting 10 years or more in over 90 percent of patients. Treatment of localized NHL is somewhat less effective. The 5-year remission rates average 60 to 80 percent for all types; success rates are higher in nodular forms than for diffuse forms.

Although these are encouraging results, it should be noted that few lymphoma patients are diagnosed while in the early stages of the disease. For example, only 10-15 percent of NHL patients are diagnosed at stages I or II. For lymphomas at stages III and IV, treatment most often involves chemotherapy. Combination chemotherapy, in which two or more drugs are administered simultaneously, is the most effective treatment. For

---

Table A-17 Cancer Staging in Lymphoma

Stage I: Involvement of a single node or region (or of a single extranodal site).

Typical treatment: surgical removal and/or localized irradiation; in slowly progressing forms of NHL, treatment may be postponed indefinitely.

Stage II: Involvement of nodes in two or more regions (or of an extranodal site and nodes in one or more regions) on the same side of the diaphragm.

Typical treatment: surgical removal and localized irradiation that includes an extended area around the cancer site (the extended field).

Stage III: Involvement of lymph node regions on both sides of the diaphragm. This is a large category that is subdivided on the basis of the organs or regions involved. For example, in stage III, the spleen contains cancer cells.

Typical treatment: combination chemotherapy, with or without radiation; radiation treatment may involve irradiating all of the thoracic and abdominal nodes plus the spleen (*total axial nodal irradiation,* or *TANI*).

Stage IV: Widespread involvement of extranodal tissues above and below the diaphragm.

Treatment is highly variable, depending on the circumstances. Combination chemotherapy is always used; it may be combined with whole-body irradiation. The "last resort" treatment involves massive chemotherapy followed by a bone marrow transplant.

15

Hodgkin's disease, a four-drug combination with the acronym **MOPP** (nitrogen **M**ustard, **O**ncovin [*vincristine*], **P**rednisone, and **P**rocarbazine) produces lasting remission in 80 percent of patients.

Bone marrow transplantation is a treatment option for acute, late-stage lymphoma. When suitable donor marrow is available, the patient receives whole-body irradiation, chemotherapy, or some combination of the two sufficient to kill tumor cells throughout the body. This treatment also destroys normal bone marrow cells. Donor bone marrow is then infused, and over the next 2 weeks the donor cells colonize the bone marrow and begin producing red blood cells, granulocytes, monocytes, and lymphocytes.

Potential complications of this treatment include the risk of infection and bleeding while the donor marrow is becoming established. The immune cells of the donor marrow may also attack the tissues of the recipient, a response called **graft-versus-host disease, (GVH)**. For a patient with stage I or II lymphomas, without bone marrow involvement, bone marrow can be removed and stored (frozen) for over 10 years. If other treatment options fail, or the patient comes out of remission at a later date, an autologous marrow transplant can be performed. This eliminates the need for donor typing and the risk of GVH disease.

⚕ Disorders of the Spleen EAP *p. 392*

The spleen responds like a lymph node to infection, inflammation, or invasion by cancer cells. The enlargement that follows is called **splenomegaly** (splen-ō-MEG-a-lē; megas, large), and splenic rupture may also occur under these conditions. One relatively common condition causing splenomegaly is **mononucleosis**. This condition, also known as the "kissing disease," results from chronic infection by the *Epstein-Barr virus (EBV)*. In addition to splenic enlargement, symptoms of mononucleosis include fever, sore throat, widespread swelling of lymph nodes, increased numbers of atypical lymphocytes in the blood, and the presence of circulating antibodies to the virus. The condition most often affects young adults (age 15 to 25) in the spring or fall. Treatment is symptomatic, as there are no drugs that are effective against this virus. The most dangerous aspect of the disease is the risk of rupturing the enlarged spleen, which becomes fragile. Patients are therefore cautioned against heavy exercise or other activities that increase abdominal pressures. If the spleen does rupture, severe hemorrhaging may occur; death will follow unless transfusion and an immediate splenectomy can be performed.

An individual whose spleen is missing or nonfunctional has **hyposplenism** (hī-pō-SPLĒN-ism). Hyposplenism usually does not pose a serious problem, but such individuals are more prone to some bacterial infections, such as *Streptococcus pneumoniae*, than individuals with normal spleens, and immunization against this pathogen is recommended. In **hypersplenism** the spleen becomes overactive, and the increased phagocytic activities lead to anemia (low number of RBCs), leukopenia (low number of WBCs), and thrombocytopenia (low number of platelets). Splenectomy is the only known cure for hypersplenism.

℞ Immunization EAP *p. 392*

Immunization is the manipulation of the immune system by providing antigens under controlled conditions, or by providing antibodies that can combat an existing infection. In **active immmunization** a primary response to a particular pathogen is intentionally stimulated before an individual encounters the pathogen in the environment. The result is lasting immunity against that pathogen. Immunization is accomplished by administering a **vaccine** (vak-SĒN), a preparation of antigens derived from a specific pathogen. The vaccine may be given orally or via intramuscular or subcutaneous injection. Most vaccines consist of the pathogenic organism, in whole or in part, living (but weakened) or dead. In some cases a vaccine contains one of the metabolic products of the pathogen.

Before live bacteria or viruses are administered they are weakened, or **attenuated** (a-TEN-u-ā-ted), to lessen or eliminate the chance of a serious infection developing from exposure to the vaccine. The rubella, mumps, measles, smallpox, yellow fever, and oral polio vaccines are examples of vaccines using live attenuated viruses. Despite attenuation, the administration of live microorganisms may produce mild symptoms comparable to those of the actual disease. However, the risks of serious illness developing as a result of vaccination are very small compared with the risks posed by pathogen exposure *without* prior vaccination.

Inactivated or "killed" vaccines consist of bacterial cell walls or viral protein coats. These vaccines have the advantage that they cannot produce even mild symptoms of the disease. Unfortunately, inactivated vaccines do not stimulate as strong an immune response and so do not confer as long-lasting immunity as do live-organism vaccines. In the years following exposure, the antibody titer declines and the system eventually fails to produce an adequate secondary response. As a result, the immune system must be "reminded" of the antigen periodically by the administration of *boosters*. Influenza, typhoid, cholera, typhus, plague, and injected polio vaccines use inactivated viruses or bacteria. In some cases fragments of the bacterial or viral walls, or their toxic products, can be used to produce a vaccine. The tetanus, diphtheria, and hepatitis B vaccines are good examples. Data concerning attenuated and inactivated vaccines are presented in Table A-18.

Gene-splicing techniques can now be used to incorporate antigenic compounds from pathogens into the cell walls of harmless bacteria. When exposed to these bacteria, the immune system responds by producing antibodies and memory B cells that are equally effective against the engineered bacterium and the pathogen.

Table A-18   Immunizations Currently Available

Immunization Target	Type of Immunity Provided	Vaccine Type	Remarks
VIRUSES			
Poliovirus	Active	Live, attenuated	Oral
	Active	Killed	Boosters every 2-3 years
Rubella	Active	Live, attenuated	
	Passive	Human antibodies (pooled)	
Mumps	Active	Live, attenuated	
Measles (rubeola)	Active	Live, attenuated	May need second booster
Hepatitis A	Passive	Human antibodies (pooled)	
Hepatitis B	Active	Killed	May need periodic boosters
	Passive	Human antibodies (pooled)	
Smallpox	Active	Live, related virus	Boosters every 3-5 years (no longer required as disease appears to have been eliminated)
Yellow fever	Active	Live, attenuated	Boosters every 10 years
Herpes zoster	Passive	Human antibodies (pooled)	
Hemophilus influenza B (HIB)	Active	Killed	May need periodic boosters
Rabies	Passive	Human antibodies (pooled)	
	Passive	Horse antibodies	
	Active	Killed	Boosters required
BACTERIA			
Typhoid	Active	Killed	Boosters every 10 years
Tuberculosis	Active	Live, Attenuated	
Plague	Active	Killed	Boosters every 2 years
Cholera	Active	Killed	Boosters every 6 months
Tetanus	Active	Toxins only	Boosters every 5–10 years
	Passive	Human antibodies (pooled)	
Diphtheria	Active	Toxins only	Boosters every 10 years
	Passive	Horse antibodies	
Streptococcal pneumonia	Active	Bacteria and cell wall components	
Botulism	Passive	Horse antibodies	
Rickettsia: Typhus	Active	Killed	Boosters yearly
OTHER TOXINS			
Snake bite	Passive	Horse antibodies	
Spider bite	Passive	Horse antibodies	
Venomous fish spine	Passive	Horse antibodies	

Passive immunization is usually selected if the individual has already been exposed to a dangerous pathogen or toxin, so that there is not enough time for active immunization to take effect. In passive immunization the patient receives a dose of antibodies that will attack the pathogen and overcome the infection, even without the help of the host's own immune system. Passive immunization provides only short-term resistance to infection, for the antibodies are gradually removed from circulation and are not replaced.

The antibodies provided during passive immunization have traditionally been acquired by collecting and combining antibodies from the sera of many other individuals. This *pooled sera* is used to obtain large quantities of antibodies, but the procedure is very expensive, and improper treatment of the sera carries the risk of accidental transmission of an infectious agent, such as the hepatitis or AIDS virus. Antibodies can also be obtained from the blood of a domesticated animal (usually a horse) exposed to the same antigen. Unfortunately, recipients may suffer allergic reactions to horse serum proteins.

At present, antibody preparations are available to treat hepatitis A, hepatitis B, diphtheria, tetanus, rabies, measles, rubella, botulism, and the venoms of certain fishes, snakes, and spiders. Gene-splicing technology can also be used to reproduce pure antibody preparations free from antigenic or viral contaminants, and this should eventually eliminate the need for pooled or foreign plasma.

15

It should also be noted that passive immunity occurs naturally during fetal development, because maternal IgG antibodies can cross the placental barriers and enter the fetal circulation.

✝ Transplants and Graft Rejection

Organ transplantation is a treatment option for patients with severe disorders of the kidneys, liver, heart, lungs, or pancreas. Finding a suitable donor is the first major problem. In the United States, each day 6 people die while awaiting an organ transplant and another 48 people are added to the transplant waiting list. After surgery has been performed, the major problem is **graft rejection.** In graft rejection, T cells are activated by contact with MHC proteins on cell membranes in the donated tissues. The cytotoxic T cells that develop then attack and destroy the foreign cells.

Significant improvements in transplant success can be made by reducing the sensitivity of the immune system. Until recently the drugs used to produce this **immunosuppression** did not selectively target the immune system. For example, **prednisone** (PRED-ni-sōn), a corticosteroid, was used because it has anti-inflammatory effects that reduce the number of circulating white blood cells and depress the immune response. However, corticosteroid use also caused undesirable changes in glucose metabolism.

An understanding of the communication among T cells, macrophages, and B cells has now led to the development of drugs with more selective effects. **Cyclosporin A (CsA),** a compound derived from a fungus, was the most important *immunosuppressive drug* developed in the 1980s. This compound depresses all aspects of the immune response, primarily by suppressing helper T cell activity while leaving suppressor T cells relatively unaffected.

✝ AIDS   EAP *p. 397*

Acquired immune deficiency syndrome (AIDS), or *late-stage HIV disease*, develops following infection by the *human immunodeficiency virus* (HIV). There are at least three different types of AIDS virus, designated *HIV-1, HIV-2,* and *HIV-3.* Most AIDS patients in the United States are infected with HIV-1. HIV-2 infections are most common in West Africa. Because individuals infected with HIV-2 may not always develop AIDS, it is thought that this may be a less dangerous virus (so far nearly 100 percent of those infected with HIV-1 eventually develop AIDS). The distribution and significance of HIV-3 infection remain to be determined.

The discussion that follows, based on information pertaining to HIV-1 infection, expands upon the discussion on *p. 397* of the text.

Symptoms of HIV Disease

The initial infection may produce a flulike illness with fever and swollen lymph nodes a few weeks after exposure to the virus. This exposure usually triggers production of antibodies against the virus. These antibodies appear in the serum within 2-6 months of exposure, and antibody tests can be used to detect infection. Further symptoms may not appear for 5-10 years or more. Over this period the virus content of the blood is very low, but the viruses are at work within lymphoid tissues, especially in the lymph nodes. There is a steady decline in the number of CD4 T cells in the body, and, for reasons as yet uncertain, a destruction of dendritic cells in lymphoid tissues.

The life expectancy is short because HIV-1 selectively infects helper T cells. This impairs the immune response, and the effect is magnified because suppressor T cells are relatively unaffected by the virus. Over time, circulating antibody levels decline, cellular immunity is reduced, and the body is left without defenses against a wide variety of bacterial, viral, fungal, and protozoan invaders.

This vulnerability is what makes AIDS dangerous. The effects of HIV on the immune system are not by themselves life-threatening, but the infections that result certainly are. With the depression of immune function, ordinarily harmless pathogens can initiate lethal infections, known as opportunistic infections. In fact, the most common and dangerous pathogens for an AIDS patient are microorganisms that seldom cause illnesses in humans with normal immune systems. AIDS patients are especially prone to lung infections and pneumonia, often caused by infection with the fungus *Pneumocystis carinii.* They are also subject to a variety of other fungal infections, such as *cryptococcal meningitis*, and an equally broad array of bacterial and viral infections. Because AIDS patients are virtually defenseless, the symptoms and time course of these infections are very different from those in normal individuals.

In addition to problems with pathogenic invasion, immune surveillance is depressed, and the risk of cancer increases. One of the most common cancers seen in AIDS patients is *Kaposi's sarcoma,* a condition that is extremely rare in uninfected individuals. Kaposi's sarcoma typically begins with rapid cell divisions in endothelial cells of cutaneous blood vessels. Associated lesions, blue or a deep brown-purple in color, generally appear first in the hands or feet, and later occur closer to the trunk. In a small number of patients the lesions develop in the epithelium of the digestive or respiratory tracts. In normal individuals the tumor usually does not metastasize; in the case of AIDS patients, whose immune systems are relatively ineffective, the tumor often converts to an aggressive, invasive cancer.

If the AIDS patient survives all of these assaults, infection of the CNS by HIV eventually produces neurological disorders and a progressive dementia. So far, AIDS is invariably fatal, and it appears likely that all those who carry the virus will eventually die of the disease. There are a few long-term survivors who test as HIV positive but who have no signs of abnormal immune function. It is not certain that these individuals were infected with a less virulent form of the virus, and their cases are being closely monitored.

Incidence

During 1996, roughly 64,000 cases of AIDS were diagnosed in the United States, and the number of U.S. deaths attributed to this disorder since its discovery 12 years ago exceeded 350,000. The Centers for Disease Control (CDC) in Atlanta, which has been monitoring the spread of AIDS, estimates that the number of AIDS cases in the United States will continue to increase rapidly (Figure A-42).

Because the virus can remain in the body for years without producing clinical symptoms, the number of individuals infected and at risk is certain to be far higher. An estimated 1-2 million Americans are infected with HIV, and all will eventually develop AIDS, probably within the next decade. The numbers worldwide are even more frightening. The World Health Organization (WHO) estimates that as many as 20 million people may be infected, and the number of AIDS patients worldwide continues to climb rapidly. Several African nations are already on the verge of social and economic collapse because of devastation by AIDS. In Malawi, one-third of the population is infected, and 60 percent of pregnant women carry the virus.

Modes of Infection

Infection with HIV occurs through intimate contact with the body fluids of infected individuals. Although all body fluids, including saliva, tears, and breast milk, carry the virus, the major routes of infection involve contact with blood, semen, or vaginal secretions. The transmission pattern has been analyzed for U.S. adult and adolescent AIDS patients. Four major transmission routes have been identified, and we will consider them individually.

- *Sexual transmission:* In approximately 62 percent of all cases, exposure occurred through sexual contact with infected individuals. Male homosexual contact was most often involved (57 percent versus 5 percent for heterosexual contact), and the ratio of male to female AIDS patients is approximately 9 to 1. A comparable transmission pattern is found in Canada, Europe, South America, Australia, and New Zealand. In the Caribbean and in Africa, by contrast, AIDS began in the heterosexual community. It affects heterosexual men and women in roughly equal num-

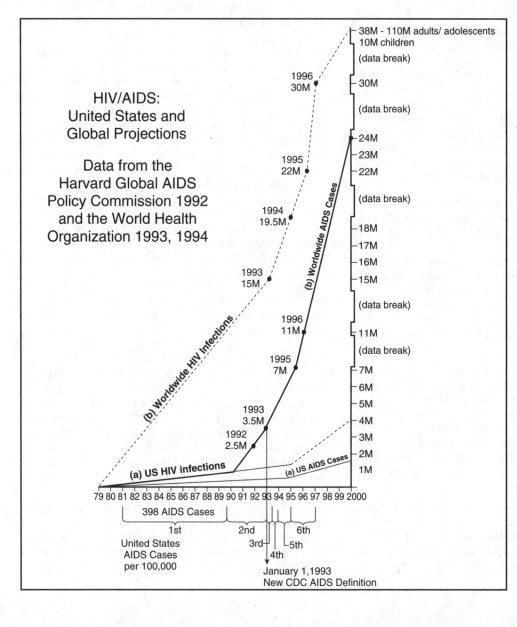

Figure A-42
United States and Global Projections for Total Number of HIV and AIDS Cases. *(From Gerald J. Stine (1996), AIDS Update 1996, © Prentice Hall, Englewood Cliffs, NJ)*

15

bers, and the sex ratio for AIDS patients in these countries is approximately 1 to 1.

Two factors may account for the pattern of transmission observed in the United States. First, the disease appears to have spread through the homosexual community before entering the heterosexual population via bisexual males. Second, there is statistical evidence that on a per-exposure basis the risk of male-to-male or male-to-female transmission is many times greater than the risk of female-to-male transmission. As a result, it may take longer to spread the virus through the heterosexual population (male 1 to female 1 to male 2) than through the homosexual population (male 1 to male 2 to male 3). Whether homosexual or heterosexual contact is involved, a sex partner whose epithelial defenses are weakened is at increased risk of infection. This accounts for the relatively higher rate of transmission through anal intercourse, which often damages the delicate lining of the anorectal canal. Genital ulcers from other sexually transmitted diseases, such as syphilis, chancroid, or herpes, also increases the risk of transmission.

Because of the predominance of homosexual transmission in this country, many people still consider AIDS to be a homosexual disease. It is not. Over time, the number of cases in the heterosexual population has been steadily increasing; since 1987, the percentage of homosexual or bisexual AIDS patients has dropped by roughly 25 percent, and the number of cases transmitted by heterosexual contact has steadily increased. It can be anticipated that over time the sex ratio will continue to shift toward the 1 to 1 male-to-female ratio typical of Africa and the Caribbean.

- *Injecting drug use:* Roughly 19 percent of AIDS patients contracted the disease through the shared use of needles. (Another 6 percent were homosexual males who shared needles with other drug users, making it unclear which factor was responsible for transmission.) Although only small quantities of blood are inadvertently transferred when needles are shared, this practice injects the AIDS virus directly into the bloodstream. It is thus a very effective way to transmit the disease.

- *Receipt of blood or tissue products:* About 3 percent of AIDS patients have become infected with the virus after receiving a transfusion of contaminated whole blood or plasma, an infusion of blood products, such as platelets or extracts of pooled sera, or an organ transplant from an infected individual. With careful screening of blood and blood products, the rate of new transmission via this route is essentially zero in the U.S.

- *Prenatal exposure:* Approximately 2000 infants are born each year already infected with AIDS. Although this is a relatively small number compared to the total number of AIDS patients in the United States, it is increasing rapidly. A pregnant woman infected with the AIDS virus has a 20-33 percent chance of infecting her baby across the placenta. As AIDS spreads through the heterosexual population, more pregnant women will become infected, and maternal-fetal transmission will become more common. These unfortunate infants will place social and financial stresses on our society for the rest of the 1990s.

Prevention of AIDS

The best defense against AIDS consists of avoiding exposure to the virus. The most important rule is to avoid sexual contact with infected individuals. All forms of sexual intercourse carry the potential risk of viral transmission. The use of synthetic (latex) condoms has been recommended when the previous history of a sex partner is not known. (Condoms that are not made of synthetic materials are effective in preventing pregnancy but do not block the passage of viruses.) Although condom use does not provide absolute protection, it drastically reduces the risk of infection.

Attempts are underway to ensure that blood and blood products are adequately screened for the presence of HIV-1. A simple blood test exists for the detection of HIV-1 antibodies, and a positive reaction indicates previous exposure to the virus. The assay, an example of an **ELISA test** (enzyme-linked immunoabsorbent assay), is now used to screen blood donors, reducing the risk of infection by transfusion or the use of blood products from pooled sera. Pooled sera can also be heat-treated by exposing it to temperatures sufficient to kill the virus but too low to permanently denature blood proteins.

Most public health facilities will perform the test on request for individuals who fear that they may have already been exposed to the AIDS virus. Unfortunately the test is not 100 percent reliable, and false positive reactions occur at a rate of about 0.4 percent. In addition, the ELISA test does not detect HIV-2 or HIV-3. In the event of a positive test result, a retest should be performed using the more sensitive western blot procedure. It should be recognized that because of the variable incubation period, a positive test for HIV infection does not mean that the individual has AIDS. It does mean that the individual is likely to develop AIDS at some time in the future and that the person is now a carrier and capable of infecting others. By the time an individual develops AIDS, he or she is obviously sick and usually has little interest in sexual activity. In terms of the spread of this disease, the most dangerous individuals are those who appear perfectly healthy and have no idea that they are carrying the virus.

Despite intensive efforts, a vaccine has yet to be developed that will provide immunity from HIV infection. Current research programs are attempting to stimulate antibody production in response to (1) killed but intact viruses, (2) fragments of the viral envelope, (3) HIV proteins on the surfaces of other, less dangerous viruses, or (4) T cell proteins that are targeted by HIV. (The last approach is based on the hypothesis that the antibodies produced will cover the binding sites, preventing viral attachment and penetration.)

Three different groups plan to test proposed vaccines in the next few years. The World Health Organization has selected Uganda, Brazil, and Thailand for initial testing. The U.S. Army is considering Thailand, and the National Institute of Allergy and Infectious Diseases is considering Zaire and unspecified sites inside the U.S. Clinical trials of several vaccines were scheduled for 1995, but plans were changed after negative preliminary results. It is now uncertain when vaccine trials will actually begin.

Treatment

There is no cure for HIV infection. However, the length of survival for AIDS patients has been steadily increasing because (1) new drugs are available that slow the progress of the disease and (2) improved antibiotic therapies have helped overcome infections that would otherwise prove fatal. This combination is extending the lifespan of patients as the search for more effective treatment continues. It should be noted, however, that overcoming an infection in an AIDS patient with antibiotics may require doses up to 10 times greater than those used to fight infections in normal individuals. Moreover, once the infection has been overcome, the patient may have to continue taking that drug for the rest of his or her life. As a result, some AIDS patients find themselves taking 50-100 pills per day just to prevent recurrent infections.

An antiviral drug, *azidothymidine* or *AZT* (sold as *Zidovudine* or *Retrovir*), can slow the progression of AIDS. Low doses of AZT have now been shown to be effective in delaying the progression of HIV infection to AIDS. Higher doses are used to treat AIDS, but this use can lead to a variety of unpleasant side effects, including anemia and even bone marrow failure. However, AZT is effective in reducing the neurological symptoms of AIDS because it can cross the blood-brain barrier to reach infected CNS tissues. By alternating AZT treatment with other antiviral drugs, *dideoxycytidine (ddC)* and *dideoxyinosine (ddI)*, side effects are reduced. Other drugs are undergoing clinical testing in the U.S. and Europe. But, although the next generation of drugs may further extend the lives of AIDS patients, none of the drugs under review promises a cure for the disease.

AIDS and the Cost to Society

Treatment of AIDS is complex and expensive. The average cost is $70,000 per patient per year, giving a projected annual cost of roughly $34 billion (1994). As more individuals become infected, and more of the 1-2 million people already infected develop AIDS, the costs will continue to escalate rapidly.

There are other, less obvious costs to society.

Individuals with depressed immune systems get sick more often and remain sick longer, and as a result, are more likely to spread the infecting pathogen to other, healthy individuals. Some diseases, such as tuberculosis, that were previously present at very low levels in the U.S. population, are now occurring with increased frequency. In part, the reason for this increase is that AIDS patients are succumbing to infection, but it also reflects the spread of these diseases via transmission from AIDS patients into the general population. Long-term use of multiple antibiotics in AIDS patients may also lead to more resistant disease organisms (see p. 130).

⚕ Lyme Disease   EAP *p. 399*

In November 1975, the town of Lyme, Connecticut experienced an epidemic of adult and juvenile arthritis. Between June and September, 59 cases were reported, 100 times the statistical average for a town of that size. Symptoms were unusually severe; in addition to joint degeneration, victims experienced chronic fever and a prominent rash that began as a bull's-eye centered around what appeared to be an insect bite (Figure A-43). It took almost 2 years to track down the cause of the problem.

Lyme disease is caused by a bacterium, *Borrelia burgdorferi*, that normally lives in white-footed mice. It is transmitted to humans through the bite of a tick. The high rate of infection among children reflects the fact that they play outdoors during their summer vacations. After 1975, the problem became regional and then national in scope. In 1996, there were approximately 50,000 people diagnosed with Lyme disease in the United States. It has also been found in Europe.

Although some of the joint destruction results from immune complex deposition, in a mechanism comparable to that involved in rheumatoid arthritis, many of the symptoms (fever, pain, skin rash) develop in response to the release of interleukin-1 (Il-1) by activated macrophages. The cell walls of *B. burgdorferi* contain lipid-carbohydrate complexes

15

Figure A-43   Lyme disease rash.

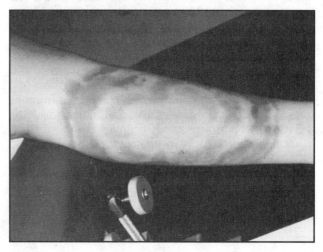

that stimulate secretion of Il-1 in large quantities. By stimulating the specific and nonspecific defense mechanisms of the body, Il-1 exaggerates the inflammation, rash, fever, pain, and joint degeneration associated with the primary infection. Treatment for Lyme disease consists of administering antibiotics and anti-inflammatory drugs.

Most cases of Lyme disease are reported from the Northeast coast, the Midwest, and the West Coast of the United States. On the West Coast, the transmission cycle is somewhat different. The bacterium is found in woodrats in the wild, and the disease reaches humans through the bite of a tick. However, few woodrats are infected, and transmission from rat to rat involves a second species of tick. As a result, the disease spreads more slowly.

A vaccine against Lyme disease is expected by 1997. Over the interim, attempts are being made to control the spread of the disease on the West Coast by attracting woodrats to baits soaked in an oral form of the vaccine.

✝ Immune Competence   EAP p. 400

The ability to demonstrate an immune response upon exposure to an antigen is called **immune competence.** Cell-mediated immunity can be demonstrated as early as the third month of fetal development, but active antibody-mediated immunity does not appear until somewhat later.

The first cells that leave the fetal thymus migrate to the skin and into the epithelia lining the mouth, digestive tract, and, in females, the uterus and vagina. These cells take up residence in these tissues as antigen-presenting cells, such as the Langerhans cells of the skin, whose primary function will be the activation of T cells. T cells that leave the thymus later in development populate lymphoid organs throughout the body.

The first B cells to be produced in the liver and bone marrow carry IgM antibodies in their membranes. Sometime after the fourth developmental month, the fetus may produce IgM antibodies if exposed to specific pathogens. Fetal antibody production is uncommon, however, because the developing fetus has natural passive immunity due to the transfer of IgG antibodies from the maternal bloodstream. These are the only antibodies that cross the placenta, and they include the antibodies responsible for the clinical problems that accompany fetal-maternal Rh incompatibility, discussed on p. 93. Because the agglutinins anti-A and anti-B are IgM antibodies, and IgM cannot cross the placenta, problems with maternal-fetal incompatibilities involving the ABO blood groups rarely occur.

The natural immunity provided by maternal IgG may not be enough to protect the fetus if the maternal defenses are overrun by a bacterial or viral infection. For example, the microorganisms responsible for syphilis and rubella ("German measles") can cross from the maternal to the fetal bloodstream, producing a congenital infection that leads to the production of fetal antibodies. IgM provides only a partial defense, and these infections can result in severe developmental problems for the fetus.

Fetal infections are rare because the developing fetus acquires passive from antibodies produced by the mother. These defenses break down if the maternal antibodies are unable to cope with a bacterial or viral infection. The fetus may then begin producing IgM antibodies. Blood drawn from a newborn infant or taken from the umbilical cord of a developing fetus can be tested for the presence of IgM antibodies. This procedure provides concrete evidence of congenital infection. For example, a newborn infant with congenital syphilis will have IgM antibodies that target the pathogenic bacterium involved (*Treponema pallidum*). Fetal or neonatal (newborn) blood may also be tested for antibodies against the rubella (German measles) virus or other pathogens.

In the case of congenital syphilis, antibiotic treatment of the mother can prevent fetal damage. In the absence of antibiotic treatment, fetal syphilis can cause liver and bone damage, hemorrhaging, and a susceptibility to secondary infections in the newborn infant. There is no satisfactory treatment for congenital rubella infection, which can cause severe developmental abnormalities. For this reason rubella immunization has been recommended for young children (to slow the spread of the disease in the population) and for women of childbearing age. The vaccination, which contains live attenuated virus, must be administered before pregnancy to prevent maternal infection during pregnancy, with resulting fetal damage.

Delivery eliminates the maternal supply of IgG, and although the mother provides IgA antibodies in the breast milk, the infant gradually loses its passive immunity. The amount of maternal IgG in the infant's bloodstream declines rapidly over the first 2 months after birth. During this period the infant becomes vulnerable to infection by bacteria or viruses that were previously overcome by maternal antibodies. The infant also begins producing IgG on its own, as the immune system begins to respond to infections, environmental antigens, and the vaccinations administered by pediatricians. It has been estimated that children encounter a "new" antigen every 6 weeks from birth to age 12. (This fact explains why parents, exposed to the same antigens when they were children, usually remain healthy while their children develop runny noses and colds.) Over this period the concentration of circulating antibodies gradually rises toward normal adult levels, and the populations of memory B cells and T cells continue to increase.

Skin tests can be used to determine whether an individual has been exposed to a particular antigen. In this procedure, small quantities of antigen are injected into the skin, usually on the anterior surface of the forearm. If resistance exists, the region will become inflamed over the next 2–4 days. Many states require a tuberculosis test, called a *tuberculin skin test,* when someone applies for a job in the food-service industry. If the test is positive, the individual has been exposed to the disease and has developed antibodies. Further tests must be performed to determine whether an infection is under way. (Anyone with tuberculosis may accidentally transmit the bacteria when preparing or serving food.) Skin tests are also used to check for allergies to environmental antigens.

☤ Immediate Hypersensitivity

EAP *p. 400*

Immediate hypersensitivity begins with the process of *sensitization.* Sensitization is the initial exposure to an allergen that leads to the production of antibodies, specifically large quantities of IgE. The tendency to produce IgE antibodies in response to specific allergens may be genetically determined. In drug reactions, such as penicillin allergies, IgE antibodies are produced in response to a hapten (partial antigen) bound to a larger molecule inside the body.

Because of the lag time needed to activate B cells, produce plasma cells, and synthesize antibodies, the first exposure to an allergen does not produce symptoms. It merely sets the stage for the next encounter. After sensitization, the IgE antibodies become attached to the cell membranes of basophils and mast cells throughout the body. When exposed to the same allergen at a later date, the bound antibodies stimulate these cells to release histamine, heparin, several cytokines, prostaglandins, and other chemicals into the surrounding tissues. The result is a sudden, massive inflammation of the affected tissues.

Basophils, eosinophils, T cells, and macrophages are drawn to the area by the cytokines and other secretions of the mast cells. These cells release chemicals of their own, extending and exaggerating the responses initiated by mast cells.

The severity of the allergic reaction depends on the sensitivity of the individual and the location involved. If allergen exposure occurs at the body surface, the response may be restricted to that area. If the allergen enters the systemic circulation, the response may be more dramatic and occasionally lethal.

ANAPHYLAXIS In **anaphylaxis** (a-na-fi-LAK-sis; *ana-*, again + *phylaxis*, protection) a circulating allergen affects mast cells throughout the body (Figure A-44). The entire range of symptoms can develop within a matter of minutes. Changes in capillary permeabilities produce swelling and edema in the dermis, and raised welts, or *hives,* appear on the surface of the skin. Smooth muscles along the respiratory passageways contract, and the narrowed passages make breathing extremely difficult. In severe cases of anaphylaxis an extensive peripheral vasodilation occurs, producing a fall in blood pressure that may lead to a circulatory collapse. This response has been called **anaphylactic shock.**

☤ Immune Complex Disorders EAP *p. 400*

Under normal circumstances, immune complexes are promptly eliminated by phagocytosis. But when an antigen appears suddenly, in high concentrations, the local phagocytic population may not be able to cope with the situation. The immune complex then enlarges further, eventually forming insoluble granules that are deposited in the affected area. The presence of these complexes triggers extensive activation of complement, leading to inflammation and tissue damage at that site. This condition is known as an **immune complex disorder**. The process of immune complex formation is further enhanced by neutrophils, which release

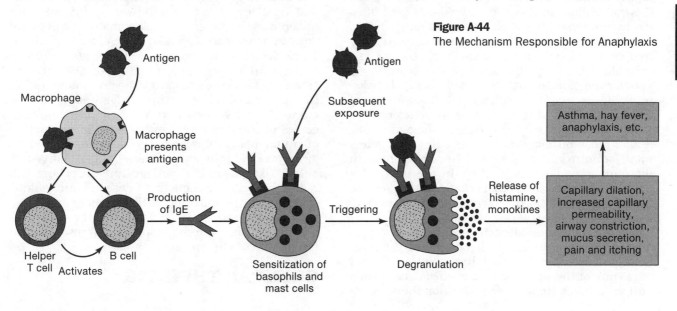

Figure A-44
The Mechanism Responsible for Anaphylaxis

15

Antigen

Macrophage

Macrophage presents antigen

Helper T cell   B cell
Activates

Production of IgE

Sensitization of basophils and mast cells

Antigen

Subsequent exposure

Triggering

Degranulation

Release of histamine, monokines

Capillary dilation, increased capillary permeability, airway constriction, mucus secretion, pain and itching

Asthma, hay fever, anaphylaxis, etc.

enzymes that attack the inflamed cells and tissues. The most serious immune complex disorders involve deposits within blood vessels and in the filtration membranes of the kidney.

⚕ Delayed Hypersensitivity EAP p. 400

Delayed hypersensitivity begins with the sensitization of cytotoxic T cells. At the initial exposure, macrophages present antigenic materials to T cells. On subsequent exposure, the T cells respond by releasing cytokines that stimulate macrophage activity and produce a massive inflammatory response in the immediate area. Examples of delayed hypersensitivity include the many types of contact dermatitis, such as poison ivy, a form of dermatitis considered on p. 38.

Skin tests can be used to check for delayed hypersensitivity. The antigen is administered by shallow injection, and the site is inspected after a period of minutes to days. Most skin tests inject antigens taken from bacteria, fungi, viruses, or parasites. If the individual has previously been exposed to the antigen, in 2-4 days the injection site will become red, swollen, and invaded by macrophages, T cells, and neutrophils. These signs are considered a positive test which indicates previous exposure, but does not necessarily indicate the presence of the disease.

⚕ Technology, Immunity, and Disease
EAP p. 403

It is possible to identify an individual B cell responsible for producing a specific antibody. That B cell can then be isolated and fused with a cancer cell. This produces a **hybridoma** (hī-bri-Dō-ma), a cancer cell that pro-duces large quantities of a single antibody. Hybridomas undergo rapid mitotic divisions, like other cancer cells, so culturing that original hybridoma cell soon produces an entire popula-tion, or clone, of genetically identical cells. This particular clone will produce large quantities of a **monoclonal** (mo-nō-KLō-nal) **antibody**.

Monoclonal antibodies are useful because they are free from impurities. One important use for this technology has been the development of antibody tests for the clinical analysis of body fluids. Labeled antibodies can be used to detect small quanti-ties of specific antigens in a sample of plasma or other body fluids. For example, a popular home pregnancy test relies on monoclonal antibodies that detect small amounts of a placental hormone, human chorionic gonadotrophin (HCG), in the urine of a pregnant woman. Other monoclonal antibodies are used in standard blood screening tests for venereal diseases and urine tests for ovulation.

Monoclonal antibodies can also be used to provide passive immunity to disease. Passive immunizations using monoclonal antibodies do not cause any of the unpleasant side effects associated with antibodies from pooled sera, for the product will not contain plasma proteins, viruses, or other contami-nants. The antibodies can be made to order by exposing a popu-lation of B cells to a particular antigen, then isolating any B cells that produce the desired antibodies.

Genetic engineering techniques can also be used to promote immunity in other ways. One interesting approach involves gene-splicing techniques. The genes coding for an antigenic protein of a viral or bacterial pathogen are identified, isolated, and inserted into a harmless bacterium that can be cultured in the laboratory. The clone that eventually develops will produce large quantities of pure antigen that can then be used to stimulate a primary immune response. Vaccines against malaria and hepatitis were developed in this manner, and a similar strategy may be used to design an AIDS vaccine.

A more controversial experimental technique involves taking a pathogenic organism and adding or removing genes to make it harmless. The modified pathogen can then be used to produce active immunity without the risk of severe illness. Fears that the engineered organism could mutate or regain its pathogenic habits have so far limited the use of this approach, even in animal trials.

Hybridomas that manufacture other products of the immune system can also be produced. Interferons are not effective against all viruses, but interferon nasal sprays appear to provide resistance against the viruses responsible for the common cold. (Unfortunately these sprays can cause nasal bleed-ing and other unpleasant side effects, and they have not been approved for sale to the public.) Interferons can also control certain forms of virally induced cancers. Interleukins may prove useful in increasing the intensity of the immune response.

As our understanding of the immune system grows, complex therapies involving combinations of lymphokines, interleukins, and monoclonal antibodies are appearing. For example, in a recent clinical trial, cytotoxic T cells were removed from patients with malignant melanoma, a particularly dangerous cancer discussed in Chapter 5. (p. 100). These lymphocytes were able to recognize tumor cells, and they had migrated to the tumor. However, for some reason they were unable to kill the tumor cells. The extracted lymphocytes were cultured, and viruses were used to insert multiple copies of the genes responsible for the production of tumor necrosis factor. The patients were then given periodic infusions of these "super-charged" T cells. To enhance T cell activity further, the researchers also administered doses of interleukin-2. Although the outcome of this particular therapy remains uncertain, it is clear that the ability to manipulate the immune response will revolutionize the treatment of many serious diseases.

CRITICAL THINKING

QUESTIONS

7-1. Matthew, a 17-year-old hemophiliac, reports to Dr. M. concerning his recent unexplained weight loss and a flu disorder that he has had "on and off" for the last month. Matthew has a fever of 100°F, cervical lymph node swelling, and a persistent cough. Dr. M. is worried about Matthew's immune state and recommends a bronchoscopic analysis to detect the cause of the respiratory problems. The cultured specimen from Matthew's lungs contains the bacteria *Pneumocystis carinii*. Matthew is diagnosed with PCP or *Pneumocystis carinii pneumonia*. What further test would you order for Matthew?

7-2. Paula's grandfather is diagnosed as having lung cancer. His physician orders biopsies of several lymph nodes from neighboring regions of the body. Paula wonders why, since his cancer is in the lungs. What would you tell her?

7-3. Anne, an 18-year-old college student, goes to her physician with symptoms that have persisted for 2 weeks. She has been fatigued and feverish, with a temperature as high as 38.6°C (101°F). She has inflamed tonsils and enlarged cervical lymph nodes. On palpation her spleen is slightly enlarged; no other signs or symptoms are noted.

Her lab results are as follows:

RBC count	4.2 million/mm^3
Hct	37
Hb	12 g/dl
MCV	80 μm^3
MCH	27 pg
Reticulocytes	1% of total RBCs
WBC count	7,500/mm3
Diff. count:	40% neutrophis
	60% lymphocytes
	2% monocytes
	1% eosinophils .
	0.5% basophils
Platelets	200,000/mm^3

Based on the lab results, what is a likely diagnosis?

NOTES

The Respiratory System

The anatomical components of the respiratory system can be divided into an *upper respiratory system*, which includes the nose, nasal cavity, paranasal sinuses, and pharynx, and a *lower respiratory system* composed of the larynx, trachea, bronchi and lungs. The *respiratory tract* consists of the airways that carry air to and from the exchange surfaces of the lungs. The respiratory tract can be divided into a *conducting portion* and a *respiratory portion*. The conducting portion begins at the entrance to the nasal cavity and extends through the pharynx and larynx, and along the trachea and bronchi to the terminal bronchioles. The respiratory portion of the tract includes the respiratory bronchioles and the alveoli that are part of the respiratory membrane, where gas exchange occurs.

SYMPTOMS OF RESPIRATORY DISORDERS

Individuals with lower respiratory disorders usually seek medical attention due to one or two major symptoms, specifically *chest pain* and *dyspnea*.

1. The chest pain associated with a respiratory disorder usually worsens when the person takes a deep breath or coughs. This pain with breathing distinguishes it from the chest pain experienced by individuals with angina (pain appears during exertion) or a myocardial infarction (pain is continuous, even at rest). Several disorders, such as those affecting the pleural membranes, cause chest pain that is localized to specific regions of the thoracic cage. A person with such a condition will usually press against the sensitive area and avoid coughing or deep breathing in an attempt to reduce the pain.

2. *Dyspnea,* or difficulty in breathing, may be a symptom of pulmonary disorders, cardiovascular disorders, metabolic disorders, or environmental factors such as hypoxia at high altitudes. It may be a chronic problem, or it may develop only upon exertion, or only when lying down.

   Dyspnea due to respiratory problems usually indicates one of the following classes of disorders:

   - *Obstructive disorders* result from increased resistance to air flow along the respiratory passageways. The individual usually struggles to breathe, even at rest, and expiration is more difficult than inspiration. Examples of obstructive disorders include *emphysema* (p. 130) and *asthma* (p. 129).

   - *Restrictive disorders* include (1) arthritis, (2) paralysis of respiratory muscles caused by trauma and scarring, muscular dystrophy, myasthenia gravis, multiple sclerosis, or polio, and (3) physical trauma or congenital structural disorders that limit lung expansion. Individuals with restrictive disorders usually experience dyspnea during exertion because pulmonary ventilation cannot increase enough to meet the respiratory demand.

   Cardiovascular disorders that produce dyspnea include *heart disease, congestive heart failure*, and *pulmonary embolism*. In *paroxysmal nocturnal dyspnea* a person awakens at night, gasping for air. The underlying cause is usually a problem with cardiac output due to advanced heart disease or heart failure. *Cheyne-Stokes respiration* consists of cycles of rapid, deep breathing separated by periods of respiratory arrest. This breathing pattern is most often seen in persons with CNS disorders or congestive heart failure.

   Dyspnea may also be related to metabolic problems, such as the acute acidosis associated with *diabetes mellitus* (p. 83) or *uremia* (p. 149). The fall in blood pH can trigger *Kussmaul breathing*, which usually consists of rapid, deep respiratory cycles.

THE PHYSICAL EXAMINATION AND THE RESPIRATORY SYSTEM

Several components of the physical examination will detect indications of respiratory disorders:

1. *Inspection* can reveal abnormal dimensions, such as the "barrel chest" that develops in emphysema or other obstructive disorders (p. 130), or *clubbing* of the fingers (p. 36). Clubbing is most often a late sign of disorders such as emphysema or congestive heart failure. *Cyanosis*, a blue color of the skin and mucous membranes, usually indicates hypoxia. Laboratory testing of arterial blood gases will assist in determining the cause and extent of the hypoxia.

2. *Palpation* of the bones and muscles of the thoracic cage can detect structural problems or asymmetry. For example, asymmetrical contraction of respiratory muscles during breathing may indicate a restrictive disorder.

3. *Percussion* on the surface of the thoracic cage over the lungs will yield sharp, resonant sounds. Dull or flat sounds may indicate structural changes in the lungs, such as those accompanying pneumonia, or collapse of a lung (atelectasis). Increased resonance may result in obstructive disorders, such as emphysema, due to hyperinflation of the lungs as the individual attempts to improve alveolar ventilation.

4. *Auscultation* of the lungs with a stethoscope yields the distinctive sounds of inspiration and expiration. These sounds vary in intensity, pitch, and duration. Abnormal breath

sounds accompany several different pulmonary disorders.

- *Rales* (rahls) are hissing, whistling, scraping, or rattling sounds associated with increased airway resistance. The sounds are created by turbulent airflow past accumulated pus or mucus, or through airways narrowed by inflammation or bronchospasms. *Moist rales* are gurgling sounds produced as air flows over fluids within the respiratory tract. They are heard in conditions such as *bronchitis, tuberculosis,* or *pneumonia. Dry rales* are produced as air flows over thick masses of mucus, through inflamed airways, or into fluid-filled alveoli. Dry rales are characteristic of *asthma* and *pulmonary edema. Rhonchi* are loud dry rales produced by mucus buildup in the air passages.

- *Stridor* is a very loud, high-pitched sound that can be heard without a stethoscope. Stridor usually indicates acute airway obstruction, such as the partial blockage of the glottis by a foreign object.

- *Wheezing* is a whistling sound that can occur with inspiration or expiration. It usually indicates airway obstruction, due to mucus buildup or bronchospasms.

- *Coughing* is a familiar sign of several respiratory disorders. Although primarily a reflex mechanism that clears the airway, coughing may also indicate irritation of the lining of the respiratory passageways. The duration, pitch, causative factors, and productivity (a productive cough ejects sputum; a nonproductive cough does not) may be important clues in the diagnosis of a respiratory disorder. If the cough is productive, the sputum ejected can be analysed. This analysis will provide information about the presence of epithelial cells, macrophages, blood cells, and/or pathogens. If pathogens are present, the sputum can be cultured to permit identification of the specific microorganism involved.

- A *friction rub* is a distinctive crackling sound produced by abrasion between abnormal serous membranes. A *pleural rub* accompanies respiratory movements and indicates problems with the pleural membranes, such as *pleurisy*. A *pericardial rub* accompanies the heartbeat and indicates inflammation of the pericardium, as in *pericarditis*.

5. During assessment of vital signs, the respiratory rate (number of breaths per minute) is recorded, along with notations concerning the general rhythm and depth of respiration. *Tachypnea* is defined as a respiratory rate faster than 20/min. in an adult; *bradypnea* is an adult respiratory rate below 12/min.

DISORDERS OF THE RESPIRATORY SYSTEM

The respiratory system provides a route for air movement in and out of the lungs and supplies a large, protected surface area for the exchange of oxygen and carbon dioxide between the air and circulating blood. Disorders affecting the respiratory system (Figure A-45) may therefore involve:

1. Interfering with the movement of air along the respiratory passageways: Internal or external factors may be involved. Within the respiratory tract, constriction of small airways, as in *asthma* (p. 129), can reduce airflow to the lungs. Blockage of major airways, as in choking, may completely shut off the air supply. External factors that interfere with air movement include (1) the introduction of air (*pneumothorax*) or blood (*hemothorax*) into the the pleural cavity, with subsequent lung collapse, (2) the buildup of fluid within the pleural cavities (a *pleural* effusion) which compresses and collapses the lungs, and (3) arthritis, muscular paralysis, or other conditions that prevent the normal skeletal or muscular activities responsible for moving air in and out of the respiratory tract.

2. Damaging or otherwise impeding the diffusion of gases at the respiratory membrane: The walls of the alveoli are part of the respiratory membrane, where gas exchange occurs. Any disease process that affects the alveolar walls will reduce the efficiency of gas exchange. In *emphysema* (p. 130) or *lung cancer*, alveoli are destroyed. Respiratory exchange can also be disrupted by the buildup of fluid or mucus within the alveoli. This disruption may occur as the result of inflammation or infection of the lungs, as in the various types of pneumonia.

3. Blocking or reducing the normal circulation of blood through the alveolar capillaries: Blood flow to portions of the lungs may be prevented by a *pulmonary embolism*, a circulatory blockage discussed in the text (*p. 414*). A pulmonary embolism not only prevents normal gas exchange in the affected regions of a lung, but also results in tissue damage and, if the blockage persists for several hours, permanent alveolar collapse. Pulmonary blood pressure may rise, a condition called *pulmonary hypertension*, leading to pulmonary edema and a reduction in alveolar function in other portions of the lungs.

✝ Cystic Fibrosis   EAP p. 409

Cystic fibrosis (CF) is the most common lethal inherited disease affecting Caucasians of Northern European descent, occurring at a frequency of 1 birth in 2500. It occurs, with less frequency, in those with Southern European ancestry, in the Ashkenazi

16

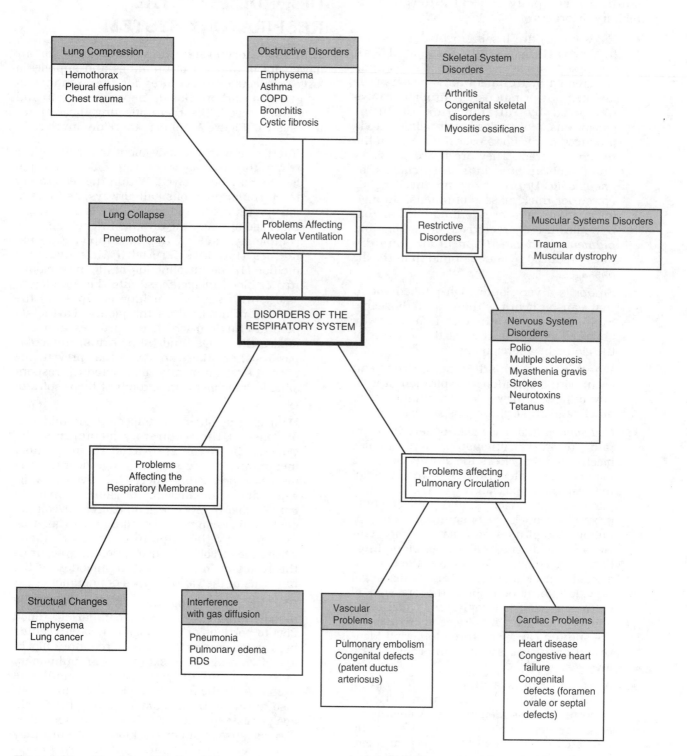

Figure A-45   Disorders of the Respiratory System.

Jewish population, and in African Americans. The condition results from a defective gene located on chromosome 7. Within the U.S. Caucasian population, 1 person in 25 carries one copy of the gene for this disorder, and an infant receiving a copy from each parent will develop CF. Each year 2000 babies are born with CF, and there are roughly 30,000 persons with this condition in the U.S. Individuals with classic CF seldom survive past age 30; death is usually the result of a massive bacterial infection of the lungs and associated heart failure.

The gene involved carries instructions for a transmembrane protein responsible for the active transport of chloride ions. This membrane protein is abundant in exocrine cells that produce watery secretions. In persons with CF, the protein does not function normally. The secretory cells cannot transport salts and water effectively, and the secretions produced are thick and gooey. Mucous glands of the respiratory tract and secretory cells of the pancreas, salivary glands, and digestive tract are affected.

The most serious symptoms appear because the respiratory defense system cannot transport such dense mucus. The mucus escalator stops working, and mucus plugs block the smaller respiratory passageways. This blockage reduces the diameter of the airways, and the inactivation of the normal respiratory defenses leads to frequent bacterial infections.

Treatment has primarily been limited to supportive care and antibiotic therapy to control infections. In a few instances, lung transplants have provided relief, but the technical and logistical problems involved with this approach are formidable. The normal and abnormal gene structure has now been determined, and the current goal is to correct the defect by inserting normal genes within the cells in critical areas of the body. In the meantime, it has been discovered that one of the factors contributing to the thick nature of the mucus is the presence of DNA released from degenerating cells within areas of inflammation. Inhaling an aerosol spray containing an enzyme that breaks down DNA has proven to be remarkably effective in improving respiratory performance.

⚕ Asthma EAP *p. 414*

Asthma (AZ-ma) affects an estimated 3-6 percent of the U.S. population. There are several different forms of asthma, but all are characterized by unusually sensitive and irritable conducting passageways. In many cases the trigger appears to be an immediate hypersensitivity reaction to an allergen in the inspired air. Drug reactions, air pollution, chronic respiratory infections, exercise, and/or emotional stress can also cause an asthmatic attack in sensitive individuals.

The most obvious and potentially dangerous symptoms include (1) the constriction of smooth muscles all along the bronchial tree, (2) edema and swelling of the mucosa of the respiratory passageways, and (3) accelerated production of mucus. The combination makes breathing very difficult. Exhalation is affected more than inhalation; the narrowed passageways often collapse before exhalation is completed. Although mucus production increases, mucus transport slows, and fluids accumulate along the passageways. Coughing and wheezing then develop.

Severe asthmatic attacks reduce the functional capabilities of the respiratory system, and the peripheral tissues become oxygen-starved. This condition can prove fatal, and asthma fatalities have been increasing in recent years. The annual death rate from asthma in the United States is approximately 4 deaths per million population (for ages 5-34). Mortality among asthmatic African-Americans is twice that of Caucasian Americans.

Treatment of asthma involves dilation of the respiratory passageways by administering **bronchodila-** **tors** (brong-kō-dī-LĀ-torz) and reducing inflammation and swelling of the respiratory mucosa. Important bronchodilators include *theophylline, epinephrine, albuterol,* and other beta-adrenergic drugs. Although the strongest beta-adrenergic drugs are very useful in a crisis, they are effective only for relatively brief periods, and the individual must be closely monitored because of the potential effects on cardiovascular function. Anti-inflammatory medication with less acute effects, such as inhaled or ingested steroids, is becoming increasingly important.

⚕ Respiratory Distress Syndrome (RDS) EAP *p. 414*

Surfactant cells begin producing surfactants at the end of the sixth fetal month. By the eighth month surfactant production has risen to the level required for normal respiratory function. **Neonatal respiratory distress syndrome (NRDS)**, also known as *hyaline membrane disease (HMD)*, develops when surfactant production fails to reach normal levels. Although there are inherited forms of HMD, the condition most often accompanies premature delivery.

In the absence of surfactants, the alveoli tend to collapse during exhalation, and although the conducting passageways remain open, the newborn infant must then inhale with extra force to reopen the alveoli on the next breath. In effect, every breath must approach the power of the first, and the infant rapidly becomes exhausted. Respiratory movements become progressively weaker, eventually the alveoli fail to expand, and gas exchange ceases.

One method of treatment involves assisting the infant by administering air under pressure, so that the alveoli are held open. This procedure, known as **positive end-expiratory pressure (PEEP)**, can keep the newborn alive until surfactant production increases to normal levels. Surfactant from other sources can also be provided; suitable surfactants can be extracted from cow lungs (*Survanta*), obtained from the liquid (amniotic fluid) that surrounds full-term infants, or synthesized using gene-splicing techniques (*Exosurf*). These preparations are usually administered in the form of a fine mist of surfactant droplets.

Surfactant abnormalities may also develop in adults as the result of severe respiratory infections or other sources of pulmonary injury. Alveolar collapse follows, producing a condition known as **adult respiratory distress syndrome (ARDS)**. PEEP is often used in an attempt to maintain life until the underlying problem can be corrected, but at least 50-60 percent of ARDS cases result in fatalities.

⚕ Tuberculosis EAP *p. 417*

Tuberculosis spreads rapidly because it can be transmitted through the air, without direct person-to-person contact. Coughing, sneezing, or speaking by an infected individual will spread the pathogen

16

through the air, in the form of tiny droplets that can be inhaled by others. Once infection has occurred, the disease progresses in stages. At the site of the infection, macrophages and fibroblasts proceed to wall off the area, forming an abscess. If the scar-tissue barricade fails, the bacteria move into the surrounding tissues and the process repeats itself. The resulting masses of fibrous tissue distort the conducting passageways, increasing resistance and decreasing airflow. In the alveoli, the attacked surfaces are destroyed. The combination severely reduces the area available for gas exchange.

Treatment for TB is complex, because (1) the bacteria can spread to many different tissues, and (2) they can develop a resistance to standard antibiotics relatively quickly. As a result, several drugs are used in combination over a period of 6–9 months. The most effective drugs now available include *isoniazid*, which interferes with bacterial replication, and *rifampicin*, which blocks bacterial protein synthesis.

The TB problem is much less severe in developed nations, such as the United States, than in third-world nations. However, tuberculosis was extremely common in the United States earlier in this century. An estimated 80 percent of Americans born around the turn of the century became infected with tuberculosis during their lives. Although many were able to meet the bacterial challenge, it was still the number one cause of death in 1906. These statistics have been drastically altered with the arrival of antibiotics and techniques for early detection of infection. Between 1906 and 1993 the death rate fell from 200 deaths per 100,000 population to 0.6 deaths per 100,000 population.

Tuberculosis today is unevenly distributed through the U.S. population, with several groups at relatively high risk for infection. For example, Hispanics, blacks, prison inmates, individuals with immune disorders, such as AIDS patients, and hospital employees who work around infected patients are more likely to be infected than any other members of the population. Although at present only 2–5 percent of young American adults have been infected, the incidence has increased each year since 1984, reaching nearly 24,361 in 1994. A growing percentage of these cases are caused by antibiotic-resistant strains. An estimated 14% of TB cases diagnosed in the U.S. each year are resistant to both isoniazid and rifampicin. The statistics for some parts of the country are worse; in New York City, 33% of new cases are drug-resistant. The frightening part about this surge in resistant TB is that the fatality rate for infections resistant to 2 or more antibiotics is 50%. Unless efforts are made to combat the spread of TB, over the next decade the U.S. may face a public health crisis as deadly as HIV infection and even more difficult to control.

Decompression Sickness   EAP *p. 421*

Decompression sickness can develop when an individual experiences a sudden change in pressure. Nitrogen is the gas responsible for this condition.

Nitrogen, which accounts for 78.6 percent of the atmospheric gas mixture, has a relatively low solubility in body fluids. Under normal atmospheric pressures there are few nitrogen molecules in the blood, but at higher than normal pressures additional nitrogen molecules diffuse across the alveolar surfaces and into the bloodstream.

As more nitrogen enters the blood it is distributed throughout the body. Over time nitrogen diffuses into peripheral tissues and into body fluids such as the cerebrospinal fluid (CSF), aqueous humor, and synovial fluids. If the pressure decreases, the change must occur slowly enough that the excess nitrogen can diffuse out of the tissues, into the blood, and across the alveolar surfaces. If the pressure falls suddenly, this gradual movement of nitrogen from the periphery to the lungs cannot occur. Instead, the nitrogen leaves solution and forms bubbles of nitrogen gas in the blood, tissues, and body fluids.

A few bubbles in peripheral connective tissues may not be particularly dangerous, at least initially. However, these bubbles can fuse together, forming larger bubbles that distort tissues, causing pain. Bubbles often develop in joint capsules first. These bubbles cause severe pain, and the afflicted individual tends to bend over or curl up. This symptom accounts for the popular name of this condition, "the bends." Bubbles in the systemic or pulmonary circulation can cause infarcts, and those in the cerebrospinal circulation can cause strokes, leading to sensory losses, paralysis, or respiratory arrest.

Treatment consists of recompression, exposing the individual to pressures that force the nitrogen back into solution and alleviate the symptoms. Pressures are then reduced gradually over a period of 1 or more days. Breathing air with a higher percentage of oxygen and a lower percentage of nitrogen than atmospheric air accelerates the removal of excess nitrogen from the blood.

Today most bends cases involve scuba divers who have gone too deep or stayed at depth for too great a time. The condition is not restricted to divers, however, and the first reported cases involved construction crews working in pressurized surroundings. Although such accidents are exceedingly rare, the sudden loss of cabin pressure in a commercial airliner can also produce symptoms of decompression sickness.

Emphysema,   EAP *p. 421*

Bronchitis (brong-KĪ-tis) is an inflammation of the bronchial lining. The most characteristic symptom is the overproduction of mucus, which leads to frequent coughing. An estimated 20 percent of adult males have *chronic bronchitis*. This condition is most often related to cigarette smoking, but it can also result from other environmental irritants, such as chemical vapors. Over time the increased mucus production can block smaller airways and reduce respiratory efficiency. This condition is called **chronic airways obstruction**.

Emphysema (em-fi-SĒ-ma) is a chronic, progressive condition characterized by shortness of breath and an inability to tolerate physical exertion. The underlying problem is destruction of respiratory exchange surfaces. In essence, respiratory bronchioles and alveoli are functionally eliminated. The alveoli gradually expand, capillaries deteriorate, and gas exchange in the affected region comes to a halt.

Emphysema has been linked to the inhalation of air containing fine particulate matter or toxic vapors, such as those found in cigarette smoke. Early in the disease, local regulation shunts blood away from the damaged areas, and the individual may not notice problems, even with strenuous activity. As the condition progresses, the reduction in exchange surface limits the ability to provide adequate oxygen. However, obvious clinical symptoms typically fail to appear until the damage is extensive.

An enzyme, *alpha-antitrypsin*, that is normally present in the lungs helps prevent degenerative changes in lung tissue. Most people requiring treatment for emphysema are adult smokers; this group includes individuals with alpha-antitrypsin deficiency and those with normal tissue enzymes. In the U.S., 1 person in 1000 carries two copies of a gene that codes for an abnormal and inactive form of this enzyme. A single change in the amino acid sequence appears responsible for this inactivation. At least 80 percent of nonsmokers with abnormal alpha-antitrypsin will develop emphysema, usually at age 45-50. *All* smokers will develop at least some emphysema, typically by age 35-40.

Unfortunately, the loss of alveoli and bronchioles in emphysema is permanent and irreversible. Further progression can be limited by cessation of smoking; the only effective treatment for severe cases is administration of oxygen. Lung transplants have helped some patients. For persons with alpha-antitrypsin deficiency who are diagnosed early, attempts are underway to provide enzyme supplements by daily infusion or periodic injection.

Two different patterns of symptoms may appear in individuals with advanced emphysema. In one group of patients other aspects of pulmonary structure and function are relatively normal. The respiratory rate in these patients increases dramatically. The lungs are fully inflated at each breath, and expirations are forced. These individuals maintain near-normal arterial PO_2. Their respiratory muscles are working hard, and they use a lot of energy just breathing. As a result, these patients are usually relatively thin. Because blood oxygenation is near normal, skin color in Caucasian patients will be pink.

In the second group, emphysema has been complicated by chronic bronchitis and chronic airways obstruction. This combination, known as **chronic obstructive pulmonary disease** (COPD), is particularly dangerous. Individuals with COPD often expand their chests permanently, in an effort to enlarge their lung capacities and make the best use of the remaining functional alveoli. This adaptation gives them a distinctive "barrel-chested" appearance. COPD sufferers also have symptoms of heart failure, including widespread edema. Blood oxygenation is low, and the skin has a bluish coloration.

CRITICAL THINKING QUESTIONS

8-1. Joe is playing in an intramural football game, when he is tackled so hard that he breaks a rib. On the way to the hospital, Joe is having a difficult time breathing. Joe may be suffering from

　a.　a collapsed trachea

　b.　an obstruction in the bronchi

　c.　a pneumothorax

　d.　decreased surfactant production

　e.　a bruised diaphragm

8-2. A patient recovering from bacterial pneumonia goes to Dr. Smith complaining of recurrent shortness of breath and pain when breathing. The patient reluctantly admits not finishing his antibiotics as prescribed. Dr. Smith auscultates the thorax with a stethoscope and detects a scratching sound during inspiration and decreased breath sounds at the base of one lung. Thoracentesis reveals cloudy fluid with an increase in the number of WBCs and the presence of bacteria in the cultured pleural fluid specimen. What condition is the patient most likely suffering from?

8-3. Ann, a respiratory care technician, complains of low-grade fever, persistent cough, night sweats, and unexplained weight loss over the last two months. Ann does not have a history of cigarette smoking. She has coughed up blood several times in the last two days. The physician orders routine lab tests that are within normal limits. Ann's chest X ray shows lesions related to prior infection. The doctor is awaiting the results of the sputum culture, and has placed a skin test on Ann's forearm. What is a reasonable preliminary diagnosis?

　a.　emphysema

　b.　lung cancer

　c.　tuberculosis

　d.　pleurisy

　e.　pneumonia

What is the significance of the skin test?

The Digestive System

The digestive system consists of the *digestive tract* and *accessory digestive organs*. The digestive tract is divided into the oral cavity, pharynx, esophagus, stomach, small intestine, and large intestine. Most of the absorptive functions of the digestive system occur in the small intestine, with smaller amounts in the stomach and large intestine. The accessory digestive organs provide acids, enzymes, and buffers that assist in the chemical breakdown of food. The accessory organs include the salivary glands, the liver, the gallbladder, and the pancreas. The salivary glands produce saliva, a lubricant that contains enzymes that aid in the digestion of carbohydrates. The liver produces bile, which is concentrated in the gallbladder and released into the small intestine for fat emulsification. The pancreas secretes enzymes and buffers important to the digestion of proteins, carbohydrates, and lipids.

The activities of the digestive system are controlled through a combination of local reflexes, autonomic innervation, and the release of gastrointestinal hormones such as *gastrin, secretin,* and *cholecystokinin.*

SYMPTOMS AND SIGNS OF DIGESTIVE SYSTEM DISORDERS

The functions of the digestive organs are varied, and the symptoms and signs of digestive system disorders are equally diverse. Common symptoms of digestive disorders include:

1. *Pain* is a common symptom of digestive disorders.

   - Widespread pain in the oral cavity may result from (1) trauma, (2) infection of the oral mucosa by bacteria, fungi, or viruses, or (3) a deficiency in vitamin C (*scurvy*, p. 10) or one or more of the B vitamins. Focal pain in the oral cavity usually accompanies (1) infection or blockage of salivary gland ducts, (2) tooth disorders such as tooth fractures, *dental caries* , *pulpitis*, abscess formation, or *gingivitis*, and (3) oral lesions, such as those produced by the *herpes* virus.

   - Abdominal pain is characteristic of a variety of digestive disorders. In most cases the pain is perceived as distressing but tolerable; if the pain is acute and severe, a surgical emergency may exist. Abdominal pain sometimes causes the abdominal muscle overlying the painful area to become rigid. This can be easily felt on palpation. The contractions (*guarding*) may be voluntary, in an attempt to protect a painful area or an involuntary spasm resulting from irritation of the peritoneal lining, as in *peritonitis*. Persons with peri-

toneal inflammation also experience *rebound tenderness*. In rebound tenderness, pain appears when fingertip pressure on the abdominal wall is suddenly removed.

Abdominal pain can result from disorders of the digestive, circulatory, urinary, or reproductive systems. Digestive tract disorders producing abdominal pain include *appendicitis* (p. 114), *gastric* or *duodenal ulcers* (p. 135), *pancreatitis, cholecystitis* (p. 137), *hepatitis* (p. 136), *intestinal obstruction* (p. 133), *diverticulitis, peritonitis*, and cancers of the digestive tract.

2. *Dyspepsia*, or indigestion, is pain or discomfort in the substernal or gastric region. Most cases are idiopathic (of no known cause). Digestive tract disorders associated with dyspepsia include *esophagitis* (EAP, p. 440), *gastric or duodenal ulcers* (p. 135), *gastroesophageal reflux*, and *cholecystitis* (p. 137).

3. *Nausea* is a sensation that usually precedes or accompanies vomiting. Nausea may result from digestive disorders or from disturbances of CNS function.

4. *Anorexia* is a decrease in appetite usually accompanied by weight loss. Digestive disorders causing anorexia include *gastric cancer*), *hepatitis* (p. 136), and several forms of *diarrhea* (p. 138). Anorexia may also accompany disorders involving other systems. *Anorexia nervosa*, an eating disorder with a psychological basis, is discussed on p. 140.

5. *Dysphagia* is difficulty in swallowing. This may result from trauma, infection, inflammation, or blockage of the posterior oral cavity, pharynx, or esophagus. For example, the infections of *tonsillitis, pharyngitis*, and *laryngitis* may cause dysphagia.

THE PHYSICAL EXAMINATION AND THE DIGESTIVE SYSTEM

Physical assessment can provide information useful in diagnosis of digestive system disorders. The abdominal region is particularly important because most of the digestive system is located within the abdominopelvic cavity.

1. *Inspection* can provide a variety of useful diagnostic clues:

   - Bleeding of the gums, as in *gingivitis*, and characteristic oral lesions can be seen on inspection of the oral cavity. Examples of distinctive lesions include those of oral herpes infections and *thrush*, lesions produced by infection of the mouth by *Candida albicans*. This fungus, a normal resident of the digestive tract, causes widespread oral infections in immunodeficient persons, such as AIDS patients or patients undergoing immunosuppressive therapies.

- Peristalsis in the stomach and intestines may be seen as waves passing across the abdominal wall in persons that do not have a thick layer of abdominal fat. The waves become very prominent during the initial stages of intestinal obstruction.

- A general yellow discoloration of the skin, a sign called *jaundice* or *icterus*, results from excessive levels of bilirubin in body fluids. Jaundice is often seen in individuals with *cholecystitis* (p. 137) or liver diseases such as *hepatitis* (p. 136) and *cirrhosis* (p. 137).

- Abdominal distention may be caused by (1) fluid accumulation in the peritoneal cavity, as in *ascites* (EAP, p. 435), (2) air or gas (flatus) within the digestive tract or peritoneal cavity, (3) obesity, (4) abdominal masses, such as tumors, or enlargement of visceral organs, (5) pregnancy, (6) presence of an *abdominal hernia* (p. 61), or (7) fecal impaction, as in severe and prolonged constipation.

- *Striae* are multiple scars, 1-6 cm in length, that are visible through the epidermis. Striae develop in damaged dermal tissues after stretching, and they are often seen in the abdominal region after a pregnancy or other rapid weight gain. Abnormal striae may develop following ascites, of subcutaneous edema, and purple striae are characteristic signs of *Cushing's syndrome* (adrenocortical hypersecretion, p. 87).

2. *Palpation* of the abdomen may reveal specific details about the status of the digestive system, including:

- the presence of abnormal masses, such as tumors, within the peritoneal cavity

- abdominal distention from (1) excess fluid within the digestive tract or peritoneal cavity, or (2) gas within the digestive tract

- herniation of digestive organs through the inguinal canal or other weak spots in the abdominal wall (p. 61)

- changes in the size, shape, or texture of visceral organs. For example, in several liver diseases the liver becomes enlarged and firm, and these changes can be detected on palpation of the right upper quadrant.

- detection of voluntary or involuntary abdominal muscle contractions

- detection of rebound tenderness

- identification of specific areas of tenderness and pain. For example, someone with acute *hepatitis* (p. 136), a liver disease, will usu-ally experience pain on palpation of the upper right quadrant. In contrast, a person with *appendicitis* (p. 114) usually experiences pain when the right lower quadrant is palpated.

3. *Percussion* of the abdomen is less instructive than percussion of the chest, because the visceral organs do not contain extensive airspaces that will reflect the sounds conducted through surrounding tissues. However, the stomach usually contains a small air bubble, and percussion over this area produces a sharp, resonant sound. The sound becomes dull or disappears when the stomach fills, the spleen enlarges, or the peritoneal cavity contains abnormal quantities of peritoneal fluid, as in *ascites*. Percussion may also detect gas accumulation within the peritoneal cavity or within portions of the colon.

4. *Auscultation* can detect gurgling abdominal sounds, or *bowel sounds*, produced by peristaltic activity along the digestive tract. Increased bowel sounds occur in persons with acute diarrhea, and bowel sounds may disappear in persons with (1) advanced intestinal obstruction, (2) *peritonitis*, an infection of the peritoneal lining, and (3) spinal cord injuries that prevent normal innervation.

Diagnostic procedures, such as endoscopy, are often used to provide additional information.

DISORDERS OF THE DIGESTIVE SYSTEM

Due to the number and diversity of digestive organs, there are many different types of digestive system disorders. One common method of categorizing these disorders (Figure A-46, p. 134) is based on a combination of anatomical and functional characteristics.

1. *Disorders of the oral cavity*: This category includes disorders affecting the oral mucosa, the teeth, and the salivary glands.

2. *Disorders of the digestive tract*: This category includes all of the conditions usually called *gastro-intestinal disorders*.

3. *Disorders of the digestive viscera*: This category includes all of the hepatic (liver), biliary (gallbladder and bile ducts), and pancreatic disorders.

4. *Nutritional and metabolic disorders*: These disorders, which affect the entire body, are often considered with disorders of the digestive system because they often accompany digestive system disorders. However, nutritional and metabolic disorders may also reflect (1) disorders involving the endocrine or nervous systems, (2) congenital metabolic problems, or (3) dietary abnormalities.

17

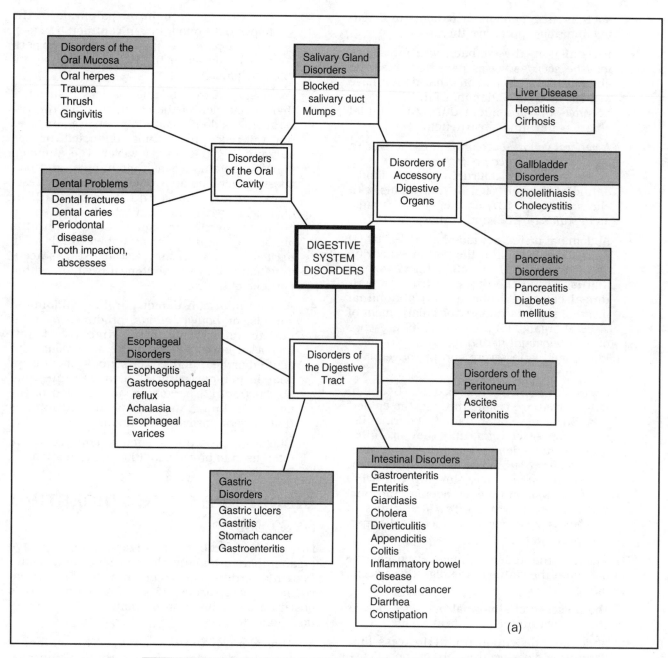

Disorders of the Oral Mucosa
Oral herpes
Trauma
Thrush
Gingivitis

Salivary Gland Disorders
Blocked
 salivary duct
Mumps

Liver Disease
Hepatitis
Cirrhosis

Disorders of the Oral Cavity

Disorders of Accessory Digestive Organs

Dental Problems
Dental fractures
Dental caries
Periodontal
 disease
Tooth impaction,
 abscesses

Gallbladder Disorders
Cholelithiasis
Cholecystitis

DIGESTIVE SYSTEM DISORDERS

Pancreatic Disorders
Pancreatitis
Diabetes
 mellitus

Esophageal Disorders
Esophagitis
Gastroesophageal
 reflux
Achalasia
Esophageal
 varices

Disorders of the Digestive Tract

Disorders of the Peritoneum
Ascites
Peritonitis

Gastric Disorders
Gastric ulcers
Gastritis
Stomach cancer
Gastroenteritis

Intestinal Disorders
Gastroenteritis
Enteritis
Giardiasis
Cholera
Diverticulitis
Appendicitis
Colitis
Inflammatory bowel
 disease
Colorectal cancer
Diarrhea
Constipation

(a)

**Figure A-46
Disorders of the Digestive System**
(a) Major categories of digestive system disorders (b) Nutritional and metabolic disorders.

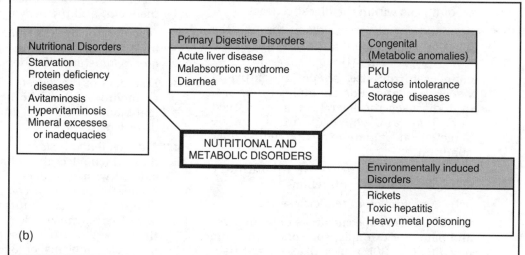

Nutritional Disorders
Starvation
Protein deficiency
 diseases
Avitaminosis
Hypervitaminosis
Mineral excesses
 or inadequacies

Primary Digestive Disorders
Acute liver disease
Malabsorption syndrome
Diarrhea

**Congenital
(Metabolic anomalies)**
PKU
Lactose intolerance
Storage diseases

NUTRITIONAL AND METABOLIC DISORDERS

Environmentally induced Disorders
Rickets
Toxic hepatitis
Heavy metal poisoning

(b)

✇ Dental Problems and Dental Implants
EAP p. 439

The mass of theplaque deposit protects the bacteria fromsalivary secretions, and as they digest nutrients the bacteria generate acids that erode the structure of the tooth. The results are dental caries, otherwise known as cavities. Streptococcus mutans is the most abundant bacteria at these sites, and vaccines are now being developed to promote resistance and prevent dental caries. If these bacteria reach the pulp and infect it, pulpitis (pul-PĪ-tis) results. Treatment usually involves the complete removal of the pulp tissue, especially the sensory innervation and all areas of decay, followed by packing the pulp cavity with appropriate materials. This procedure is called a root canal.

Brushing the exposed surfaces of the teeth after meals helps to prevent the settling of bacteria and the entrapment of food particles, but bacteria between the teeth, in the region known as the **interproximal space**, and within the gingival sulcus may elude the brush. Dentists therefore recommend the daily use of dental floss to clean these spaces and stimulate the gingival epithelium. If bacteria and plaque remain within the gingival sulcus for extended periods, the acids generated will begin eroding the connections between the neck of the tooth and the gingiva. The gums appear to recede from the teeth, and periodontal disease develops. As the disease progresses the bacteria attack the cementum, eventually destroying the periodontal ligament and eroding the alveolar bone. This deterioration loosens the tooth and it falls out or must be pulled. Periodontal disease is the most common cause for the loss of teeth.

Lost or broken teeth are often replaced with "false teeth" attached to a plate or frame inserted into the mouth. Since the 1980s an alternative has been developed using dental implants. A ridged titanium cylinder is inserted into the alveolus, and osteoblasts lock the ridges into the surrounding bone. After 4–6 months, an artificial tooth is screwed into the cylinder. Dental implants are not suitable for everyone—for example, there must be enough alveolar bone present to provide a firm attachment—and there may be complications during or after surgery. Nevertheless, as the technique evolves, dental implants will become increasingly important. Roughly 42 percent of individuals over age 65 have lost all of their teeth; the rest have lost an average of 10 teeth.

Lost or broken teeth are often replaced with "false teeth" attached to a plate or frame inserted into the mouth. Since the 1980s an alternative has been developed using dental implants. A ridged titanium cylinder is inserted into the alveolus, and osteoblasts lock the ridges into the surrounding bone. After 4–6 months, an artificial tooth is screwed into the cylinder. Dental implants are not suitable for everyone—for example, there must be enough alveolar bone

present to provide a firm attachment—and there may be complications during or after surgery. Nevertheless, as the technique evolves, dental implants will become increasingly important. Roughly 42 percent of individuals over age 65 have lost all of their teeth; the rest have lost an average of 10 teeth.

✇ Gastric Ulcers
EAP p. 442

Peptic ulcers are lesions of the gastric or duodenal mucosa caused by the action of acids and pepsin in gastric juice. Regardless of the primary cause of a gastric ulcer, once gastric juices have destroyed the epithelial layers, the virtually defenseless lamina propria will be exposed to digestive attack. Sharp abdominal pain results, and bleeding can develop. In severe cases the damage to the mucosa of the stomach may cause significant bleeding, and the acids may even erode their way through the wall of the stomach and into the peritoneal cavity. This condition, called a **perforated ulcer**, requires immediate surgical correction.

The administration of antacids can often control gastric or duodenal ulcers by neutralizing the acids and allowing time for the mucosa to regenerate. The drug *cimetidine* (sī-MET-i-dēn), or *Tagamet*, inhibits the secretion of acid by the parietal cells. Dietary restrictions limit the intake of foods that promote acid production (caffeine and pepper) or damage unprotected mucosal cells (alcohol).

The treatment of peptic ulcers has changed radically since the identification of *Helicobacter pylori* as a likely causative agent. These bacteria are able to resist gastric acids long enough to penetrate the mucus coating the epithelium. Once within the protective layer of mucus, they bind to the epithelial surfaces, where they are safe from the action of gastric acids and enzymes. Over time, the bacteria release toxins that damage the epithelial lining. These toxins ultimately result in erosion of the epithelium and the destruction of the lamina propria by gastric juices. Individuals whose stomachs harbor *H. pylori* are also at increased risk of gastric cancer, although the reason is not known.

Current treatment for *H. pylori* related ulcers consists of a combination of antibiotics (*tetracycline, amoxycillin,* and *metronidazole*), PeptoBismol™ and treatment with drugs that suppress acid production. Because strains of *H. pylori* resistant to at least one of these antibiotics have already appeared, research is underway to find alternative methods for controlling these infections.

✇ Drastic Weight-Loss Techniques
EAP p. 444

At any given moment, an estimated 20 percent of the U.S. population is dieting to promote weight loss. In addition to the appearance of "fat farms" and exercise clubs across the country, there has

17

been an increase in the use of surgery to promote weight loss. Many of these techniques involve surgically remodeling the gastrointestinal tract. **Gastric stapling** attempts to correct an overeating problem by reducing the size of the stomach. A large portion of the gastric lumen is stapled closed, leaving only a small pouch in communication with the esophagus and duodenum. After this surgery the individual will be able to eat only a small amount before the stretch receptors in the gastric wall become stimulated and the individual feels "full."

Gastric stapling is a major surgical procedure, and there are many potential complications. In addition, the smooth muscle in the wall of the functional portion of the stomach gradually becomes increasingly tolerant of distension, and the operation may have to be repeated.

A more drastic approach involves the surgical removal or bypass of a large portion of the jejunum. This procedure reduces the effective absorptive area, producing a marked weight loss. After the operation the individual must take a number of dietary supplements to ensure that all the essential nutrients and vitamins can be absorbed before the chyme enters the large intestine. Chronic diarrhea and serious liver disease are potential complications of this procedure.

⚕ Liver Disease　　　　EAP *p. 451*

The liver is the largest and most important visceral organ, and liver disorders affect almost every other vital system in the body.

- **Liver function tests** assess specific functional capabilities.

- A **serum bilirubin assay** indicates how efficiently the liver has been able to extract and excrete this compound.

- Serum and plasma protein assays can detect changes in the liver's rate of plasma protein synthesis, and serum enzyme tests can reveal liver damage by detecting intracellular enzymes in the circulating blood.

- **Liver scans** involve the injection of radioisotope-labeled compounds into the circulation. Compounds are chosen that will be selectively absorbed by either Kupffer cells, liver cells, or abnormal liver tissues.

- CT scans of the abdominal region are often used to provide information about cysts, abscesses, tumors, or hemorrhages in the liver.

- A **liver biopsy** can also be taken through the abdominal wall using a long needle, or performing a laparoscopy. Laparoscopic examination can also reveal gross structural changes in the liver or gallbladder.

The term *liver disease* includes a variety of different disorders. We will focus here on two major liver diseases, *hepatitis* and *cirrhosis*.

Hepatitis　　　　EAP *p. 449*

Hepatitis (hep-a-TĪ-tis) is defined as inflammation of the liver. Viruses specifically targeting the liver are responsible for most cases of hepatitis, although some environmental toxins can cause similar symptoms. There are at least five different forms of viral hepatitis: hepatitis A, B, C, D, and E.

1. **Hepatitis A**, or "infectious hepatitis," usually results from ingestion of food, water, milk, or shellfish contaminated by infected fecal wastes. It has a relatively short incubation period of 2-6 weeks. The disease usually runs its course in a matter of months, and fatalities are rare in individuals under 40 years of age.

2. **Hepatitis B**, or "serum hepatitis," is transmitted by intimate body fluid contact. For example, infection may occur through blood products, through a break in the skin or mucosa, or by sexual contact. The incubation period is much longer, ranging from 1 to 6 months. If a pregnant woman is infected, the newborn baby may become infected at birth

3. **Hepatitis C**, originally designated "non-A, non-B" hepatitis. Hepatitis C is most often transferred from individual to individual through the collection and transfusion of contaminated blood. As of 1990, screening procedures have been used to lower the incidence of transfusion-related hepatitis C. Hepatitis C can also be transmitted between intravenous drug users through shared needles, and there is evidence that heptatitis C may also be sexually transmitted. Chronic hepatitis C infections produce significant liver damage in roughly half the individuals infected with the virus. Interferon treatment early in the disease can slow the progression of the disease in some patients.

4. **Hepatitis D** is a virus that only causes symptoms in persons already infected with hepatitis B. Transmission of hepatitis D resembles that of hepatitis B, and in the U.S. the disease is most common among intravenous drug users. The combination of hepatitis B and hepatitis D causes progressive and severe liver disease.

5. **Hepatitis E** resembles hepatitis A in that it is transmitted by ingestion of contaminated food or water. Hepatitis E is the most common form of hepatitis worldwide, but cases seldom occur in the U.S. For reasons that are as yet unknown, hepatitis E infections are most acute and severe among pregnant women, and fatalities often result.

The hepatitis viruses disrupt liver function by attacking and destroying liver cells. The individual may develop a high fever, and the liver may become inflamed and tender. As the disease progresses, several hematological parameters change markedly. For example, enzymes usually confined to the cytoplasm of functional liver cells appear in the circulating blood. Normal metabolic regulatory activities become less effective, and blood glucose declines. Plasma protein synthesis slows, and the clotting time becomes unusually long. The injured hepatocytes stop removing bilirubin from the circulating blood, and symptoms of jaundice appear.

Hepatitis may be *acute* or *chronic*. Acute hepatitis is characteristic of hepatitis A and E. Almost everyone who contracts hepatitis A or hepatitis E eventually recovers, although full recovery may take several months. Symptoms of acute hepatitis include severe fatigue and jaundice. Fatigue is less pronounced, and jaundice is rare in chronic hepatitis. *Chronic active hepatitis* is a progressive disorder that leads to severe medical problems as liver function deteriorates and *cirrhosis* develops. Common complications include:

- the formation of *dilated esophageal veins* due to portal hypertension,

- *ascites* due to increased peritoneal fluid production at the elevated venous pressures,

- *bacterial peritonitis*, which may recur for unknown reasons, and

- *hepatic encephalopathy*. Hepatic encephalopathy is characterized by disorientation and confusion due to disruption of CNS function due to some combination of factors in the blood. High ammonia levels and the presence of abnormal concentrations of fatty acids, amino acids, and waste products have been implicated.

Hepatitis B, C, D, and E infection may produce either acute or chronic hepatitis. The chronic forms may be progressive, eventually leading to liver failure and death. Roughly 10 percent of hepatitis B patients develop potentially dangerous complications; the prognosis is significantly worse for those infected with both hepatitis B and hepatitis D.

Passive immunization with pooled immunoglobulins is available for both the hepatitis A and B viruses. (Active immunization for hepatitis B is also available.) Although there are now blood tests available that can detect any of the hepatitis viruses, there are no vaccines available to stimulate immunity to hepatitis C, D, or E.

Cirrhosis   EAP *p. 449*

The underlying problem in **cirrhosis** (sir-Ō-sis) appears to be the widespread destruction of hepatocytes by exposure to drugs (especially alcohol), viral infection, ischemia, or blockage of the hepatic ducts. Two processes are involved in producing the symptoms. Initially the damage to hepatocytes leads to the formation of extensive areas of scar tissue that branch throughout the liver. The surviving hepatocytes then undergo repeated cell divisions, but the fibrous tissue prevents the new hepatocytes from achieving a normal lobular arrangement. As a result, the liver gradually converts from an organized assemblage of lobules to a fibrous aggregation of poorly functioning cell clusters. Jaundice, ascites, and other symptoms may appear as the condition progresses.

Problems with Bile Storage and Excretion   EAP *p. 452*

If bile becomes too concentrated, crystals of insoluble minerals and salts begin to appear. These deposits are called **gallstones.** Merely having them, a condition termed **cholelithiasis** (kō-lē-li-THĪ-a-sis; *chole,* bile), does not represent a problem as long as the stones remain small. Small stones are normally flushed down the bile duct and excreted. In *cholecystitis,* the gallstones are so large that they can damage the wall of the gallbladder or block the cystic or common bile duct. A recent therapy for cholecystitis involves immersing the individual in water and shattering the stones with focused sound waves. The apparatus used is called a *lithotripter.* The particles produced are then small enough to pass through the bile duct without difficulty. In severe cases of cholecystitis, the gallbladder may become infected, inflamed, or perforated. Under these conditions it may be surgically removed, in a procedure known as a *cholecystectomy.* This loss does not seriously impair digestion, for bile production continues at normal levels. However, the bile is more dilute, and its entry into the small intestine is not as closely tied to the arrival of food in the duodenum.

An estimated 16-20 million people in the United States have gallstones that are unnoticed, and small stones are often flushed down the bile duct and excreted. If gallstones enter and jam in the cystic or bile duct, the painful symptoms of **cholecystitis** (kō-lē-sis-TĪ-tis) appear. Approximately one million people develop acute symptoms of cholecystitis each year. The gallbladder becomes swollen and inflamed, infections may develop, and symptoms of *obstructive jaundice* develop.

If the blockage does not work its way down the duct to the duodenum, it must be removed or destroyed. Small gallstones can be chemically dissolved. One chemical now under clinical review is *methyl tert-butyl ether* (MTBE). When introduced into the gallbladder it dissolves gallstones in a matter of hours; testing to date has not shown many undesirable side effects. (You may have heard of this compound in another context, as it is a gasoline additive used to prevent engine knocking.)

Surgery is usually required to remove large gallstones. (The gallbladder is also removed to prevent recurrence.) Many surgeons are now using a laparoscope, inserted through 2-3 small abdominal incisions, to perform this surgery. A *lithotripter* may also be used to shatter the gallstones with focused sound waves.

17

✝ Colon Inspection and Cancer

EAP *p. 452*

Each year there are roughly 150,000 new cases of colorectal cancer diagnosed in the United States, and 55,000 deaths from this condition. This deadly form of cancer is usually diagnosed in persons over 50 years of age. Primary risk factors for colorectal cancer include: (1) a diet rich in animal fats and low in fiber, (2) *inflammatory bowel disease* (p. 138), and (3) a number of inherited disorders that promote epithelial tumor formation along the intestines.

Successful treatment of colorectal cancer depends on the early identification of the condition. It is believed that colorectal cancers begin as small, localized mucosal tumors, or **polyps** (POL-ips), that grow from the intestinal wall. The prognosis improves dramatically if cancerous polyps are removed before metastasis has occurred. If the tumor is restricted to the mucosa and submucosa, the 5-year survival rate is over 90 percent. If it extends into the serosa, the survival rate drops to 70-85 percent, and after metastasis to other organs the rate drops to around 5 percent.

One of the early signs of polyp formation is the appearance of blood in the feces. Unfortunately, people often ignore small amounts of blood in fecal materials because they attribute the bleeding to "harmless" hemorrhoids. This offhand diagnosis should always be professionally verified.

When blood is detected in the feces, X-ray techniques are often used as a first step in diagnosis. In the usual procedure a large quantity of a liquid barium solution is introduced by enema. Because this solution is radiopaque, the X-rays will reveal any intestinal masses, such as a large tumor, blockages, or structural abnormalities.

Often the most precise surveys can be obtained with the aid of a flexible **colonoscope** (ko-LON-o-skōp). This instrument permits direct visual inspection of the lining of the large intestine. Variations on this procedure permit the collection of tissue samples. The colonoscope can also be used to remove polyps, and this removal avoids the potential complications of traditional surgery.

✝ Inflammatory Bowel Disease

EAP *p. 454*

The general term **colitis** (ko-LĪ-tis) may be used to indicate a condition characterized by inflammation of the colon. The **irritable bowel syndrome** is characterized by diarrhea, constipation, or an alternation between the two. When constipation is the primary problem, this condition may be called a *spastic colon*, or *spastic colitis*. The irritable bowel syndrome may have a partly psychological basis. **Inflammatory bowel disease**, such as *ulcerative colitis*, involves chronic inflammation of the digestive tract, most often affecting the colon. The mucosa becomes inflamed and ulcerated, extensive areas of scar tissue develop, and colonic function deteriorates. Acute bloody diarrhea with cramps are common symptoms. Fever and weight loss are also frequent complaints. Treatment of inflammatory bowel disease often involves anti-inflammatory drugs and corticosteroids that reduce inflammation. In severe cases, oral or intravenous fluid replacement is required. In cases that do not respond to other therapies, immunosuppressive drugs, such as *cyclosporine* may be used to good effect.

Treatment of severe inflammatory bowel disease may also involve a **colectomy** (kō-LEK-to-mē), the removal of all or a portion of the colon. If a large part or even all of the colon must be removed, normal connection with the anus cannot be maintained. Instead, the end of the intact digestive tube is sutured to the abdominal wall, and wastes then accumulate in a plastic pouch or sac attached to the opening. If the attachment involves the colon, the procedure is a **colostomy** (ko-LOS-to-mē); if the ileum is involved it is an **ileostomy** (il-ē-OS-to-mē).

✝ Diarrhea

EAP *p. 455*

Diarrhea is characterized as the production of copious, watery stools. There are many different conditions that will result in diarrhea, and we will consider only a representative sampling. Most infectious diarrhea involves organisms shed in the stool, that are then spread person to person (or person to food to person). Proper hand washing after defecation and good sewage disposal are the best preventative measures.

GASTROENTERITIS. An irritation of the small intestine may lead to a series of powerful peristaltic contractions that eject the contents of the small intestine into the large intestine. An extremely powerful irritating stimulus will produce a "clean sweep" of the absorptive areas of the digestive tract. Vomiting clears the stomach, duodenum, and proximal jejunum, and peristaltic contractions evacuate the distal jejunum and ileum. Bacterial toxins, viral infections, and various poisons will sometimes produce these extensive gastrointestinal responses. Conditions affecting primarily the small intestine are usually referred to as **enteritis** (en-ter-Ī-tis) of one kind or another. If both vomiting and diarrhea are present, the term **gastroenteritis** (gas-trō-en-ter-Ī-tis) may be used instead.

TRAVELER'S DIARRHEA. *Traveler's diarrhea*, a form of infectious diarrhea usually caused by a bacterial or viral infection, develops because the irritated or damaged mucosal cells are unable to maintain normal absorption levels. The irritation stimulates the production of mucus, and the damaged cells and mucus secretions add to the volume of feces produced. Despite the inconvenience, this type of diarrhea is usually temporary, and mild diarrhea is probably a reasonably effective method of rapidly removing an intestinal irritant. Drugs, such as *Lomotil*, that prevent peristaltic contractions in the colon relieve the diarrhea but leave the irritant intact, and the symptoms may return with a vengeance when the drug effects fade.

GIARDIASIS. Giardiasis is an infection caused by the protozoan *Giardia lamblia*. This pathogen colonizes the duodenum and jejunum and interferes with the normal absorption of lipids and carbohydrates. Many people do not develop acute symptoms, and these individuals act as carriers who can spread the disease. When acute symptoms develop, they usually appear within 3 weeks of initial exposure. Violent diarrhea, abdominal pains, cramps, nausea, and vomiting are the primary complaints. These symptoms persist for 5-7 days, although some patients are subject to relapses. Treatment usually consists of the oral administration of drugs, such as *quinacrine* or *metronidazole*, that can kill the protozoan.

Transmission of giardiasis requires the contamination of food or water with feces containing cysts, resting stages of the protozoan that are produced during passage through the large intestine. Rates of infection are highest (1) in Third World countries with poor sanitation, (2) among campers drinking surface water, (3) among individuals with impaired immune systems (as in AIDS), and (4) among toddlers and young children. The cysts can survive in the environment for months, and they are not killed during the chlorine treatment used to kill bacteria in drinking water. Travelers are advised to boil or ultrafilter water and boil food before consumption, as these preventative measures will destroy the cysts.

CHOLERA EPIDEMICS. Cholera epidemics are most common in areas with poor sanitation and where drinking water is contaminated by fecal wastes. After an incubation period of 1-2 days, the symptoms of nausea, vomiting, and diarrhea persist for 2-7 days. Fluid loss during the worst stage of the disease can approach 1 liter per hour. This dramatic loss causes a rapid drop in blood volume, leading to acute hypovolemic shock and damage to the kidneys and other organs.

Treatment consists of oral or intravenous fluid replacement while the disease runs its course. Antibiotic therapy may also prove beneficial. There is a vaccine available, but its low success rate (40-60 percent) and short duration (4-6 month protection) make it relatively ineffective in preventing or controlling cholera outbreaks. In 1991, a cholera epidemic began in Peru and has since spread throughout South America. More than 500,000 cases have been reported, with a death rate of 0.5 percent. This outbreak has had a remarkably low mortality rate; death rates in other outbreaks in this century have been as high as 60 percent.

Metabolic Problems

LACTOSE INTOLERANCE EAP *p. 456*

Lactose intolerance is a malabsorption syndrome that results from the lack of the enzyme *lactase* at the brush border of the intestinal epithelium. This condition poses more of a problem than would be expected if the only important outcome were the inability to make use of this particular disaccharide. The clinical symptoms result because undigested lactose provides a particularly stimulating energy source for the bacterial inhabitants of the colon. Increased gas generation, cramps, and diarrhea often bring extreme discomfort and distress after the individual drinks a single glass of milk or eats small amounts of any other dairy product.

There appears to be a genetic basis for lactose intolerance. Infants produce lactase to digest milk, but older children and adults may stop producing this enzyme. In certain populations lactase production continues throughout adulthood. Only around 15 percent of Caucasians develop lactose intolerance, while estimates ranging from 80 to 90 percent have been suggested for the adult African American and Oriental populations. These differences have an obvious effect on dietary preferences within these groups. Such differences must also be kept in mind when planning relief efforts, for shipping powdered milk to starvation areas in Africa may actually make matters worse if supplies are distributed to adults rather than to children.

Phenylketonuria EAP *p. 472*

Phenylketonuria (fen-il-kē-tō-NU-rē-a), or **PKU**, is one of about 130 disorders that have been traced to the lack of a specific enzyme. Individuals suffering from PKU are deficient in a key enzyme, *phenylalanine hydroxylase*, responsible for the conversion of the amino acid phenylalanine to tyrosine. This reaction is a necessary step in the synthesis of tyrosine, an important component of many proteins and the structural basis for a pigment (melanin), two hormones (epinephrine and norepinephrine), and two neurotransmitters (dopamine and norepinephrine). This conversion must also occur before the carbon chain of a phenylalanine molecule can be recycled or broken down in the TCA cycle.

If PKU is undetected and untreated, plasma concentrations of phenylalanine gradually escalate from normal (about 3 mg/dl) to levels above 20 mg/dl. High plasma concentrations of phenylalanine affect overall metabolism, and a number of unusual byproducts are excreted in the urine. The synthesis and degradation of proteins and other amino acid derivatives are affected. Developing neural tissue is most strongly influenced by these metabolic alterations, and severe brain damage and mental retardation results.

Fortunately, this condition is detectable shortly after birth because there are elevated levels of phenylalanine in the blood and phenylketone, a metabolic byproduct, in the blood and urine of the newborn infant. (During pregnancy, the normal mother metabolizes phenylalanine for the PKU fetus, so fetal levels are normal prior to delivery.) Treatment consists of controlling the amount of phenylalanine in the diet while monitoring plasma concentrations. This treatment is most important in infancy and childhood, when the nervous system is developing. Once grown, dietary restriction of phenylalanine can

17

be relaxed, except during pregnancy. A pregnant woman with PKU must protect the fetus from high levels of phenylalanine by following a strict diet that must actually begin before the pregnancy occurs.

Although the dietary restrictions are more relaxed for adults than for children, those with PKU must still monitor the ingredients used in the preparation of their meals. For example, one popular artificial sweetener, *Nutrasweet*, consists of phenylalanine and aspartic acid. Consumption of food or beverages containing this sweetener can therefore cause problems for PKU sufferers. Because tyrosine cannot be synthesized from dietary phenylalanine, the diet of these patients must also contain adequate amounts of tyrosine.

In its most severe form PKU affects approximately 1 infant in 20,000. Individuals who carry only a single gene for PKU will produce the affected enzyme but in lesser amounts. These individuals are asymptomatic but have slightly elevated phenylalanine levels in their blood. Statistical analysis of the incidence of fully developed PKU indicates that as many as 1 person in 70 may carry a gene for this condition.

⚕ Protein Deficiency Diseases

EAP *p. 473*

Regardless of the energy content of the diet, if it is deficient in essential amino acids, the individual will be malnourished to some degree. In a **protein deficiency disease**, protein synthesis decreases throughout the body. As protein synthesis in the liver fails to keep pace with the breakdown of plasma proteins, plasma osmolarity falls. This reduced osmolarity results in a fluid shift, as more water moves out of the capillaries and into interstitial spaces, the peritoneal cavity, or both. The longer the individual remains in this state, the more severe the ascites and edema that results.

This clinical picture is relatively common outside of the developed countries, where dietary protein is often scarce or prohibitively expensive. Growing infants suffer from **marasmus** (ma-RAZ-mus) when deprived of adequate proteins and calories. **Kwashiorkor** (kwash-ē-OR-kor) occurs in children whose protein intake is inadequate, even if the caloric intake is acceptable (Figure A-47). In each case, additional complications include damage to the developing brain. It is estimated that over 100 million children worldwide are suffering from protein deficiency diseases.

⚕ Gout

EAP *p. 473*

At concentrations above 7.4 mg/dl, body fluids are supersaturated with uric acid. Although symptoms may not appear at once, uric acid crystals begin to precipitate in body fluids. The condition that results is called **gout**. The severity of the symptoms depends on the amount and location of the crystal deposits.

Initially the joints of the extremities, especially the metatarsal/phalangeal joint of the great toe,

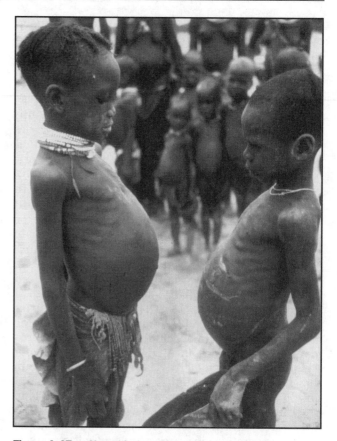

Figure A-47   Kwashiorkor

are likely to be affected. This exquisitely painful condition, called *gouty arthritis*, may persist for several days and then disappear for a period of days to years. Recurrences often involve other joints and may produce generalized fevers. Precipitates may also form within cartilages, in synovial fluids, tendons, or other connective tissues, or in the kidneys. At serum concentrations of over 12-13 mg/dl, half of the patients will develop kidney stones, and kidney function may be affected to the point of kidney failure.

The incidence of gout is much lower than that of hyperuricemia, with estimates ranging from 0.13 to 0.37 percent of the population. Only about 5 percent of the sufferers are women, and most affected males are over 50. Foods high in purines, such as meats or fats, may aggravate or initiate the onset of gout. These foods usually cost more than carbohydrates, and "rich foods" have often been associated with this condition.

⚕ Eating Disorders

EAP *p. 474*

Eating disorders are psychological problems that result in either inadequate or excessive food consumption. The most common conditions are *anorexia nervosa*, characterized by self-induced starvation, and *bulimia*, characterized by feeding binges followed by vomiting, laxative use, or both. Adolescent females account for most cases. These conditions are less common in males, who account for only 5–10 percent of anorexia or bulimia cases. A common thread in the two conditions is an obsessive concern about food and body weight.

Current estimates are that the incidence of **anorexia nervosa** in the United States ranges from 0.5 to 1.0 percent of young women age 12–18. A typical person with this condition is an adolescent Caucasian woman whose weight is roughly 30 percent below normal levels. Although underweight, she is usually convinced that she is still too fat and refuses to eat normal amounts of food.

The psychological factors responsible for anorexia are complex. Young women with this condition tend to be high achievers who are attempting to reach an "ideal" weight that will be envied and admired, and thereby achieve a sense of security and accomplishment. The factors thus tend to be a combination of their view of society ("thin is desirable or demanded"), their view of themselves ("I am not yet thin enough"), and a desire to be able to control their fate ("I can decide when to eat"). Female models and theater and arts majors at any age may feel forced to drop weight to remain competitive. The few male anorexics diagnosed often face comparable stresses. They tend to be athletes, such as jockeys or wrestlers, who need to maintain a minimal weight to maintain their careers.

Young anorexic women will often continue to starve themselves down to a weight of 30–35 kg. Dry skin, peripheral edema, an abnormally low heart rate and blood pressure, a reduction in bone and muscle mass, and a cessation of menstrual cycles are relatively common symptoms. Some of the changes, especially in bone mass, may be permanent. Treatment is difficult, and only 50–60 percent of patients who regain normal weight stay there for 5 years or more. Death rates from severe anorexia nervosa range from 10 to 15 percent.

Bulimia is more common than anorexia. In this condition the individual goes on an "eating binge" that may involve a meal that lasts 1–2 hours and may include 20,000 or more calories. The meal is followed by induced vomiting, often accompanied by the use of laxatives (to promote movement of the material through the digestive tract) and diuretics (drugs that promote fluid loss in the urine). These often expensive binges may occur several times each week, separated by periods of either normal eating or fasting.

Bulimia most often involves women of the same age class as anorexia, and recent estimates of the incidence of bulimia in young women range from 5 to 20 percent or more. However, many bulimics are not diagnosed until they are age 30–40. Because of the amount of food ingested, bulimics may have normal body weight, and therefore the condition is harder to diagnose than anorexia nervosa.

The health risks of bulimia result from (1) cumulative damage to the stomach, esophagus, oral cavity, and teeth by repeated exposure to stomach acids; (2) electrolyte imbalances resulting from the loss of sodium and potassium ions in the gastric juices, diarrhea, and urine; (3) edema; and (4) cardiac arrhythmias.

The underlying cause of bulimia remains uncertain. Societal factors are certainly involved, but bulimia has also been strongly correlated with depression and with elevated CSF levels of ADH.

Regulatory obesity results from a failure to regulate food intake so that appetite, diet, and activity are in balance. Most instances of obesity fall within this category. Usually there is no obvious organic cause, although in rare cases the problem may arise because of some disorder, such as a tumor, affecting the hypothalamic centers dealing with appetite and satiation. Typically, chronic overeating is thought to result either from psychological or sociological factors, such as stress, neurosis, long-term habits, family or ethnic traditions, or from inactivity. Genetic factors may also be involved, but because the psychological and social environment plays such an important role in human behavior the exact connections have been difficult to assess. In short, individuals suffering from regulatory obesity are overeating for some reason and thereby extending the duration and magnitude of the absorptive state.

In **metabolic obesity** the condition is secondary to some underlying organic malfunction that affects cell and tissue metabolism. These cases are relatively rare and typically involve chronic hypersecretion or hyposecretion of metabolically active hormones, such as glucocorticoids.

Recent genetic discoveries have shed light on another metabolic aspect of obesity. In 1995, researchers identified a weight-control hormone, *leptin*, and its gene. *Leptin* (*leptos*, slender) is produced by the body's fat cells and relased into the blood. After passing through the blood-brain barrier, leptin (1) binds and surppresses the appetite centers of the hypothalamus and (2) sppeds up metabolic processes that burn energy, thereby promoting weight loss. Interestingly, many individuals may have become insensitive to this hormone for unknown reasons. For these people, simply supplying the body with more leptin (which still has not undergone testing in humans) will not necessarily promote weight loss. In 1996, researchers identified the cell membrane receptor for leptin and its gene. Inherited forms of obesity may be caused by genetic mutations that produce receptors unable to bind leptin. There is also evidence that leptin may play a role in the maintenance of normal reproductive function. Leptin may stimulate GnRH release in the hypothalamus. This could explain why women who have extremely low percent body fat ratios (under 10%) have reduced fertility and may cease menstruating.

Perspectives on Dieting   EAP *p. 479*

Dieting has become a national pastime, and magazines and bookstores are flooded with "how-to" guides for losing weight. Despite the unusual, astounding, and often preposterous claims made on the covers, none has proved to be ideal and many have actually turned out to be dangerous. Now that

18

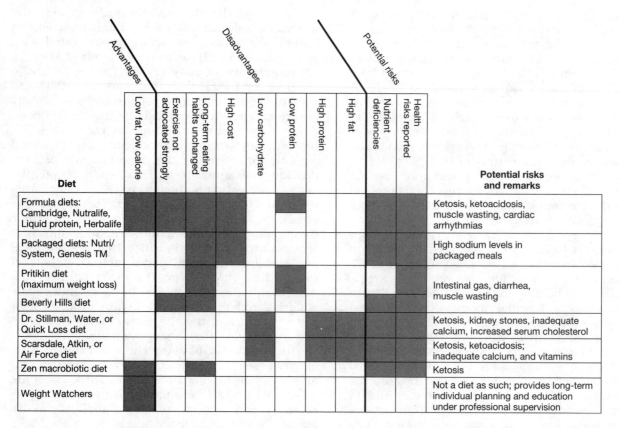

The table columns, grouped under **Advantages**, **Disadvantages**, and **Potential risks**, are: Low fat, low calorie; Exercise not advocated strongly; Long-term eating habits unchanged; High cost; Low carbohydrate; Low protein; High protein; High fat; Nutrient deficiencies; Health risks reported.

Diet	Potential risks and remarks
Formula diets: Cambridge, Nutralife, Liquid protein, Herbalife	Ketosis, ketoacidosis, muscle wasting, cardiac arrhythmias
Packaged diets: Nutri/System, Genesis TM	High sodium levels in packaged meals
Pritikin diet (maximum weight loss)	Intestinal gas, diarrhea, muscle wasting
Beverly Hills diet	Intestinal gas, diarrhea, muscle wasting
Dr. Stillman, Water, or Quick Loss diet	Ketosis, kidney stones, inadequate calcium, increased serum cholesterol
Scarsdale, Atkin, or Air Force diet	Ketosis, ketoacidosis; inadequate calcium, and vitamins
Zen macrobiotic diet	Ketosis
Weight Watchers	Not a diet as such; provides long-term individual planning and education under professional supervision

Figure A-48 **A Critical Review of Popular Fad Dieting Methods**

Data Sources: "Fad Diet Summary," *Kaiser Permanente;* "Rating the diets," *Consumer Reports;* "A Practical Guide to Fad Diets," *Clinical Sports Medicine* 3 (3): 723-729; "The Sense and Nonsense of Best-Selling Diet Books," *Can. Med. Assoc. Journal* 126 (6):F 696-701 (1982)

you have a basic familiarity with metabolic processes and interconversions, we can briefly review several dietary myths. Analyzing the most popular fad diets is beyond the scope of this text, but you might find the information summarized in Figure A-48 useful if you are considering an intensive diet program.

Identifying fad diets can be quite easy, once you have learned the warning signs. They invariably promise almost immediate results, and most claim to be virtually effortless. The "immediate results" turn out to be temporary water losses, and if dieting in any form were easy, Americans would not have a weight problem in the first place.

Carried to extremes, some fad diets can produce fatal alterations in blood chemistry and physiological systems. Many of the trendy diets advocate the elimination of most or all of the members of one or more food groups, making it difficult to obtain dietary essentials. Several of the diets intentionally produce ketosis, a condition that can cause several unpleasant and even disastrous side effects. Equally significant, such restrictions are artificial and usually are intended to be followed only while dieting. Because the eating patterns that caused the original problem are not addressed, once the diet has ended, the individual immediately begins putting on weight. This weight gain leads to another dieting cycle 6 months later, often using a different fad diet. This cycle of dieting delights publishing companies, who thrive on a combination of diet books and cookbooks, but it can be very frustrating for the individual. Only about 5 percent of dieters sustain their weight loss for 1-5 years or more.

The advocacy of "secret ingredients" should also ring warning bells, and you should be on the alert for exaggerated claims concerning either the amount of weight loss or the potential fringe benefits of the program. Paid testimonials from successful dieters are highly suspect and hardly a suitable ground for decision making. Such testimonials often take the place of scientific evidence, and pseudoscientific terms are often used to convince the public that scientific proof exists to support the inflated statements. In some cases the "scientific justifications" are totally unrelated to the weight loss experienced by those following a particular diet. For example, eating fruits and vegetables, coupled with reduced quantities of beef, will lower caloric intake and bring the diet in closer agreement with current dietary recommendations. Most people will lose weight on such a diet regardless of whether it is accompanied by complicated mumbo-jumbo about acupressure, cosmic contemplation, or toxic combinations in the stomach.

The most effective weight-loss program does not involve a crash diet, miracle drugs, elimination of all foods beginning with the letter B, or subscriptions to the diet-of-the-month club.

The weight loss should be gradual, rather than sudden. Dietary modifications should be moderate, not sweeping, and intended to alter long-term eating habits and ensure that the weight loss will be permanent.

Given an opportunity, the human body will preserve homeostasis. By following the advice "everything in moderation," we can avoid testing our homeostatic limits, but many people tend to follow the "more must be better" philosophy when it comes to meeting nutritional needs. For example, it has been said that Americans have the most valuable urine in the world, thanks to our tendency to overdose ourselves on water soluble vitamins. *It should be kept in mind that no amount of vitamins, minerals, dietary restrictions, or attention to nutritional recommendations can counteract the negative effects of even a single serious behavioral health risk such as smoking, excessive alcohol consumption, or drug abuse.*

✝ Heat Exhaustion and Heat Stroke
EAP *p. 482*

Heat exhaustion and *heat stroke* represent malfunctions of the thermoregulatory system. In **heat exhaustion,** also known as heat prostration, the individual experiences difficulties with the maintenance of blood volume. The heat-loss center is stimulating sweat glands whose secretions moisten the surface of the skin to provide evaporative cooling. As fluid losses mount, blood volume decreases. The resultant decline in blood pressure is not countered by peripheral vasoconstriction, because the heat-loss center is actively stimulating peripheral vasodilation. As blood flow to the brain declines, headache, nausea, and eventual collapse follow. Treatment is straightforward: Provide fluids, salts, and a cooler environment.

Heat stroke is more serious and may occur following an untreated case of heat exhaustion. Predisposing factors include any preexisting condition, such as heart disease or diabetes, that affects peripheral circulation. The thermoregulatory center ceases to function, the sweat glands are inactive, and the skin becomes hot and dry. Unless the situation is recognized in time, body temperature may climb to 41°–45°C (106°–113°F). Temperatures in this range will quickly disrupt a variety of vital physiological systems and destroy brain, liver, skeletal muscle, and kidney cells. Proper treatment involves lowering body temperature as rapidly as possible.

✝ Hypothermia
EAP *p. 482*

Hypothermia may be intentionally produced during surgery to reduce the metabolic rate of a particular organ or of the entire body. In controlled hypothermia, the individual is first anesthetized to prevent the shivering that would otherwise fight the process.

During open-heart surgery the body is often cooled to 25°–32°C (79°–89°F). This cooling reduces the metabolic demands of the body, which will be receiving blood from an external pump or oxygenator. The heart must be stopped completely during the operation, and it cannot be well supplied with blood over this period. So the heart is exposed to an *arresting solution* at 0°–4°C (32°–39°F) and maintained at a temperature below 15°C (60°F) for the duration of the operation. At these temperatures the cardiac muscle can tolerate several hours of ischemia without damage.

When cardiac surgery is performed on infants, a deep hypothermia may be produced by cooling the entire body to temperatures as low as 11°C (52°F) for an hour or more. In effect this procedure duplicates the conditions experienced by the accidental drowning victims.

✝ Accidental Hypothermia
EAP *p. 482*

If body temperature declines significantly below normal levels, the thermoregulatory system begins to lose sensitivity and effectiveness. Cardiac output falls, respiratory rate decreases, and if the core temperature falls below 28°C (82°F), cardiac arrest is likely. The individual then has no heartbeat, no respiratory rate, and no response to external stimuli, even painful ones. Body temperature continues to decline, and the skin turns blue and cold.

At this point we would probably assume that the individual has died. But because of the systemwide decrease in metabolic activities, the victim may still be saved, even after several hours have elapsed. Treatment consists of cardiopulmonary support and gradual rewarming, both external and internal. The skin can be warmed up to 45°C (110°F) without damage; warm baths or blankets can be used. One effective method of raising internal temperatures involves the introduction of warm saline into the peritoneal cavity.

Hypothermia is a significant risk for those engaged in water sports, and its presence may complicate treatment of a drowning victim. Water absorbs heat roughly 27 times faster than air, and the heat-gain mechanisms are unable to keep pace over long periods or when faced with a large temperature gradient. But hypothermia in cold water does have a positive side. On several occasions small children who have drowned in cold water have been successfully revived after periods of up to 4 hours. Children lose body heat quickly, and their systems stop functioning very quickly as body temperature declines. This rapid drop in temperature prevents the oxygen-starvation and tissue damage that would otherwise occur when breathing stops.

Resuscitation attempts are not attempted if the individual has actually frozen. Water expands roughly 7 percent during ordinary freezing, and cell membranes throughout the body are destroyed in the process. Very small organisms can be frozen and subsequently thawed without ill effects because their surface-to-volume ratio is enormous, and the freezing process occurs so rapidly that ice crystals never form.

18

⚕ Thermoregulatory Problems of Infants

EAP *p. 482*

Infants have problems with thermoregulation because of their relatively high surface-to-volume ratios. During embryonic development, temperature regulation is no concern of theirs, as the maternal surroundings are at normal body temperature. At birth, the temperature-regulating mechanisms are not fully functional. With such high surface-to-volume ratios, newborns must be dried quickly and kept bundled up; for those born prematurely, a thermally regulated incubator is required. Infants' body temperatures are more unstable than those of adults, and their metabolic rates decline when they are sleeping, then rise after arousal.

Infants cannot shiver, but they have a different mechanism for raising body temperature rapidly. The adipose tissue between the shoulder blades, around the neck, and possibly elsewhere in the upper body is histologically and functionally different from most of the adipose tissue in the adult. The tissue is highly vascularized, and the individual adipocytes contain numerous mitochondria. Together these characteristics give the tissue a deep, rich color responsible for the name **brown fat.** The individual adipocytes are innervated by sympathetic autonomic fibers. When these nerves are stimulated, lipolysis accelerates in the adipocytes. The cells do not capture the energy released through fatty acid catabolism, and it radiates into the surrounding tissues as heat. This heat quickly warms the blood passing through the surrounding network of vessels, and it is then distributed throughout the body. In this way an infant can accelerate metabolic heat generation by 100 percent very quickly, while nonshivering thermogenesis in the adult will raise heat production by only 10–15 percent after a period of weeks.

With increasing age and size, body temperature becomes more stable, and the importance of this thermoregulatory mechanism declines. There is little if any brown fat in the adult; with increased body size, skeletal muscle mass, and insulation, shivering thermogenesis is significantly more effective in elevating body temperature.

CRITICAL THINKING QUESTIONS

9-1. Tony is a chronic alcoholic who suffers from cirrhosis of the liver. Which of the following symptoms would you expect to observe in Tony?

    a. prolonged clotting time

    b. jaundice

    c. ascites

    d. portal hypertension

    e. all of the above

9-2. Paul, a 60-year-old man, complains of rectal bleeding for the last 24 hours. He has experienced four bloody stools in the last several hours and is beginning to weaken. Prior to this episode, he claims he was constipated. He is admitted to the hospital, where lab and diagnostic studies results are all within normal limits except for the following:

Hemoglobin: 9 g/dl

Hematocrit: 25

Barium enema: reveals narrowing of a small portion of the descending colon

Colonoscopy: mass detected in the descending colon

If the physician took a tissue biopsy during the colonoscopy, what results would you expect from the pathologic analysis of the specimen?

9-3. Mary, a 25-year-old school teacher, tells her physician she has been experiencing bouts of painful abdominal cramping, gas, and diarrhea for the last six months for unknown reasons. The episodes are short-lived and sporadic and do not have an associated fever. When questioned about her diet, Mary cannot remember a specific food related to the episodes. A stool specimen is negative for bacteria and parasites. A barium enema is negative for any abnormalities. What might be Mary's problem? How would you go about checking this diagnosis?

9-4. Barry is participating in a nutrition study in which he must follow a prescribed diet and have his blood tested regularly. On one of his visits, his test results indicate low levels of insulin, elevated levels of glucagon and epinephrine, and moderate levels of cortisol and growth hormone. What would you expect the levels of glucose, ketones, and fatty acids to be?

NOTES

The Urinary System

The urinary system consists of the kidneys, where urine production occurs, and the conducting system that transports and stores urine prior to its elimination from the body. The conducting system includes the ureters, the urinary bladder, and the urethra. Although the kidneys perform all the vital functions of the urinary system, problems with the conducting system can have direct and immediate effects on renal function.

THE PHYSICAL EXAMINATION AND THE URINARY SYSTEM

The primary symptoms of urinary system disorders are pain and changes in the frequency of urination. The nature and location of the pain can provide clues to the source of the problem (see Figure 10-1, EAP *p. 251*). For example,

- Pain superior to the pubic region may be associated with urinary bladder disorders.

- Upper lumbar back pain radiating to the flank and the right upper quadrant or left upper quadrant can be caused by kidney infections such as *glomerulonephritis* and *pyelonephritis*.

- *Dysuria* (painful or difficult urination) may occur with *cystitis* and *urinary obstructions*. In males, prostatic enlargement can lead to compression of the urethra and dysuria.

Individuals with urinary system disorders may urinate more or less often and produce normal or abnormal amounts of urine.

- Irritation of the lining of the ureters or urinary bladder can lead to the desire to urinate with increased frequency, although the total amount of urine produced each day remains normal. When these problems exist, the individual feels the urge to urinate when the urinary bladder volume is very small. The irritation may result from trauma, urinary bladder infection or tumors, or increased acidity of the urine.

- *Incontinence*, an inability to control urination voluntarily, may involve periodic involuntary urination, or a continual, slow trickle of urine from the urethra. Incontinence may result from urinary bladder or urethral problems, damage or weakening of the muscles of the pelvic floor, or interference with normal sensory or motor innervation in the region. Renal function and daily urinary volume are normal.

- In *urinary retention* renal function is normal, at least initially, but urination does not occur. Urinary retention in males often results from prostatic enlargement and compression of the prostatic urethra. In both sexes urinary retention may result from obstruction of the outlet of the urinary bladder, or from CNS damage, such as a stroke or damage to the spinal cord.

- Changes in the volume of urine produced indicate that there are problems either at the kidneys or with the control of renal function. **Polyuria,** the production of excessive amounts of urine, may result from hormonal or metabolic problems, such as those associated with *diabetes* (pp. 80, 83), or damage to the glomeruli, as in *glomerulonephritis* (*p. 148*). **Oliguria** (urine volume 50-500 ml/day) and **anuria** (0-50 ml/day) are conditions that indicate serious kidney problems and potential renal failure. Renal failure can occur with *heart failure* (p. 109), renal ischemia, *circulatory shock* (*p. 106*), *burns* (*p. 44*), and a variety of other disorders.

Important clinical signs of urinary system disorders include

- *Hematuria*, the presence of red blood cells in the urine, indicates bleeding at the kidneys or conducting system. Hematuria producing dark red urine usually indicates bleeding in the kidney, and hematuria producing bright red urine indicates bleeding in the lower urinary tract. Hematuria most often occurs with trauma to the kidneys, calculi, tumors, or urinary tract infections.

- *Hemoglobinuria* is the presence of hemoglobin in the urine. Hemoglobinuria indicates increased RBC hemolysis within the circulation, due to cardiovascular or metabolic problems. Examples of conditions resulting in hemoglobinuria include the *thalassemias* (p. 91), *sickle cell anemia* (p. 92), *hypersplenism* (p. 116), and some autoimmune disorders.

- Changes in urine color may accompany some renal disorders. For example, the urine may become (1) cloudy, due to the presence of bacteria, lipids, or epithelial cells; (2) red or brown from hemoglobin or myoglobin; (3) blue-green from bilirubin; or (4) brown-black from excessive concentration. Not all color changes are abnormal, however. Some foods and several prescription drugs can cause changes in urine color. A serving of beets can give urine a reddish color, whereas eating rhubarb can give the urine an orange tint.

- Renal disorders often lead to a generalized edema in peripheral tissues. Facial swelling, especially around the eyes, is often seen.

- A fever often develops when the urinary system is infected by pathogens. Urinary bladder infections *(cystitis)* often result in a low-grade fever; kidney infections, such as *pyelonephritis*, usually produce very high fevers.

During the physical assessment, palpation can be used to check the status of the kidneys and urinary bladder. The kidneys lie in the *costovertebral area*, the region bounded by the lumbar spine and the 12th ribs on either side. To detect tenderness due to kidney inflammation, the examiner places one hand over the costovertebral area and taps it with the other hand held in a fist. This usually does not cause pain unless the adjacent kidney is inflamed.

The general shapes and sizes of the kidneys can be determined by palpation in the flank and abdominal area on either side. The right kidney can be palpated more easily than the left, because the spleen normally presses against the left kidney. Enlargement of the left kidney can therefore be difficult to distinguish from enlargement of the spleen.

The urinary bladder can be palpated just superior to the pubic symphysis. It can be difficult to distinguish between urinary bladder enlargement due to urine retention and the presence of an abdominal mass on the basis of palpation alone.

There are many procedures and laboratory tests that may be used in the diagnosis of urinary system disorders. The functional anatomy of the urinary system can be examined using a variety of sophisticated procedures. For example, administering a radiopaque compound that will enter the urine permits the creation of an **intravenous pyelogram** (PĪ-el-ō-gram), or **IVP,** by taking an X- ray of the kidneys. This procedure permits detection of unusual kidney, ureter, or bladder structures and masses. Computerized tomography (CT) scans may also provide useful information concerning localized abnormalities.Figure A-49 is a simple concept map outlining the major classes of disorders of the urinary system.

☤ Conditions Affecting Filtration

EAP *p. 492*

Changes in filtration pressure (P_f) can result in significant alterations in kidney function. Examples of factors that can disrupt normal filtration rates include physical damage to the filtration apparatus and interference with normal filtrate or urine flow.

PHYSICAL DAMAGE TO THE FILTRATION APPARATUS

The lamina densa and podocytes can be injured by mechanical trauma, such as a blow to the kidneys, bacterial infection, circulating immune complexes, or exposure to metabolic poisons, such as mercury. The usual result is a sudden increase in the permeability of the glomerulus. When damage is severe, plasma proteins and even blood cells enter the capsular spaces. The loss of plasma proteins has two immediate effects: (1) it reduces the osmotic pressure of the blood, and (2) it increases the osmotic pressure of the filtrate. The result is an increase in the net filtration pressure and an increased rate of filtrate production.

Blood cells entering the filtrate will not be reabsorbed. The presence of blood cells in the urine is called **hematuria** (hēm-a-TŪR-ē-a). Although small

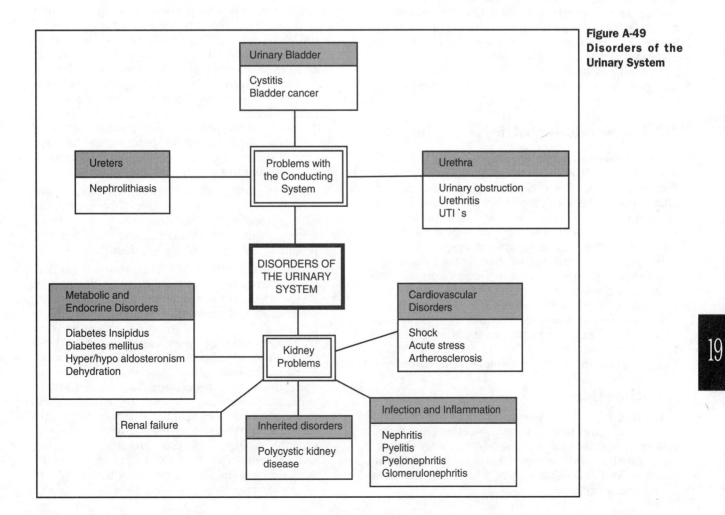

**Figure A-49
Disorders of the
Urinary System**

amounts of protein can be reabsorbed, when glomeruli are severely damaged the nephrons are unable to reabsorb all the plasma proteins entering the filtrate. Plasma proteins then appear in the urine, a condition termed **proteinuria** (prō-tēn-ŪR-ē-a). Proteinuria and hematuria indicate that kidney damage has occurred.

INTERFERENCE WITH FILTRATE OR URINE FLOW

If the tubule, collecting duct, or ureter becomes blocked and urine flow cannot occur, capsular pressures gradually rise. When the capsular hydrostatic pressure and blood osmotic pressure equal the glomerular hydrostatic pressure, filtration stops completely. The severity of the problem depends upon the site of the blockage. If it involves a single nephron, only a single glomerulus will be affected. If the blockage occurs within the ureter, filtration in that kidney will come to a halt. If the blockage occurs in the urethra, both kidneys will become nonfunctional. Examples of factors involved in urinary blockage are discussed later (see Problems with the Conducting System.)

Elevated capsular pressures can also result from inflammation of the kidneys, a condition called **nephritis** (nef-RĪ-tis). A generalized nephritis may result from bacterial infections or exposure to toxic or irritating drugs. One of the major problems in nephritis is that the inflammation causes swelling, but the renal capsule prevents an increase in the size of the kidney. The result is an increase in the hydrostatic pressures in the peritubular fluid and filtrate. This pressure opposes the glomerular hydrostatic pressure, lowering the net filtration pressure and the GFR.

✚ Glomerulonephritis   EAP *p. 492*

Glomerulonephritis (glo-mer-ū-lō-nef-RĪ-tis) is an inflammation of the renal cortex that affects the filtration mechanism. This condition, which may develop after an infection involving *Streptococcus* bacteria, is an example of an *immune complex disorder*. The primary infection may not occur in or near the kidneys. However, as the immune system responds to the infection, the number of circulating antigen-antibody complexes skyrockets. These complexes are small enough to pass through the lamina densa but too large to fit between the slit pores. As a result, the filtration mechanism clogs up, and filtrate production declines. Any condition that leads to a massive immune response can cause glomerulonephritis, including viral infections and autoimmune disorders.

℞ Diuretics   EAP *p. 497*

Diuretics (dī-ū-RET-iks) are drugs that promote the loss of water in the urine. Diuretics have many different mechanisms of action, but all affect transport activities and/or water reabsorption along the nephron and collecting system. Important diuretics in use today include

- *Osmotic diuretics:* Osmotic diuretics are metabolically harmless substances that are filtered at the glomerulus and ignored by the tubular epithelium. Their presence in the urine increases its osmolarity and limits the amount of water reabsorption possible. **Mannitol** (MAN-i-tol) is the most frequently administered osmotic diuretic. It is used to accelerate fluid loss and speed the removal of toxins from the blood and to elevate the GFR after severe trauma or other conditions have impaired renal function.

- *Drugs that block sodium and chloride transport:* A class of drugs called **thiazides** (THĪ-a-zīdz) reduce sodium and chloride transport in the proximal and distal tubules. Thiazides such as *chlorothiazide* are often used to accelerate fluid losses in the treatment of hypertension and peripheral edema.

- *High-ceiling diuretics:* The **high-ceiling diuretics,** such as *furosemide* and *bumetanide,* inhibit transport along the loop of Henle, reducing the osmotic gradient and the ability to concentrate the urine. They are called "high-ceiling" diuretics because they produce a much higher degree of diuresis than other drugs. They are fast-acting and are often used in a clinical crisis—for example, in treating acute pulmonary edema. In both the thiazide and the furosemide diuretics, water, Na^+, and K^+ are lost in the urine.

- *Aldosterone blocking agents:* Blocking the action of aldosterone prevents the reabsorption of sodium along the DCT and collecting tubule, and so accelerates fluid losses. The drug *spironolactone* is an example of this type of diuretic. It is often used in conjunction with other diuretics because blocking the aldosterone-activated exchange pumps helps reduce the potassium ion loss. These drugs are also known as *potassium-sparing diuretics.* Atrial natriuretic peptide may be used as a diuretic because it counteracts the effects of both aldosterone and ADH at the kidneys.

- *ACE inhibitors:* **ACE inhibitors** (*angiotensin converting enzyme inhibitors*) prevent the conversion of angiotensin I to angiotensin II by converting enzyme. This prevents stimulation of aldosterone production and promotes water loss.

- *Drugs with diuretic side effects:* Many drugs prescribed for other conditions promote diuresis as a side effect. For example, drugs that block carbonic anhydrase activity, such as acetazolamide *(Diamox),* have an indirect effect on sodium transport. Although they cause diuresis, these drugs are seldom prescribed with that in mind. (Because carbonic anhydrase is also involved in aqueous human secretion, Diamox is used to reduce intraocular pressure in glaucoma patients.) Two more familiar drugs, caffeine and alcohol, have pronounced diuretic effects. Caffeine produces diuresis directly, by reducing sodium reabsorption along the tubules. Alcohol works indirectly, by suppressing the release of ADH at the posterior pituitary gland.

Urinalysis

EAP *p. 497*

Several basic screening tests can be performed by recording changes in the color of test strips that are dipped in the sample. Urine pH and urinary concentrations of glucose, ketones, bilirubin, urobilinogen, plasma proteins, and hemoglobin can be monitored using this technique. In addition, the density or *specific gravity* of the urine is usually determined, using a simple device known as a **urinometer** (ū-ri-NOM-e-ter) or **densitometer** (den-si-TOM-e-ter). The sample may also be spun in a centrifuge and any sediment examined under the microscope. Mineral crystals, bacteria, red or white blood cells, and deposits, known collectively as **casts,** can be detected in this way. Figure A-50 provides an overview of the major categories of urinary casts. During a urinary tract infection, bacteria may be cultured to determine their specific identities.

More comprehensive analyses can determine the total osmolarity of the urine and the concentration of individual electrolytes and minor metabolites, metabolic wastes, vitamins, and hormones. A test for one hormone in the urine,

human chorionic gonadotrophin (hCG), provides an early and reliable indication of pregnancy.

The information provided by urinalysis can be especially useful when correlated with the data obtained from blood tests. The term **azotemia** (a-zō-TĒ-mē-a) refers to the presence of excess metabolic wastes in the blood. This condition may result from overproduction of urea or other nitrogenous wastes by the liver ("pre-renal syndrome"). In **uremia** (ū-RĒ-mē-a), by contrast, all normal kidney functions are adversely affected. The symptoms of uremia, which are those of kidney failure, include hypertension, anemia (because of a decline in the production of erythropoietin), and central nervous problems that may lead to sleeplessness, seizures, delirium, and even coma.

The total volume of urine produced in a 24-hour period may also be of interest. **Polyuria** (pol-ē-U-rē-a) refers to excessive production of urine, well over 2 liters per day. Polyuria most often results from endocrine disorders, such as the various forms of *diabetes,* metabolic disorders, or damage to the filtration apparatus, as in *glomerulonephritis.* **Oliguria** (o-li-GŪ-rē-a) refers to inadequate urine production (50-500 ml/day). In **anuria** (a-NŪ-rē-a), a negligible amount of urine is produced (0-50 ml/day), and a potentially fatal problem exists.

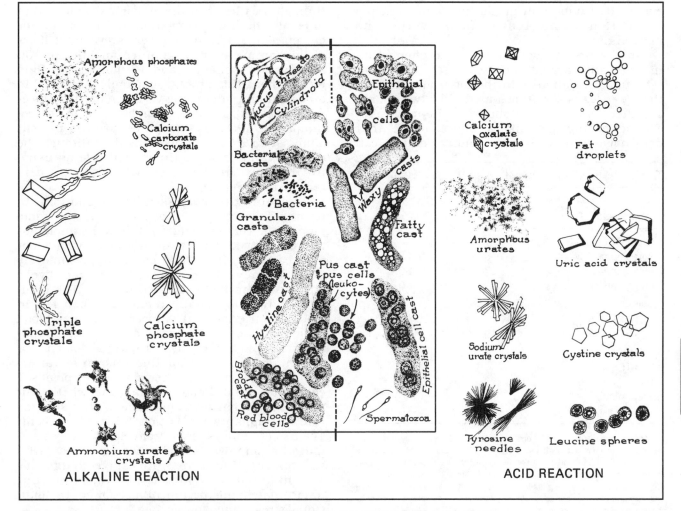

Figure A-50   Microscopic examination of urine sediment (redrawn after Todd and Sanford)

⚚ Advances in the Treatment of Renal Failure

EAP *p. 501*

In dialysis, the functions of damaged kidneys are partially compensated for by diffusion between the patient's blood and a *dialysis fluid* whose coomposition is carefully regulated. In one technique, called *hemodialysis* (hē-mō-dĪ-AL-i-sis), an artificial membrane is used to regulate the composition of the blood. The basic principle involved in this process, called *dialysis*, involves passive diffusion across a semipermeable membrane. The patient's blood flows across an artificial *dialysis membrane* that contains pores large enough to permit the diffusion of small ions, but small enough to prevent the loss of plasma proteins. On the other side of the membrane.

The composition of dialysis fluid is indicated in Table A-19. As diffusion takes place across the membrane, the composition of the blood changes. Potassium ions, phosphate ions, sulfate ions, urea, creatinine, and uric acid diffuse across the membrane into the dialyzing fluid. Bicarbonate ions and glucose diffuse into the bloodstream. In effect, diffusion across the dialysis membrane takes the place of normal glomerular filtration, and the characteristics of the dialysis fluid ensure that important metabolites remain in the circulation, rather than diffusing across the membrane.

In practice, silastic tubes, called *shunts*, are inserted into a medium-sized artery and vein. (The usual location is in the forearm, although the lower leg is sometimes used.) The two shunts are then connected as shown in Figure A-51b. The connection acts like a "short circuit" that does not impede blood flow, and the shunts can be used like taps in a wine barrel, to draw a blood sample or to connect the individual to a *dialysis machine* (Figure A-51a). When connected to the dialysis machine, the individual sits quietly while blood circulates from the arterial shunt, through the machine, and back via the venous shunt. Inside the machine, the blood flows across a dialyzing membrane, where diffusion occurs.

Use of a dialysis machine is suggested when a patient's *BUN* (*blood urea nitrogen*) exceeds 100 mg/dl (the normal value is 30 mg/dl). Dialysis techniques are useful because they can maintain patients awaiting a transplant or those whose kidney function has been temporarily disrupted. Hemodialysis does have a number of drawbacks, however: (1) the patient must sit by the machine about 15 hours per week; (2) between treatments the symptoms of uremia will gradually develop; (3) hypotension can develop as a result of fluid loss during dialysis; (4) air bubbles in the tubing can cause embolism formation in the bloodstream; (5) anemia often develops; and (6) the shunts can serve as sites for recurring infections.

One alternative to the use of a dialysis machine is *peritoneal dialysis*, in which the peritoneal lining is used as a dialysis membrane. Dialyzing fluid is introduced into the peritoneum through a catheter in the abdominal wall, and at intervals the fluid is removed and replaced. For example, one procedure involves cycling 2 liters of fluid in an hour-15 minutes for infusion, 30 minutes for exchange, and 15 minutes for fluid reclamation. This process is usually performed in a hospital. An interesting variation on this procedure is called *continuous ambulatory peritoneal dialysis (CAPD)*. In this procedure the patient administers 2 liters of dialyzing fluid through the catheter and then continues with life as usual until 4 to 6 hours later, when the fluid is removed and replaced.

Probably the most satisfactory solution, in terms of overall quality of life, is *kidney transplantation*. This procedure involves the implantation of a new kidney obtained from a living donor or a cadaver. One-third of the approximately 8000 kidneys transplanted last year were obtained from living, related donors. The damaged kidney is usually removed. When left in place, an arterial graft is inserted to carry blood from the iliac artery or the aorta to the transplant, located in the pelvis or lower abdomen.

The success rate for this procedure varies, depending on how aggressively the recipient's T cells attack the donated organ and whether or not an infection develops. The 1-year success rate for implantation is now 85 to 95 percent. The use of kidneys taken from close relatives significantly improves the chances for a successful transplant. Immunosuppressive drugs are administered to reduce tissue rejection, but unfortunately this treatment also lowers the individual's resistance to infection.

TABLE A-19 The Composition of Dialysis Fluid

Constituent	Normal Plasma	Dialyzing Fluid
ELECTROLYTES (mEq/1)		
Potassium	5	3
Bicarbonate	27	36
Phosphate	3	0
Sulfate	0.5	0
NUTRIENTS (mg/dl)		
Glucose	100	125
NITROGENOUS WASTES (mg/dl)		
Urea	26	0
Creatinine	1	0
Uric acid	0.3	0

Note: Only the significant variations are noted; values for other electrolytes are usually similar. Although these values are representative, the precise composition can be tailored to meet the specific clinical needs. For example, if plasma potassium levels are too low, the dialyzing fluid concentration can be elevated to remedy the situation. Changes in the osmolarity of the dialyzing fluid can also be used to adjust an individual's blood volume, usually by adjusting the glucose content of the dialyzing fluid.

✝ Problems with the Conducting System EAP *p. 501*

Local blockages of the collecting tubules, collecting ducts, or ureter may result from the formation of *casts,* small blood clots, epithelial cells, lipids, or other materials. Casts are often excreted in the urine and visible in microscopic analysis of urine samples. **Calculi** (KAL-kū-li), or "kidney stones," form from calcium deposits, magnesium salts, or crystals of uric acid. This condition is called *nephrolithiasis* (nef-rō-li-THĪ-a-sis). The blockage of the urinary passage by a stone or other factors, such as external compression, results in **urinary obstruction.** Urinary obstruction is a serious problem because, in addition to causing pain, it will reduce or eliminate filtration in the affected kidney by elevating the capsular hydrostatic pressure.

Kidney stones are usually visible on an X-ray, and if peristalsis and fluid pressures are insufficient to dislodge them they must be surgically removed or destroyed. One interesting nonsurgical procedure involves breaking kidney stones apart with a *lithotripter,* the same apparatus used to destroy gallstones. Another nonsurgical approach entails the insertion of a catheter armed with a laser that can shatter kidney stones with intense light beams.

✝ Urinary Tract Infections EAP *p. 500*

Urinary tract infections, or **UTIs,** result from the colonization of the urinary tract by bacterial or fungal invaders. The intestinal bacterium *Escherichia coli* is most often involved, and women are particularly susceptible to urinary tract infections because of the proximity of the external urethral orifice to the anus. Sexual intercourse may also push bacteria into the urethra, and, since the female urethra is relatively short, the urinary bladder may become infected.

The condition may be asymptomatic (without symptoms), but it can be detected by the presence of bacteria and blood cells in the urine. If inflammation of the urethral wall occurs, the condition may

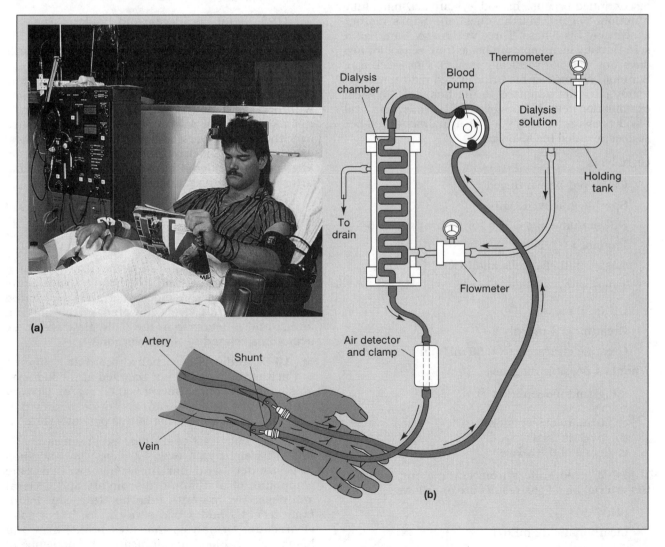

(a)

(b)

FIGURE A-51

Hemodialysis. (a) Preparation for hemodialysis typically involves implantation of a pair of shunts connected by a loop that permits normal blood flow when the patient is not hooked up to the dialysis machine. **(b)** A diagrammatic view of the dialysis procedure.

19

be termed *urethritis*, while inflammation of the lining of the bladder is called *cystitis*. Many infections, including sexually transmitted diseases such as gonorrhea, cause a combination of urethritis and cystitis. These conditions cause painful urination, a symptom known as *dysuria* (dis-Ū-rē-a), and the bladder becomes tender and sensitive to pressure. Despite the discomfort produced, the individual feels the urge to urinate frequently. Urinary tract infections usually respond to antibiotic therapies,

although subsequent reinfections may occur.

In untreated cases the bacteria may proceed along the ureters to the renal pelvis. The resulting inflammation of the walls of the renal pelvis produces *pyelitis* (pī-e-LĪ-tis), and if the bacteria invade the renal cortex and medulla as well, *pyelonephritis* (pī-e-lō-nef-RĪ-tis) results. Signs and symptoms include a high fever, intense pain on the affected side, vomiting, diarrhea, and the presence of blood cells and pus in the urine.

CRITICAL THINKING QUESTIONS

10-1. Martha's first patient on the pediatric floor is a 12-year-old boy named Sam who is undergoing kidney dialysis treatments. Sam has been on dialysis for the last week, and the treatments will probably be discontinued because his kidneys are becoming fully functional again. Martha questions Sam's mother concerning his illness. Three weeks ago, Sam had a sore throat which kept him home from school for two days but was not medically treated. One week ago, Sam told his mother his urine looked red-brown and foamy. He was admitted to the hospital. Physical examination revealed costovertebral tenderness and blood pressure of 135/90. The urinalysis and other studies revealed the following:

Urinalysis

  Color: red-brown tinged

  Specific gravity: 1.030

  RBCs: numerous RBCs per high-power field

  Protein: >1 g/24 hr

  Casts: epithelial cells and RBCs

  Other Values:

  BUN: 90 mg/dl

  Creatinine: 4 mg/dl

  Creatine clearance test: 50 ml/min

What is a possible diagnosis?

  a. glomerulonephritis
  b. cystitis
  c. urinary obstruction
  d. pyelitis
  e. none of the above

10-2. Fred suffers from chronic emphysema. His arterial blood gas results are as follows:

  pH: 7.30

  bicarbonate: 32 mEq/l

  P_{CO2}: 50 mm Hg

Which of the following best describes Fred's condition?

  a. metabolic acidosis with respiratory compensation
  b. respiratory acidosis with metabolic compensation
  c. metabolic alkalosis with respiratory compensation
  d. respiratory alkalosis with metabolic compensation

10-3. Peter, a 50-year-old banker, finds he has high blood pressure (150/110) when his nursing student daughter is practicing her skills on him. Peter's physician confirms her findings and begins some laboratory and diagnostic tests to determine the cause. Routine lab work was within normal limits. Adrenal gland function is normal. Because a prominent bruit was heard over the right upper abdomen during the physical exam, renal angiography was ordered. The renal angiogram revealed a narrowing of the right renal artery. How is this related to the hypertension?

10-4. A young college student is struck by a car. On arriving at the ER, she is unconscious with internal hemorrhaging. How will this affect her GFR? Why?

10-5. Koji collapses in the final leg of a marathon, and is rushed to a hospital with acute renal dysfunction. A blood test shows elevated serum potassium levels, lowered serum sodium levels, and a decrease in the GFR. How are these symptoms related to the marathon?

10-6. Inulin is a carbohydrate that is filtered but neither reabsorbed nor secreted at the kidneys. Plasma inulin concentration is 0.05 mg/dl. Urinary concentration is 0.1 mg/ml, and urine is formed at a rate of 1.0 ml/min. What is this patient's GFR?

10-7. Lucy, 10 years old, experiences a significant weight gain over 12 days. Her abdomen is distended, and her limbs are swollen. She complains of abdominal discomfort and general achiness. Her parents take her to a physician. Lucy recently had a bad sore throat, but it soon cleared up. Urinalysis indicates proteinuria and hematuria. A blood test indicates the presence of antibodies to toxins produced by Streptococcus bacteria. What may be the cause of Lucy's problems?

10-8. Mr Smith, 68 years old, suffers from Type II diabetes mellitus (NIDDM). One afternoon he complains of abdominal pain and fatigue. He then becomes unconscious, and he is comatose on arrival at the hospital. Make predictions about his blood pH, urine pH, plasma P_{CO_2}, and plasma potassium levels.

The Reproductive System and Development

This section considers applied topics related to the continuation of the species and the life histories of individuals. In the process we will consider aspects of the male and female reproductive systems, pregnancy, development, aging, and death.

THE REPRODUCTIVE SYSTEM

The male reproductive system includes the gonads (testes), a series of specialized ducts (the epididymis, ductus deferens, ejaculatory duct, and urethra), accessory glands (the seminal vesicles, prostate, and bulbourethral glands), and the external genitalia (penis and scrotum).

The female reproductive system includes the gonads (ovaries), derivatives of an embryonic system of ducts (the uterine tubes, uterus, and vagina), accessory glands (the greater and lesser vestibular glands), the external genitalia (the clitoris, labia majora and labia minora), and secondary sexual organs, the mammary glands of the breasts.

Assessment of the Male Reproductive System

An assessment of the male reproductive system begins with a physical examination. Common signs and symptoms of male reproductive disorders include

- *Testicular pain* may result from a variety of infections, including *gonorrhea* or other sexually transmitted diseases (p. 159), *mumps* (EAP, *p. 437*), *typhoid* (p. 24), *rheumatic fever* (p. 98), or *influenza*. Testicular pain can also result from *testicular torsion*, testicular cancer, *cryptorchidism* (EAP, *p. 517*), or the presence of a *hernia* (p. 61). The pain may also originate elsewhere along the reproductive tract, such as along the ductus deferens or within the prostate, or in other systems, as in *appendicitis* (p. 114) or a urinary obstruction.

- *Urethral discharge* and dysuria are often associated with sexually transmitted diseases. These symptoms also accompany disorders, such as *epididymitis* or *prostatitis*, that may be infectious or noninfectious.

- *Impotence* an inability to achieve or maintain an erection. It may occur as the result of psychological factors, such as fear or anxiety, medications, or alcohol abuse. It may also develop secondary to cardiovascular problems that affect blood pressure or blood flow to the penile arteries.

- *Male infertility* may be caused by a low sperm count, abnormally shaped sperm, or abnormal semen composition. Analysis of the semen can often yield important diagnostic information.

Inspection of the male reproductive system usually involves the examination of the external genitalia and palpation of the prostate gland. Inspection of the external genitalia entails the following observational steps:

1. Inspection of the penis and scrotum for skin lesions such as *vesicles, chancres, warts,* and *condylomas* (wartlike growths). For example, painful vesicles often appear in clusters following infection with the herpes simplex (type 2) virus. A chancre is a painless ulceration often associated with early-stage *syphilis* (p. 160). These skin lesions usually indicate the presence of sexually transmitted diseases (p. 159). In the course of the examination of uncircumcised males, the foreskin is retracted to observe the preputial lining. **Phimosis,** an inability to retract the foreskin in an uncircumcised male, usually indicates inflammation of the prepuce and adjacent tissues.

2. Palpation of each testis, epididymis, and ductus deferens to detect the presence of abnormal masses, swelling, or tumors. Possible abnormal findings include:

- *Scrotal swelling* due to distortion of the scrotal cavity by blood (a *hematocele),* lymph (a *chylocele),* or serous fluid.

- *Testicular swelling* due to enlargement of the testis or formation of a nodular mass. **Orchitis** is a general term for inflammation of the testis. This can be the result of an infection, such as *syphilis* (p. 160), *mumps,* or *tuberculosis* (p. 129). Testicular swelling may also accompany testicular cancer.

- *Epididymal swelling* due to cyst formation *(spermatocoele),* tumor formation, or infection. **Epididymitis** is an acute inflammation of the epididymis that may indicate an infection of the reproductive or urinary tracts. This condition may also develop due to irritation caused by the backflow, or *reflux,* of urine into the ductus deferens.

- *Swelling of the spermatic cord* may indicate (1) inflammation of the ductus deferens *(deferentitis),* (2) serous fluid accumulation in a pocket of the peritoneal cavity (a *hydrocele),* (3) bleeding within the spermatic cord, (4) *testicular torsion,* or (5) the formation of *varicose veins* (p. 105) within the pampiniform plexus, a condition known as a *varicocele.*

3. A *digital rectal examination* (DRE) is usually performed as a screening test for prostatitis or inflammation of the seminal vesicles. In this procedure, a gloved finger is inserted into the rectum and pressed against the anterior rectal wall to palpate the posterior walls of the prostate gland and seminal vesicles.

If urethral discharge is present, or if discharge occurs in the course of any of these procedures, the fluid can be cultured to check for the presence of pathogenic microorganisms.

Assessment of the Female Reproductive System

Important signs and symptoms of female reproductive disorders include the following:

- *Acute pelvic pain* is a symptom that may accompany a variety of different disorders. For example, it may be associated with pelvic inflammatory disease (PID), ruptured tubal pregnancy, a ruptured ovarian cyst, or inflammation of the uterine tubes *(salpingitis).*

- *Bleeding between menstrual cycles* can result from oral contraceptive use, hormonal fluctuation, *pelvic inflammatory disease* (EAP, *p. 527*), or *endometriosis.*

- *Amenorrhea* (EAP, *p. 529*) may occur in women with *anorexia nervosa* p. 140, women who overexercise and are underweight, in extremely obese women, and in post menopausal women.

- *Vaginal discharge* may be the result of a bacterial infection, such as an STD.

- Although the female reproductive and urinary tracts are distinct, *dysuria* may accompany an infection of the reproductive system due to migration of the pathogen to the urethral entrance.

- Infertility may be related to hormonal disturbances, a variety of ovarian disorders, or anatomical problems along the reproductive tract.

A physical examination usually includes the following steps:

1. Inspection of the external genitalia as in the male, for skin lesions, trauma, or related abnormalities. Swelling of the labia majora may result from (a) regional lymphedema, (b) a *labioinguinal hernia,* (c) bleeding within the labia, as the result of local trauma or cellulitis or (4) *Bartholinitis,* an abscess within one of the greater vestibular glands *(Bartholin's glands).*

2. Inspection and/or palpation of the perineum, vaginal opening, labia, clitoris, urethral meatus, and vestibule to detect lesions, abnormal masses, or discharge from the vagina or urethra. Samples of any discharge present can be cultured to detect and identify any pathogens involved.

3. Inspection of the vagina and cervix can be performed using a *speculum.* A speculum is an instrument that retracts the vaginal walls to permit direct visual inspection. Changes in the color of the vaginal wall may be important diagnostic clues. For example:

- Cyanosis of the vaginal mucosa normally occurs during pregnancy (see below) but it

may also occur when a pelvic tumor exists or in persons with congestive heart failure.

- Reddening of the vaginal walls occurs in *vaginitis,* bacterial infections, such as gonorrhea, protozoan infection by *Trichomonas vaginalis,* yeast infections or postmenopausally in some women (a condition known as *atrophic vaginitis).*

The cervix is inspected to detect lacerations, ulceration, polyps, or cervical discharge. A spatula or brush is then used to collect cells from the cervical os and transfer them to a glass slide. After fixation by a chemical spray, cytological examination is performed. This is the best-known example of a *Papanicolauo (PAP) test* (Table A-2 p. 5), and the process is commonly called a Pap smear. A Pap smear is a screening test for the presence of cervical cancer.

4. *Bimanual examination* is a method for the palpation of the uterus, uterine tubes, and ovaries. The physician inserts two fingers vaginally and places the other hand against the lower abdomen to palpate the uterus and surrounding structures. The contour, shape, size and location of the uterus can be determined, and any swellings or masses will be apparent. Abnormalities in other reproductive organs, such as ovarian cysts, endometrial growths, or tubal masses, can also be detected in this way.

Normal and Abnormal Signs Associated with Pregnancy

Pregnancy imposes a number of stresses on the maternal body systems. The major physiological changes are discussed in Chapter 21 (EAP *p. 552).* There are several clinical signs that may be apparent in the course of a physical examination.

- *Chadwick's sign* is a normal cyanosis of the vaginal wall during pregnancy.

- There are drastic changes in the size of the uterus during pregnancy, with the uterus at fullterm extending almost to the level of the xiphoid process of the sternum.

- Significant uterine bleeding, causing vaginal discharge of blood, most often occurs in *placenta previa* (p. 164), when the placenta forms near the cervix. Subsequent cervical stretching leads to tearing and bleeding of the vascular channels of the placenta. Vaginal bleeding may also occur prior to miscarriage.

- Nausea and vomiting often occur in pregnancy, especially during the first three months.

- Edema of the extremities, especially the legs, often occurs due to increased blood volume and weight of the uterus compressing the inferior vena cava and its tributaries. As venous pressures rise in the lower limbs and inferior trunk, *varicose veins* and *hemorrhoids* (p. 105) may develop.

20

- Back pain is common, due to increased stress on muscles of the lower back. These muscles balance the weight of the uterus over the lower limbs by accentuating the lumbar curvature.

- A weight gain of 25-30 pounds is now considered normal. Failure to gain adequate weight during a pregnancy can indicate serious problems.

- In some cases, a dangerous combination of hypertension, proteinuria, and edema may occur. This condition, called *eclampsia*, is considered in a later section (p. 165).

DISORDERS OF THE REPRODUCTIVE SYSTEM

Disorders of the reproductive system can be organized by sex and region, as in Figure A-52a. This arrangement has the advantage of linking disorders that primarily affect the reproductive system to their target organs. An alternative scheme groups the disorders as primary or secondary, depending on whether the reproductive system is the focus of the disorder. This arrangement is indicated in Figure A-52b.

✝ Prostatitus, Prostatic Hypertrophy and Prostate Cancer

Prostatic inflammation, or **prostatitis** (pros-ta-TI-tis), can occur at any age, but it most often afflicts older men. Prostatitis may result from bacterial infections, but the condition may also develop in the apparent absence of pathogens. Individuals with prostatitis complain of pain in the lower back, perineum, or rectum, sometimes accompanied by painful urination and the discharge of mucous secretions from the urethral meatus. Antibiotic therapy is usually effective in treating cases resulting from bacterial infection, but in other cases antibiotics may not provide relief. Prostatitis is taken seriously because the symptoms can resemble those of *prostate cancer*. Prostate cancer is the second most common cancer in men and it is the second leading cause of cancer deaths in males.

Prostatic enlargement, or **benign prostatic hypertrophy,** usually occurs spontaneously in men over age 50. The increase in size occurs at the same time that hormonal changes are under way within the testes. Androgen production by the interstitial cells decreases over this period, and at the same time these endocrine cells begin releasing small quantities of estrogens into the circulation. The combination of lower testosterone levels and the presence of estrogen probably stimulates prostatic growth. In severe cases, prostatic swelling can constrict and block the urethra and even the rectum. The urinary obstruction can cause permanent kidney damage if not corrected. Partial surgical removal is the most effective treatment at present. In the procedure known as a **TURP** (*transurethral prostatectomy*), an instrument

pushed along the urethra restores normal function by cutting away the swollen prostatic tissue. Most of the prostate remains in place, and there are no external scars.

Prostate cancer is the second most common cancer in men, and it is the second most common cause of cancer deaths in males. In 1996 approximately 317,000 new cases of prostate cancer were diagnosed in the United States, and there were approximately 41,400 deaths. Most patients are elderly (average age 72 at diagnosis). There are racial differences in susceptibility that are poorly understood. At age 50–54 the prostate cancer rates are twice as high for African-Americans as for Caucasian Americans. (The rates at all ages are about one-third higher for African-Americans.) The prostate cancer rates for Asian males are relatively low compared with either Caucasian Americans or African-Americans. For all age groups and all races, the rates of prostate cancer are rising sharply. The incidence of new cases increased 65% between 1980 and 1990. Much of this increase was due to improved detection.

Prostate cancer usually originates in one of the secretory glands, and as it progresses it produces a nodular lump or swelling on the prostatic surface. Palpation of the prostate gland through the rectal wall, a procedure known as a *digital rectal exam,* or DRE, is the easiest diagnostic screening procedure. *Transrectal prostatic ultrasound* (TRUS) can be used to obtain more detailed information about the status of the prostate, but at significantly higher cost to the patient.

If the condition is detected before the cancer cells have spread to other organs, the usual treatment is either localized radiation or the surgical removal of the prostate gland. This operation, called a **prostatectomy** (prosta-TEK-to-mē), is often effective in controlling the condition, but undesirable side effects may include a loss of sexual function and urinary incontinence. Modified surgical procedures can reduce these risks and maintain normal sexual function in almost three out of four patients.

One recent screening method involves a blood test for *prostate-specific antigen (PSA)*. Elevated levels of this antigen, normally present in low concentrations, may indicate the presence of prostate cancer. This test is more sensitive than the serum enzyme assay previously used for screening purposes. That enzyme test, which checks levels of the isozyme *prostatic acid phosphatase,* detects prostate cancer in comparatively late stages of development. Screening with periodic PSA tests is now being recommended for men over age 50.

Early detection is important because metastasis from the prostate soon involves the lymphatic system, lungs, bone marrow, liver, or adrenal glands. The survival rates at this stage become relatively low. Potential treatments for metastatic prostate cancer include more intensive radiation dosage, hormonal

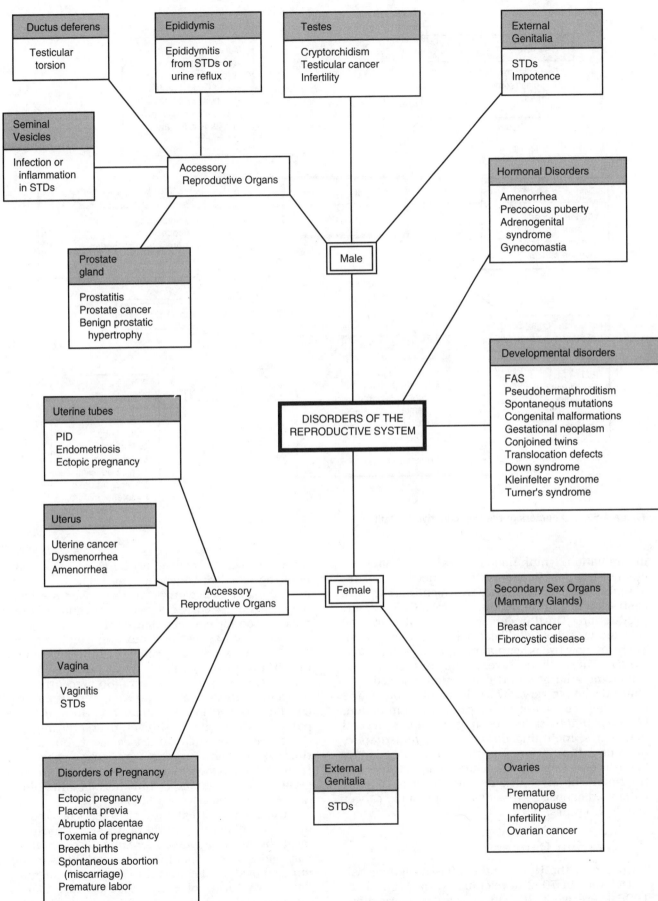

Figure A-52a   Disorders of the Reproductive System

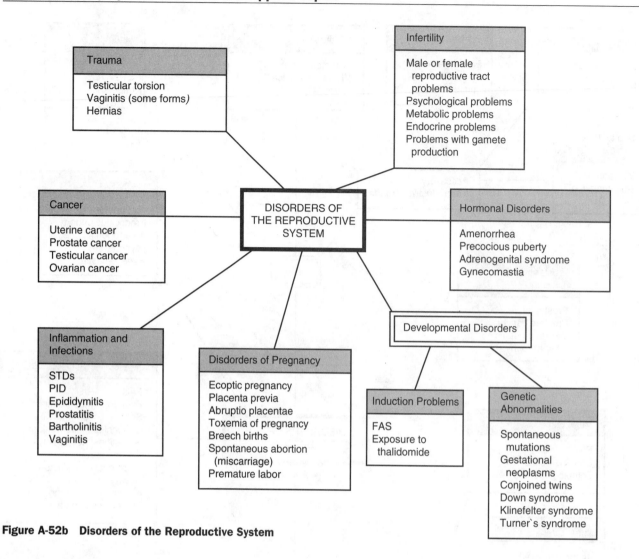

Figure A-52b   Disorders of the Reproductive System

manipulation, lymph node removal, and aggressive chemotherapy. Because the cancer cells are stimulated by testosterone, treatment may involve castration or hormones that depress GnRH or LH production. Until recently the usual hormone selected was *diethylstilbestrol* (DES), an estrogen. There are now two other options: (1) *Drugs that mimic GnRH:* These drugs are given in high doses, producing a surge in LH production followed by a sharp decline to very low levels, presumably as the endocrine cells adapt to the excessive stimulation. (2) *Drugs that block the action of androgens:* Several new drugs, including *flutamide* and *finasateride,* prevent stimulation of the cancer cells by testosterone. Despite these interesting advances in treatment, however, the average survival time for patients diagnosed with advanced prostatic cancer is only 2.5 years.

✝ Ovarian Cancer   EAP p. 525

A woman in the United States has a lifetime risk of 1 chance in 70 of developing ovarian cancer. In 1996 there were 26,700 ovarian cancers diagnosed, and 14,800 deaths from this condition.

Although ovarian cancer is the third most common reproductive cancer among women, it is the most dangerous because ovarian cancer is seldom diagnosed in its early stages. The prognosis is relatively good for cancers that originate in the general ovarian tissues, or from abnormal oocytes. These cancers respond well to some combination of chemotherapy, radiation, and surgery. However, most ovarian cancers (85 percent) develop from epithelial cells, and sustained remission can be obtained in only about one-third of these patients. Early diagnosis would greatly improve the chances for successful treatment, but as yet there is no standardized screening procedure. *(Transvaginal sonography* can detect ovarian cancer at Stage I or Stage II, but there is a high incidence of false positive results.)

The minimal treatment of Stage I or Stage II involves unilateral removal of an ovary and uterine tube (a *salpingo-oophorectomy),* or *bilateral salpingo-oophorectomy* (BSO) and *total hysterectomy* (removal of the uterus). Treatment of more dangerous forms of early stage ovarian cancer includes radiation and chemotherapy in addition to surgery.

Treatment of Stage III or Stage IV ovarian cancer often involves removal of the omentum, in addition to a BSO and total hysterectomy and aggressive chemotherapy. Bone marrow transplantation may be required, due to destruction of stem cells in the bone marrow by these chemicals. Some chemotherapy agents may be introduced into the peritoneal cavity, because higher concentrations can be administered without the systemic effects that would accompany infusion of the drugs into the bloodstream. This procedure is called *intraperitoneal therapy.*

✠ Premenstrual Syndrome   EAP *p. 529*

Several physical and physiological changes occur in women 7–10 days before the start of menses. Fluid retention, breast enlargement, headaches, pelvic pain, and an uncomfortable "bloated" feeling are common symptoms. These sensations may be associated with psychological changes producing irritability, anxiety, and depression. This combination has been called **premenstrual syndrome (PMS).**

The mechanism responsible for PMS has yet to be determined. Changes in sex hormone levels may be involved directly, by action on peripheral organ systems, or indirectly, by modifying neurotransmitter release in the CNS. There are no laboratory tests or procedures to diagnose PMS, but tracking the appearance of symptoms over a 2–3 month period can reveal characteristic patterns. Treatment at present is symptomatic and may involve exercise, dietary change, or medication, depending on the nature of the primary symptom. For example, if headache is the major problem, analgesics are prescribed; diuretics may be used to combat bloating and fluid retention. For severe PMS, drugs can be administered that block GnRH secretion and stop uterine cycles completely for 6 months or more. Over the interim, estrogens can be administered to prevent symptoms of premature menopause.

✠ Endometriosis   EAP *p. 529*

In *endometriosis* (en-dō-mē-trē-Ō-sis) an area of endometrial tissue begins to grow outside the uterus. The severity of the condition depends on the size of the abnormal mass and its location. Abdominal pain, bleeding, pressure on adjacent structures, and infertility are common symptoms. As the island of endometrial tissue enlarges, the symptoms become more severe.

Diagnosis can usually be made by using a laparoscope inserted through a small opening in the abdominal wall. Using this device a physician can inspect the outer surfaces of the uterus and uterine tubes, the ovaries, and the lining of the pelvic cavity. Treatment of endometriosis may involve hormonal therapy or surgical removal of the endometrial mass. If the condition is widespread, a hysterectomy or *oophorectomy* (removal of the ovaries) may be required.

✠ Vaginitis   EAP *p. 529*

There are several different forms of vaginitis, and minor cases are relatively common. **Candidiasis** (kan-di-DĪ-a-sis) results from a fungal (yeast) infection. The organism responsible appears to be a normal component of the vaginal ecosystem in 30-80 percent of normal women. Antibiotic administration, immun- osuppression, stress, pregnancy, and other factors that change the local environment can stimulate the unrestricted growth of the fungus. Symptoms include itching and burning sensations, and a lumpy white discharge may also be produced. Topical antifungal medications are used to treat this condition.

Bacterial (nonspecific) vaginitis results from the combined action of several bacteria. The bacteria involved are normally present in about 30 percent of adult women. In this form of vaginitis the vaginal discharge contains epithelial cells and large numbers of bacteria. The discharge has a homogeneous, sticky texture and a characteristic odor sometimes described as fishy or aminelike. Antibiotics are often effective in controlling this condition.

Trichomoniasis (trik-ō-mō-NĪ-a-sis) involves infection by a parasite, *Trichomonas vaginalis*, introduced by sexual contact with a carrier. Because it is a sexually transmitted disease, both partners must be treated to prevent reinfection.

A more serious vaginal infection by *Staphylococcus* bacteria is responsible for symptoms of **toxic shock syndrome** (TSS). Symptoms include high fever, sore throat, vomiting and diarrhea, and a generalized rash. As the condition progresses, shock, respiratory distress, and kidney or liver failure may develop, and 10-15 percent of all cases prove fatal. These symptoms result from the entry of bacterial toxins and even bacteria into the bloodstream. This disorder was first recognized in 1978, when it affected a group of children. Since that time roughly 3,000 cases have been diagnosed, 95 percent of them adult women. Most of these women (over 90 percent) have developed the condition while using a tampon, but the precise link between tampon use and TSS remains uncertain. The use of superabsorbent tampons was initially thought to be responsible, and the incidence of TSS did decline as those items were removed from the market. However, TSS continues to occur at a low but significant rate (6.2 per 100,000 menstruating women per year) in those using ordinary tampons and vaginal sponges, and in people of either sex after abrasion or burn injuries that promote bacterial infection. Treatment for TSS involves fluid administration, removal of the focus of infection (such as removal of a tampon or cleansing of a wound), and antibiotic therapy.

✠ Sexually Transmitted Diseases
EAP *p. 535*

Sexually transmitted diseases, or **STDs,** are transferred from individual to individual, usually or

20

exclusively by sexual intercourse. A variety of bacterial, viral, and fungal infections are included in this category. At least two dozen different STDs are currently recognized. All are unpleasant. *Chlamydia* can cause pelvic inflammatory disease and infertility. Other types of STDs are quite dangerous, and a few, including AIDS, are deadly. Here we will discuss four of the most common sexually transmitted diseases: *gonorrhea, syphilis, herpes,* and *chancroid.*

GONORRHEA. The bacterium *Neisseria gonorrhoeae* is responsible for gonorrhea, one of the most common sexually transmitted diseases in the United States. Nearly 2 million cases are reported each year. These bacteria usually invade epithelial cells lining the male or female reproductive tracts. In relatively rare cases they will also colonize the pharyngeal or rectal epithelium.

The symptoms of genital infection vary, depending on the sex of the individual concerned. It has been estimated that up to 80 percent of women infected with gonorrhea experience no symptoms, or symptoms so minor that medical treatment is thought to be unnecessary. As a result these individuals act as carriers, spreading the infection through their sexual contacts. An estimated 10-15 percent of women infected with gonorrhea experience more acute symptoms because the bacteria invade the epithelia of the uterine tubes. This probably accounts for many of the cases of pelvic inflammatory disease (PID) in the U.S. population; as many as 80,000 women may become infertile each year as the result of scar tissue formation along the uterine tubes after gonorrheal infections.

Diagnosis in males seldom poses as great a problem, for all but 20-30 percent of infected males develop symptoms recognized as requiring immediate medical attention. The urethral invasion is accompanied by pain on urination *(dysuria)* and often a viscous urethral discharge. A sample of the discharge can be cultured to permit positive identification of the organism involved. Treatment of gonorrhea involves the administration of antibiotics.

SYPHILIS. Syphilis (SIF-i-lis) results from infection by the bacterium *Treponema pallidum.* The first reported syphilis epidemics occurred in Europe during the sixteenth century. The death rate from the "Great Pox" was appalling, far greater than today even after taking into account the absence of antibiotic therapies at that time. It appears likely that the syphilis organism has mutated over the interim. These changes have reduced the mortality rate but prolonged the period of illness and increased the likelihood of successful transmission. Despite these relative improvements, syphilis still remains a life-threatening disease. Untreated syphilis can cause serious cardiovascular and neurologic illness years after infection, or it can be spread to the fetus during pregnancy to produce congenital malformations. The annual reported incidence of this disease has now risen to roughly 20 cases per 100,000 population, the highest rate in 40 years. An equivalent or greater number probably went unrecognized or unreported.

Primary syphilis begins as the bacteria cross the mucous epithelium and enter the lymphatics and bloodstream. At the invasion site the bacteria multiply, and after an incubation period ranging from 1.5-6 weeks their activities produce a painless raised lesion, or **chancre** (SHANG-ker). This lesion remains for several weeks before fading away, even without treatment. In heterosexual men the chancre usually appears on the penis; in women it may develop on the labia, vagina, or cervix. Lymph nodes in the region usually enlarge and remain swollen even after the chancre has disappeared.

Symptoms of *secondary syphilis* appear roughly 6 weeks later. Secondary syphilis usually involves a diffuse, reddish skin rash. Like the chancre, the rash fades over a period of 2-6 weeks. These symptoms may be accompanied by fever, headaches, and uneasiness. The combination is so vague that the disease may easily be overlooked or diagnosed as something else entirely. In a few instances more serious complications such as *meningitis* (p. 67), *hepatitis* (p. 136), or *arthritis* (p. 53) may develop.

The individual then enters the *latent phase.* The duration of the latent phase varies widely. Fifty to 70 percent of untreated individuals with latent syphilis fail to develop the symptoms of *tertiary syphilis,* or *late syphilis,* although the bacterial pathogens remain within their tissues. Those destined to develop tertiary syphilis may do so 10 or more years after infection.

The most severe symptoms of tertiary syphilis involve the CNS and the cardiovascular system. **Neurosyphilis** may result from bacterial infection of the meninges or the tissues of the brain and/or spinal cord. **Tabes dorsalis** (TĀ-bēz dor-SAL-is) results from the invasion and demyelination of the posterior columns of the spinal cord and the sensory ganglia and nerves. In the cardiovascular system the disease affects the major vessels, leading to *aortic stenosis* (p. 98), *aneurysms* (p. 103), or *focal calcification* (EAP, *p. 104*).

Equally disturbing are the effects of transmission from mother to fetus across the placenta. These cases of congenital syphilis are marked by infections of the developing bones and cartilages of the skeleton and progressive damage to the spleen, liver, bone marrow, and kidneys. The risk of transmission may be as high as 80-95 percent, so maternal blood testing is recommended early in pregnancy. Treatment of syphilis involves the administration of *penicillin* or other antibiotics.

HERPES VIRUS. Genital herpes results from infection by herpes viruses. Two different viruses are involved. Eighty to 90 percent of genital herpes cases are caused by a specific virus known as HSV-2 (herpes simplex virus Type 2), a virus usually associated with the external genitalia. The remaining cases are caused by HSV-1, the same virus responsible for cold sores on the mouth. Typically within a week of the initial infection the individual develops a number of painful, ulcerated lesions on the external genitalia. In women, ulcers may also appear on the cervix. These ulcerations gradually heal over the next 2-3 weeks. Recurring lesions are common, although subsequent incidents are less severe.

Infection of the newborn infant during delivery with herpes viruses present in the vagina can lead to serious illness, because the infant has few immunological defenses. Recent development of the antiviral agent *acyclovir* has helped treatment of initial infections.

CHANCROID. Chancroid is an STD caused by the bacterium *Haemophilus ducreyi*. Chancroid cases were rarely seen inside the United States before 1984, but since then the number of cases has risen dramatically, reaching 4,000-5,000 cases per year. The primary sign of this disease is the development of *soft chancres*, soft lesions otherwise resembling those of syphilis. The majority of chancroid patients also develop prominent inguinal lymphadenopathy.

⚕ Breast Cancer   EAP *p. 531*

The mammary glands are cyclically stimulated by the changing levels of circulating reproductive hormones that accompany the menstrual cycle. Usually the effects go unnoticed, but there can be occasional discomfort and even inflammation of mammary gland tissues late in the cycle. If inflamed lobules become walled off with scar tissue, **cysts** are created. Clusters of cysts can be felt in the breast as discrete masses, a condition known as *fibrocystic disease*. Because the symptoms are similar, biopsies may be needed to distinguish between this benign condition and breast cancer.

Despite repeated studies, there are no proven links between oral contraceptive use, estrogen therapy, fat consumption, or alcohol use and breast cancer. It appears likely that multiple factors are involved; most women never develop breast cancer, even women in families with a history of this disease. Adequate amounts of nutrients and vitamins, and a diet rich in fruits and vegetables, appear to offer some protection against the development of breast cancer. Women who have breast-fed babies have a 20 percent lower incidence of breast cancer after menopause than mothers who had not nursed their infants. The reason for this effect is not known. (Adding to the mystery, nursing does not appear to affect the incidence of premenopausal breast cancer.)

Early detection of breast cancer is the key to reducing mortalities. *Most breast cancers are found through self-examination*, but the use of clinical screening techniques has increased in recent years. *Mammography* involves the use of X-rays to examine breast tissues; the radiation dosage can be restricted because only soft tissues must be penetrated. This procedure gives the clearest picture of conditions within the breast tissues.

Ultrasound can provide some information, but the images lack the detail of standard mammograms. *Thermography* maps the surface temperatures on the skin of the breasts. Because cancer cells have abnormally high metabolic rates and increased vascularization, tumors are significantly warmer than the surrounding tissues. The heat can be detected with this technique, but unfortunately the results are subject to considerable variation.

For treatment to be successful the cancer must be identified while it is still relatively small and localized. Once it has grown larger than 2 cm (0.78 in.) the chances for long-term survival worsen. A poor prognosis also follows if the cancer cells have spread through the lymphatic system to the axillary lymph nodes. If the nodes are not yet involved, the chances of 5-year survival are about 82 percent, but if four or more nodes are involved, the survival rate drops to 21 percent.

Treatment of breast cancer begins with the removal of the tumor. Because the cancer cells usually begin spreading before the condition is diagnosed, surgical treatment involves the removal of part or all of the affected breast.

- In a *segmental mastectomy*, or "lumpectomy," only a portion of the breast is removed.

- In a *total mastectomy* the entire breast is removed, but other tissues are left intact.

- In *radical mastectomy* the pectoralis muscles, the breast, and the axillary lymph nodes are removed. In a *modified radical mastectomy*, the most common operation, the breast and nodes are removed but the muscular tissue remains intact.

A combination of chemotherapy, radiation treatments, and hormone treatments may be used to supplement the surgical procedures. *Tamoxifen* is a drug that may be used to treat breast cancer. It is more effective than conventional chemotherapy for treating breast cancer in women over 50, and it has fewer unpleasant side effects. It can also be used in addition to regular chemotherapy when treating advanced stage disease. As an added bonus, tamoxifen prevents and even reverses the osteoporosis of aging. There are down sides, however. When given to premenopausal women, tamoxifen can cause amenorrhea and hot flashes similar to those of menopause. Tamoxifen has also been linked to an increased risk of endometrial cancer and perhaps liver cancer as well. It has been proposed that this drug be used to *prevent* breast cancer, rather than treat it. Large scale trials are underway to determine whether the benefits of chronic low-dose tamoxifen therapy outweigh the risks of complications.

New treatment options are also under development. For example, a tumor-suppressor gene that inhibits breast cancer development has been isolated from normal breast tissue. The protein has been identified, and researchers are now experimenting to see if the activity of the gene can be stimulated to fight existing breast cancers.

20

Experimental Contraceptive Methods

EAP *p. 539*

A number of experimental contraceptive methods are being investigated. For example, researchers are attempting to determine whether low doses of inhibin will suppress GnRH release and prevent ovulation. Another approach is to develop a method of blocking human chorionic gonadotropin (hCG) receptors at the corpus luteum. HCG, produced by the placenta, maintains the corpus luteum for the first three months of pregnancy. If the corpus luteum were unable to respond to hCG, normal menses would occur despite implantation of a blastocyst.

Male contraceptives are also under development.

- *Gossypol*, a yellow pigment extracted from cottonseed oil, produces a dramatic decline in sperm count and sperm motility after 2 months. It can be administered topically, as it is readily absorbed through the skin. Fertility returns within a year after treatment is discontinued. Unfortunately, gossypol has not been approved as yet because of a relatively high risk of permanent sterility (around 10 percent) and the potential for development of hypokalemia.

- Weekly doses of testosterone suppress GnRH secretion over a period of 5 months. The result is a drastic reduction in the sperm count. The combination of a testosterone implant, comparable to that used in the Norplant system, with a GnRH antagonist, *cetrorelix*, effectively suppresses spermatogenesis. A new synthetic form of testosterone, *alpha-methyl-nortestosterone (MENT)*, appears even more effective than testosterone in suppressing GnRH production.

- A drug used to control blood pressure appears to cause a temporary, reversible sterility in males. This drug is now being evaluated to see if low dosages will affect fertility in normal males without affecting blood pressure.

If contraceptive methods fail, options exist to either prevent implantation or terminate the pregnancy. The "morning-after pills" contain estrogens or progestins. They may be taken within 72 hours of intercourse, and they appear to act by altering the transport of the zygote or preventing its attachment to the uterine wall. The drug known as *RU-486 (Mifepristone)* blocks the action of progesterone at the endometrial lining. The result is a normal menses, and the degeneration of the endometrium whether or not a pregnancy has occurred. RU486 is used in France and China, but because it can also be used to induce an abortion later in pregnancy it has not been licensed for use in the United States.

Technology and the Treatment of Infertility

EAP *p. 546*

An infertile, or sterile, woman is unable to produce functional eggs or support a developing embryo. An infertile man is incapable of providing a sufficient number of motile sperm for successful fertilization. Because sterility of either sexual partner will have the same result, diagnosis and treatment of infertility must involve evaluation of both sexual partners. Approximately 60 percent of infertility cases can be attributed to problems with the female reproductive system.

Recent advances in our understanding of reproductive physiology are providing new solutions to fertility problems. These approaches, called assisted reproductive technologies (ART) are diagrammed in Figure A-53:

- **Low sperm count.** In cases of male infertility due to low sperm counts, semen from several ejaculates can be pooled, concentrated, and introduced into the female reproductive tract. This technique, known as artificial insemination, may lead to normal fertilization and pregnancy. In special cases, where an individual's spermatozoa are unable to accomplish oocyte penetration, single-sperm fertilization has been accomplished with micromanipulation of the oocyte and corona radiata.

- **Abnormal spermatozoa.** If the man cannot produce functional sperm, sperm can be obtained from a "sperm bank" that stores donor sperm.

- **Hormonal problems.** If the problem involves the woman's inability to ovulate due to low gonadotropin or estrogen levels, or to maintain adequate progesterone levels after ovulation, these hormones can be provided.

   So-called *fertility drugs*, such as clomiphene (Clomid), stimulate ovarian egg production. Clomiphene works by blocking the feedback inhibition of estrogen on the hypothalamus and pituitary gland. As a result, circulating FSH levels rise, and more follicles are stimulated to complete their development. The chance of a single egg being fertilized through normal sexual intercourse is around 1 in 3. Increasing the number of eggs released increases the odds of a pregnancy. Unfortunately, it is not easy to determine just how much ovarian stimulation will be needed, so multiple births have often resulted from treatment with fertility drugs.

- **Problems with oocyte transport.** When there are problems with the transport of the egg from the ovary to the uterine tube, due to scarring of the fimbriae or other problems, a procedure called GIFT can be used. GIFT is short for *gamete intrafallopian tube transfer* (*Fallopian tube* is another name for the uterine tube or oviduct). In this procedure the ovaries

Figure A-53
The Treatment of Infertility

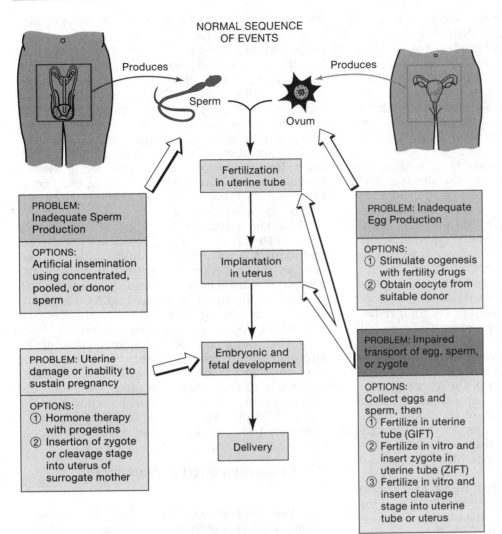

NORMAL SEQUENCE
OF EVENTS

Produces → Sperm

Produces → Ovum

Fertilization
in uterine tube

Implantation
in uterus

Embryonic and
fetal development

Delivery

PROBLEM:
Inadequate Sperm
Production

OPTIONS:
Artificial insemination
using concentrated,
pooled, or donor
sperm

PROBLEM: Uterine
damage or inability to
sustain pregnancy

OPTIONS:
① Hormone therapy
with progestins
② Insertion of zygote
or cleavage stage
into uterus of
surrogate mother

PROBLEM: Inadequate
Egg Production

OPTIONS:
① Stimulate oogenesis
with fertility drugs
② Obtain oocyte from
suitable donor

PROBLEM: Impaired
transport of egg, sperm,
or zygote

OPTIONS:
Collect eggs and
sperm, then
① Fertilize in uterine
tube (GIFT)
② Fertilize in vitro and
insert zygote in
uterine tube (ZIFT)
③ Fertilize in vitro and
insert cleavage
stage into uterine
tube or uterus

are stimulated with injected hormones, and a large "crop" of mature oocytes is removed from tertiary follicles. The the individual eggs are examined for defects, inserted into the uterine tubes, and exposed to high concentrations of sperm from the husband or donor. The success rate for this procedure is less than that of natural fertilization (33 percent), and not every pregnancy produces an infant. The cost of a single procedure (successful or not) averages $5000.

• **Blocked uterine tubes.** In the GIFT procedure, fertilization occurs in its normal location, within the uterine tube. This site is not essential, and fertilization can also take place in a test tube or petri dish. This process is called *in vitro fertilization* (*vitro*, glass). If a carefully controlled fluid environment is provided, early development will proceed normally. One variation on the GIFT procedure, called ZIFT (zygote intrafallopian tube transfer), exposes selected eggs to sperm outside the body and inserts zygotes or early cleavage-stage embryos, rather than oocytes, into the uterine tubes. If multiple zygotes are available, some can be frozen and

stored for later insertion in case the initial procedure fails to produce a successful pregnancy. The cost for a single ZIFT procedure ranges between $8000 and $10,000.

Alternatively, the zygote can be maintained in an artificial environment through the first 2 to 3 days of development. This procedure is often selected if the uterine tubes are damaged or blocked. The cleavage-stage embryo is then placed directly into the uterus, rather than into one of the uterine tubes. The cost of this procedure is comparable to that for ZIFT.

Abnormal oocytes. If the oocytes released by the ovaries are abnormal in some way, or if menopause has already occurred, viable oocytes can be obtained from a suitable donor. The donor may be anonymous or known; if anonymous, the donor usually receives a fee for the donation. After treatment with fertility drugs, the donor's ovaries are stimulated to produce a large crop of oocytes. These are collected and fertilized in vitro, usually bt y the husband's sperm. After cleavage has begun, the pre-embryo is placed in the recipient's uterus, which has been "primed" by progesterone therapy. The pregnancy rate for this procedure is

21

roughly 35 percent for women over age 40, using oocytes donated by women in their early twenties. Oocyte donation has a much higher success rate for these women than ZIFT or GIFT, with either of which the odds of a successful pregnancy are only about 4 percent. This difference suggests that age-related changes in the characteristics and quality of the oocytes, rather than changes in hormone levels or uterine responsiveness, are often the primary cause of infertility in older women.

Abnormal uterine environment. If fertilization and transport occur normally but the uterus cannot maintain a pregnancy, the problem may involve low levels of progestin secretion by the corpus luteum. Hormone therapy may solve this problem. If the maternal uterus simply cannot support development, the zygote or cleavage-stage embryo can be introduced into the uterus of a substitute mother, or surrogate mother. If the embryo survives and makes contact with the endometrium, development will proceed normally even though the mother has no genetic relationship with the embryo.

Surrogate motherhood, which sounds relatively simple and straightforward, has proven to be one of the most explosive solutions in terms of ethics and legality. Since 1990, several court cases have resulted from disputes over surrogate motherhood and who merits legal custody of the infant. Legal battles have also broken out over a variety of complex questions, and some of them will take years to sort out. To understand the problem, just take a moment and consider:

Do parents share property rights over frozen and stored zygotes? Can a husband have any of the stored zygotes implanted into the uterus of his second wife without the consent of his first wife, who provided the eggs?

If both donor egg and donor sperm are used, do adoption laws apply?

If the husband provided the sperm that fertilized the egg of a donor who is not his wife, for implantation into a surrogate mother, can the wife, the surrogate mother, or the egg donor sue for custody of the child after a divorce?

If you use your imagination, you can probably think of even more complex problems, many of which will probably be debated in a courtroom within the next decade.

DISORDERS OF DEVELOPMENT

⚕ Ectopic Pregnancies   EAP p. 547

Implantation usually occurs at the endometrial surface lining the uterine cavity. The precise location within the uterus varies, although most often implantation occurs in the body of the uterus. This is not an ironclad rule, and in an ectopic pregnancy implantation occurs somewhere other than within the uterus.

The incidence of ectopic pregnancies is approximately 0.6 percent. Women douching regularly have a

4.4 times higher risk of experiencing an ectopic pregnancy, presumably because the flushing action pushes the zygote away from the uterus. If the uterine tube has been scarred by a previous episode of pelvic inflammatory disease, there is also an increased risk of an ectopic pregnancy. Although implantation may occur within the peritoneal cavity, in the ovarian wall, or in the cervix, 95 percent of ectopic pregnancies involve implantation within a uterine tube. The tube cannot expand enough to accommodate the developing embryo, and it usually ruptures during the first trimester. At this time the hemorrhaging that occurs in the peritoneal cavity may be severe enough to pose a threat to the woman's life.

In a few instances the ruptured uterine tube releases the embryo with an intact umbilical cord, and further development can occur. About 5 percent of these abdominal pregnancies actually complete full-term development; normal birth cannot occur, but the infant can be surgically removed from the abdominopelvic cavity. Because abdominal pregnancies are possible, it has been suggested that men as well as women could act as surrogate mothers if a zygote were surgically implanted in the peritoneal wall. It is not clear how the endocrine, cardiovascular, nervous, and other systems of a man would respond to the stresses of pregnancy. However, the procedure has been tried successfully in mice, and experiments continue.

⚕ Problems with Placentation

EAP p. 551

In a **placenta previa** (PRĒ-vē-a; "in the way") implantation occurs in or near the cervix. This condition causes problems as the growing placenta approaches the internal cervical orifice. In a **total placenta previa** the placenta actually extends across the internal orifice, while a partial placenta previa only partially blocks the os. The placenta is characterized by a rich fetal blood supply, and the erosion of maternal blood vessels within the endometrium. Where the placenta passes across the internal orifice the delicate complex hangs like an unsupported water balloon. As the pregnancy advances, even minor mechanical stresses can be enough to tear the placental tissues, leading to massive fetal and maternal hemorrhaging.

Most cases are not diagnosed until the seventh month of pregnancy, as the placenta reaches its full size. At this time the dilation of the cervical canal and the weight of the uterine contents are pushing against the placenta where it bridges the internal orifice. Minor, painless hemorrhaging usually appears as the first sign of the condition. The diagnosis can usually be confirmed by ultrasound scanning. Treatment in cases of total placenta previa usually involves bed rest for the mother until the fetus reaches a size at which cesarean delivery can be performed with a reasonable chance of neonatal (newborn) survival.

In an **abruptio placentae** (ab-RUP-shē-ō pla-SEN-tē) part or all of the placenta tears away from the uterine wall sometime after the fifth month of gestation. The bleeding into the uterine cavity and the pain that follows usually will be noted and reported, although in some cases the shifting placenta may block the passage of blood through the cervical canal. In severe cases the hemorrhaging leads to maternal anemia, shock, and kidney failure. Although maternal mortality is low, the fetal mortality rate from this condition ranges from 30 to 100 percent, depending on the severity of the hemorrhaging.

⚕ Problems with the Maintenance of a Pregnancy
EAP p. 552

The rate of maternal complications during pregnancy is relatively high. Pregnancy stresses maternal systems, and the stresses can overwhelm homeostatic mechanisms. The term **toxemia** (tok-SĒ-mē-a) **of pregnancy** refers to disorders affecting the maternal cardiovascular system. Chronic hypertension is the most characteristic symptom, but fluid balance problems and CNS disturbances, leading to coma or convulsions, may also occur. Some degree of toxemia occurs in 6-7 percent of third-trimester pregnancies. Severe cases account for 20 percent of maternal deaths and contribute to an estimated 25,000 neonatal (newborn) deaths each year.

Toxemia of pregnancy includes **preeclampsia** (prē-ē-KLAMP-sē-a) and **eclampsia** (ē-KLAMP-sē-a). Preeclampsia most often occurs during a woman's first pregnancy. Systolic and diastolic pressures become elevated, reaching levels as high as 180/110. Other symptoms include fluid retention and edema, along with CNS disturbances and alterations in kidney function. Roughly 4 percent of individuals with preeclampsia develop eclampsia.

Eclampsia, or *pregnancy-induced hypertension (PIH)*, is heralded by the onset of severe convulsions lasting 1-2 minutes followed by a variable period of coma. Other symptoms resemble those of preeclampsia, with additional evidence of liver and kidney damage. The mortality rate from eclampsia is approximately 5 percent; to save the mother the fetus must be delivered immediately. Once the fetus and placenta are removed from the uterus, symptoms of eclampsia disappear over a period of hours to days.

⚕ Common Problems with Labor and Delivery
EAP p. 558

There are many potential problems during labor and delivery. Two relatively common types of complications are *forceps deliveries* and *breech births*.

By the end of gestation the fetus has usually rotated within the uterus so that it will enter the birth canal head first, with the face turned toward the sacrum. In around 6 percent of deliveries the fetus faces the pubis rather than the sacrum. Although these infants can eventually be delivered normally, risks to infant and mother increase the longer the fetus remains in the birth canal. Often the clinical response is the removal of the infant through a **forceps delivery.** The forceps used resemble a large, curved set of salad tongs that can be separated for insertion into the vaginal canal one side at a time. Once in place they are reunited and used to grasp the head of the infant. An intermittent pull is applied, so that the forces on the head resemble those encountered during normal delivery.

In 3-4 percent of deliveries, the legs or buttocks of the fetus enter the vaginal canal first. Such deliveries are known as **breech births.** Risks to the infant are relatively higher in breech births because the umbilical cord may become constricted, and placental circulation cut off. Because the head is normally the widest part of the fetus, the cervix may dilate enough to pass the legs and body but not the head. Entrapment of the fetal head compresses the umbilical cord, prolongs delivery, and subjects the fetus to severe distress and potential damage. If the fetus cannot be repositioned manually, a cesarian section is usually performed.

📋 Monitoring Postnatal Development
EAP p. 560

Each newborn infant gets a close scrutiny after delivery. The maturity of the newborn may also be determined prior to delivery via ultrasound or amniocentesis. Potential complications of premature delivery include neonatal *respiratory distress syndrome* (p. 129), the absence of some reflexes, hyperbilirubinemia, and a predisposition to infection.

A physical examination of the newborn focuses on the status of vital systems. Inspection of the infant usually focuses on the following:

- The head of a newborn infant may be misshapen following vaginal delivery, but it usually assumes its normal shape over the next few days.

- The abdomen is palpated to detect abnormalities of internal organs.

- The external genitalia are inspected. The scrotum of a male infant is checked for the presence of descended testes.

- Cyanosis of the hands and feet is normal in the newborn, but the rest of the body should be pink. A generalized cyanosis may indicate congenital circulatory disorders, such as *erythroblastosis fetalis* (*EAP p. 316*), a patent foramen ovale or ductus arteriosus, tetralogy of Fallot (p. 110), or other problems.

Measurements of body length, head circumference and body weight are taken. A weight loss in the first 48 hours is normal, due to fluid shifts that occur as the infant adapts to the change from weightlessness (floating in amniotic fluid) to normal gravity. (Comparable fluid shifts occur in astronauts returning to earth after extended periods in space.)

The excretory systems of the newborn infant are assessed by examination of the urine and feces. The first urination may have a pink coloration, due to the presence of urates. The first bowel movement consists of a mixture of epithelial cells and mucus. This *meconium* is greenish-black in color.

The **Apgar rating** considers heart rate, respiratory rate, muscle tone, response to stimulation, and color at 1 and 5 minutes after birth. In each category the infant receives a score ranging from 0 (poor) to 2 (excellent), and the scores are then totaled. An infant's Apgar rating (0-10) has been shown to be an accurate predictor of newborn survival and the presence of neurological damage. For example, newborn infants with *cerebral palsy* (EAP, *p. 238*) usually have a low Apgar rating.

In the course of this examination, the breath sounds, the depth and rate of respirations, and the heart rate are noted. Both the respiratory rate and the heart rate are considerably higher in the infant than the adult (see Table A-1, p. 3). The neuromuscular system is assessed for normal reflexes and muscle tone. Reflexes often tested include:

- The *Moro reflex* is a reflex triggered when support for the head of a supine infant is suddenly removed. The reflex response consists of trunk extension and a rapid cycle of extension-abduction and flexion-adduction of the limbs. This reflex normally disappears at an age of about 3 months.

- The *stepping reflex* consists of walking movements triggered by holding the infant upright, with a forward slant, and placing the soles of the feet against the ground. This reflex normally disappears at an age of around 6 weeks.

- The *placing reflex* can be triggered by holding the infant upright and drawing the top of one foot across the bottom edge of a table. The reflex response is to flex and then extend the leg on that side. This reflex also disappears at an age of around six weeks.

- The *sucking reflex* is triggered by stroking the lips. The associated *rooting reflex* is initiated by stroking the cheek, and the response is to turn the mouth toward the site of stimulation. These reflexes persist until age 4-7 months.

- The *Babinski reflex* is positive, with fanning of the toes in response to stroking of the side of the sole of the foot. This reflex disappears at around age 3, as descending motor pathways become established.

These procedures check for the presence of anatomical and physiological abnormalities. They also provide baseline information useful in assessing postnatal development. In addition, newborn infants are often screened for genetic and/or metabolic disorders, such as *phenylketonuria* (PKU) (p. 139), congenital *hypothyroidism* (p. 82), *sickle cell anemia* (p. 92) and *galactosemia*. Individuals with galactosemia lack the enzyme that converts galac-tose sugar in milk into glucose. Chronic high levels of galactose during childhood can cause abnormalities in nervous system development, jarndice, liver enlargement, and cataracts.

Pediatrics is a medical specialty focusing on postnatal development from infancy through adolescence. Because infants and young children cannot clearly describe the problems they are experiencing, pediatricians and parents must be skilled observers. Standardized testing procedures are also used to assess an individual's developmental progress. In the **Denver Developmental Screening Test** (DDST) infants and children are checked repeatedly during their first five years. The test checks gross motor skills, such as sitting up or rolling over, language skills, fine motor coordination, and social interactions. The results are compared with normal values determined for individuals of similar age. These screening procedures assist in identifying children who may need special teaching and attention.

Too often parents tend to focus on a single ability or physical attribute, such as the age at first step or the rate of growth. This kind of one-track analysis has little practical value, and the parents may become overly concerned with how their infant compares with the norm. *It should be realized that normal values are statistical averages,* not absolute realities. For example, an infant usually begins walking between 11 and 14 months of age. But around 25 percent start before then, and another 10 percent do not start walking by the fourteenth month. Walking early does not indicate true genius, and walking late does not mean that the infant will need physical therapy. The questions on such screening tests are intended to determine if there are *patterns* of developmental deficits. Such patterns appear only when a broad range of abilities and characteristics are considered.

Death and Dying EAP p. 561

Despite exaggerated claims, there have been no substantiated cases of individuals living more than 120 years. Estimates for the lifespan of individuals born in the United States during 1994 are 72 years for males and 78 years for females. Interestingly enough, the causes of death vary depending on the age group under discussion. Consider the graphs shown in Figure A-54, indicating the mortality statistics for various age groups. Accidents are the major cause of death in young people, and cardiovascular diseases in those over 40-45. More specific information concerning the major causes of death can be found in Table A-20. Many of the characteristic differences in mortality figures result from changes in the functional capabilities of the individuals linked to development or senescence. It should also be realized that these figures would differ significantly if tabulated for countries and cultures with different genetic and environmental pressures.

Table A–20   The Five Major Causes of Death in the U.S. Population

Rank	Ages 1-14 Male	Female	Ages 15-34 Male	Female	Ages 35-54 Male	Female	Ages 55-74 Male	Female	Ages 75+ Male	Female
1	Accidents		Accidents		Heart disease	Cancer	Heart disease		Heart disease	
2	Cancer		Homicide	Cancer	Cancer	Heart disease	Cancer		Cancer	CVD
3	Congenital anomalies		Suicide		Accidents		CVD		CVD	Cancer
4	Homicide		Cancer	Homicide	Cirrhosis of the liver	CVD	Accidents	Diabetes	Pneumonia, influenza	
5	Pneumonia, influenza		Heart disease		Suicide	Cirrhosis of the liver	COPD	Accidents	COPD	Arteriosclerosis

Note: CVD=cerebrovascular disease; COPD = chronic obstructive pulmonary disease.

The differences in mortality figures for male and female are related to differences in the accident rates for young people and in the rates of heart disease and cancer for older individuals. The upswing in female cancer rates reflects a rising breast cancer incidence for those over age 34, while lung cancer is the primary cancer killer of older men. In women the incidence of lung cancers and related killers, including pulmonary disease, heart disease, and pneumonia, has been steadily increasing as the number of women smokers has increased. This change has narrowed the difference between male and female life expectancies.

Experimental evidence and calculations suggest that the human lifespan has an upper limit of around 150 years. As medical advances continue, research must focus on two related issues: (1) extending the average lifespan toward that maximum and (2) improving the functional capabilities of long-lived individuals. The first objective may be the easiest from a technical and moral standpoint. It is already possible to reduce the number of deaths attributed to specific causes. For example, new treatments promote remission in a variety of cancer cases, and anticoagulant therapies may reduce the risks of death or permanent damage following a stroke or heart attack. Many defective organs can be replaced with functional transplants, and the use of controlled immunosuppressive drugs will increase the success rates for these operations. Artificial hearts have been used with limited success, and artificial kidneys and endocrine pancreases are under development.

The second objective poses more of a problem. Few people past their mid-90s lead active, stimulating lives, and most would find the prospect of living another 50 years rather horrifying unless the quality of their lives could be significantly improved. Our abilities to prolong life now involve making stopgap corrections in systems on the brink of complete failure. Reversing the process of senescence would entail manipulating the biochemical operations and genetic programming of virtually every organ system. Although investigations continue, breakthroughs cannot be expected in the immediate future.

Over the interim, we are left with some serious ethical and moral questions. Now that we can postpone the moment of death almost indefinitely, how do we decide when it is appropriate to do so? How can medical and financial resources be fairly allocated? Who gets the limited number of hearts, livers, kidneys, and corneas available for trans-

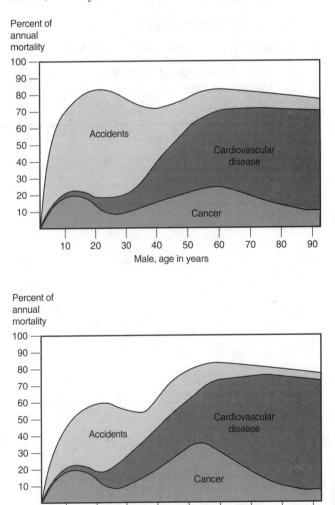

Figure A-54   Major Causes of Postnatal Mortality

21

plant? Who should be selected for experimental therapies of potential significance? Should we take into account that care of an infant or child may add decades to a lifespan, while the costly insertion of an artificial heart in a 60-year-old will add only months to years? How shall we allocate the costs for sophisticated procedures that may run to hundreds of thousands of dollars over the long run? Are these individual or family responsibilities? In either case, will only the rich survive?

Should the funds be provided by the government? If yes, what will happen to tax rates as the baby boomers become elderly citizens? And what about the role of the individual involved? If you decline treatment, are you mentally and legally competent? Can your survivors bring suit because you are forced to survive, or because you were allowed to die? These and other difficult questions will not go away, and in the years to come we will have to find answers we are content to live and die with.

CRITICAL THINKING QUESTIONS

11-1. Sally is an avid runner, and she trains incessantly. She has slimmed down so much that she is now underweight for her height and has very little fat tissue. Some of her laboratory values are below.

FSH: early cycle, 2 mIU/ml; midcycle, 3 mU/ml

LH: early cycle, 2 mIU/ml; midcycle, 15 mIU/ml

Estrogen (serum): midcycle, 60 pg/ml

From this information you would expect Sally to

a.   have heavy menstrual flows
b.   double ovulate
c.   be amenorrheic
d.   have painful menstrual cramps

11-2. Male athletes using anabolic steriods often show low sperm counts and some feminizing characteristics, such as enlarged breasts. Why do these symptoms occur?

11-3. Diane has peritonitis, which she is told resulted from a urinary tract infection. Why does this occur in females but not males?

11-4. David is brought to the hospital after a severe blow to the scrotal region during a baseball game. It later develops that the trauma has caused a breakdown in the blood-testis barrier. What effect would this have on David's immune system?

NOTES

NOTES

Answer Key

Critical Thinking Questions

THE INTEGUMENTARY SYSTEM

1-1. (d).

1-2. (b).

1-3. Carrie is a vegetarian, and her diet includes an abundant amount of carrots, oranges, apricots, and green vegetables, as well as a pint of carrot juice a day. These food sources contain carotene, a yellow-orange pigment stored in lipids, notably in the skin. The skin discoloration is not dangerous, but Carrie may elect to go a bit lighter on the carrot juice for aesthetic reasons.

1-4. The organic solvents that Sam uses for cleaning dissolve the lipids that help provide a water barrier in the skin and help keep the skin flexible. Without the lipids, the skin will become dry and lose more water to the surroundings.

1-5. You would probably assume that the hormonal changes accompanying pregnancy are responsible for the increased pigmentation, because hormones can affect melanin production by melanocytes. It is actually the hormone progesterone that is responsible for the changes noted. In fact, these skin changes are so consistent that they are referred to as the "mask of pregnancy."

THE SKELETAL SYSTEM

2-1. (c).

2-2. Trauma. Arthroscopy revealed torn ligaments, which caused bleeding in the synovial cavity, and a damaged medial meniscus that restricts movement.

2-3. (b). Tooth decay occurs when the acid produced by bacteria erodes the harder tooth structure. This particular tooth infection had spread through the tooth and into the root. The bony structure surrounding the tooth root became infected as well, producing osteomyelitis and an erosion of the adjacent bone. Antibiotics were given and the decay was removed from the tooth followed by a root canal procedure.

2-4. Since osteopenia usually begins between the ages of 30 and 40, these individuals were probably older than 40 when they died. The characteristics of the arm bones indicate that the men used their arms a great deal— perhaps lifting, pulling, chopping, or shoveling. Since the leg bones do not indicate large or well-developed muscles, these men probably did not do a great deal of walking or running.

2-5. The temperomandibular joint is found where the mandible articulates with the temporal bone. As a result of this condition you would expect to observe pain in the region of the joint, as well as in the muscles associated with stabilizing the joint and movement of the joint. Since this is a synovial joint, you might expect to hear clicking or popping sounds when the joint moves, the result of synovial fluid being squeezed through tight spaces. Tension, stress, or anxiety would produce muscle tension that would not only affect the joint but could also cause muscle pain and possibly headaches in the temporal region of the head.

2-6. The first fracture probably involved the greater trochanter or the shaft of the femur inferior to the neck. The second fracture was probably a fracture of the femoral neck, which disrupted the circulation and led to avascular necrosis at the joint. The most likely outcome is the insertion of an artificial hip joint.

THE MUSCULAR SYSTEM

3-1. (c).

3-2. (a). CeCe is concerned about the paralysis of her respiratory muscles following the injection of this drug. Dr. R. reassures her that she will be on a respirator until either the drug wears off or the drug effects are counteracted by the administration of an antagonistic drug, *neostigmine methylsulfate*.

3-3. (b).

3-4. As a result of training, we would expect the tennis player to have larger reserves of glycogen and creatine phosphate than the nonathlete. These would help supply energy during short bursts of strenuous activity. Thus we would expect to find higher levels of the enzymes required for the breakdown of glycogen and the conversion of creatine phosphate to creatine in the muscle of the tennis star. In addition, training increases the level of enzymes that hydrolyze ATP during contraction, so we would also expect to see more of this enzyme in the tennis star than in the nonathlete.

3-5. Calvin has probably injured his tibialis posterior muscle, which is the prime mover for inversion of the foot.

THE NERVOUS SYSTEM

4-1. (e).

4-2. (b).

4-3. The expected response would be plantar flexion. This is an example of a stretch reflex: (1) tapping on the tendon stretches intrafusal and extrafusal fibers in the gastrocnemius and soleus muscles, (2) the sensory processes at the muscle spindles are distorted, (3) the rate of action potentials along afferent fibers from the muscle spindles increases, (4) motor neurons in the spinal cord are stimulated, and (5) the extrafusal fibers in the stretched muscles contract. Stretch reflexes include important postural reflexes and reflexes involved with the automatic control of muscle tension during a contraction. Examples include this example (ankle jerk reflex) and the patellar reflex, the biceps reflex, and the triceps reflex. Testing stretch reflexes provides information about the status of simple reflex arcs whose components are clearly identifiable. These are monosynaptic reflexes, and each reflex involves a limited number of spinal nerves and spinal cord segments. As a result, testing stretch reflexes is one way to check for damage to specific spinal nerves or to the related segments of the spinal cord.

4-4. It appears that Mrs. Glenn is suffering from Parkinson's disease. Since the motor neurons are more excitable than they should normally be, we would expect to see exaggerated spinal reflexes due to increased muscle tone.

4-5. Chelsea probably suffered a subdural or subarachnoid hemorrhage. These can result from a blow or other injury to the head that ruptures the blood vessels in the meninges. As blood accumulates in the subdural or subarachnoid space, it puts pressure on the brain, leading to impaired neural function and (at high enough pressures) cessation of function. The pattern of sensory and motor impairment is consistent with a lateral hemorrhage in the area of the temporal lobe that gradually spreads downward to the brainstem, ultimately causing Chelsea's death.

4-7. Fifty percent of the fibers decussate (cross over) at the optic chiasm. As a result, Dave will lose vision from the temporal field of the right eye and the nasal field of the left eye. After a period of time, some accommodation will occur, helping to fill in the missing parts of the visual field.

4-8. Bone conduction tests can be used to discriminate between conductive and nerve deafness. In this type of test, the physician places a vibrating tuning fork against the patient's skull. If the patient hears the tuning fork when it touches the skull but not when it is held next to the ear, the problem must lie within the external or middle ear. If the patient does not respond to either stimulus, the problem must be at the receptors or the auditory pathway.

THE ENDOCRINE SYSTEM

5-1. (e).

5-2. (b)

5-3. (c).

5-4. Probably not, since there is no real medical reason for John to receive GH (human growth hormone), and the long-term effects of such treatments are not well known. GH affects many different tissues and can have widespread metabolic effects, such as a decline in body fat content. This in turn can affect the metabolism of many organs.

5-5. Angie's diabetes was due to excess growth hormone from the anterior pituitary. Chronic hypersecretion of GH produces chronic hyperglycemia (high blood sugar) and the same problems associated with insulin-dependent diabetes. Removal of the anterior pituitary takes away the source of the problem, allowing blood sugar levels to return to normal. As long as the retinal damage is not too extensive, the tissue can repair itself and vision can be restored.

5-6. One could inject a large dose of TRH into the patient and then monitor the blood for elevated TSH. If there was an increase in TSH following the injection, it would appear that the problem was with the hypothalamus. On the other hand, if there were no increase in TSH in response to the exogenous TRH, then it would appear that the problem was with the pituitary cells.

THE CARDIOVASCULAR SYSTEM

6-1. (b).

6-2. (c).

6-3. The most obvious possibility is patent ductus arteriosus, which occurs when the ductus arteriosus fails to close off. When the baby is not being stressed (bathing creates heat loss and thermal stress) or eating (less air is entering the lungs), she appears normal. Because some of the blood flow is shunted to the aorta during stress and eating, there is not enough blood being oxygenated and the infant becomes cyanotic.

THE LYMPHATIC/IMMUNE SYSTEMS

7-1. This type of pulmonary infection most commonly occurs in patients with AIDS. Matthew could be tested for AIDS by an ELISA test, which would detect antibodies to HIV in his blood. The Western blot test is another test for HIV and is more sensitive.

7-2. A key characteristic of cancer cells is their ability to break free from a tumor and migrate to other tissues of the body forming new tumors. This process is called metastasis. The primary route for the spread of cancer cells is the lymphatic system, and cancer cells invade lymph nodes on their way to other tissues. Examination of regional lymph nodes for the presence of cancer cells can help the physician determine if the cancer was caught in the early stage or whether it has started to spread to other tissues. It can also give the physician an idea of what other tissues may be affected by the cancer to help decide on the proper treatment.

7-3. Teresa is suffering from hypovolemic shock secondary to acute hemorrhage. A ruptured spleen is the most likely bleeding site. The accumulation of blood in the peritoneal cavity causes irritation of the diaphragm and phrenic nerves, causing pain, and this leads to abdominal rigidity.

THE RESPIRATORY SYSTEM

8-1. (c).

8-2. The patient's forgetfulness in not finishing the antibiotics has allowed the bacteria to propagate and enter the pleural cavity through the lymphatic system. The patient has pleurisy accompanied by an increase of fluid in the pleural space. This increase in fluid is called a *pleural effusion*.

8-3. (c). The tuberculin skin test is a confirming test for the presence of antibodies to the tuberculosis bacteria. Ann will have a positive reaction if she has been exposed to the TB bacterium at some time in her life. The sputum culture may take up to six weeks; it will indicate whether or not she is currently infected with the bacteria, *Mycobacterium tuberculosis*.

THE DIGESTIVE SYSTEM

9-1. (e).

9-2. Cytological examination of the tissue specimen would most likely reveal malignant cells. The mass growing in Paul's colon is cancerous, and it is protruding into the colon lumen. The rectal bleeding originates at the tumor.

9-3. Mary could have lactose intolerance due to the deficiency of the enzyme, lactase. A diet analysis over a period of several weeks might reveal that Mary's episodes of diarrhea followed the ingestion of dairy products. The physician could also order a lactose tolerance test.

9-4. You would expect low blood glucose levels and elevated ketone bodies and fatty acids.

THE URINARY SYSTEM

10-1. (a).

10-2. (b).

Use the following three steps to answer the type of question in 10-2.

1. Determine if the pH is within normal limits (7.35-7.45):

pH > 7.45 is an alkalosis.

pH < 7.35 is an acidosis.

In our example, Fred has a pH of 7.30, which indicates an acidosis.

2. Determine the primary cause of the acidosis/alkalosis by checking the values for P_{CO2} and bicarbonate.

P_{CO2} = 35-45 mm Hg

bicarbonate = 22-28 mEq/l

For the respiratory system to be the primary cause of an acidosis, the P_{CO2} value must be above 45 mm Hg.

For the respiratory system to be the primary cause of an alkalosis, the P_{CO2} value must be below 35 mm Hg.

For there to be a metabolic primary cause of an acidic pH, the bicarbonate value must be less than 22 mEq/l.

For there to be a metabolic primary cause of an alkaline pH, the bicarbonate value must be greater than 28 mEq/l.

In Fred's case the primary cause of his acidic pH cannot be of a metabolic nature because the bicarbonate level is greater than 28 mEq/l.

The primary cause of Fred's acidic pH is the respiratory system indicated by the increased P_{CO2} above the normal range.

3. Determine if the system which is not the primary cause is compensating for the imbalance in pH. If so, the value should be as follows:

P_{CO2}: < 35 mm Hg is a compensation for metabolic acidosis.

P_{CO2}: >45 mm Hg is a compensation for metabolic alkalosis.

bicarbonate: <22 mEq/l is a compensation for respiratory alkalosis.

bicarbonate: >28 mEq/l is a compensation for respiratory acidosis.

In Fred's case, his respiratory acidosis is being compensated for by the increased level of bicarbonate. This process of compensation involves the kidneys where some of the excess carbon dioxide is converted into carbonic acid and the carbonic acid is allowed to dissociate. The hydrogen ions are secreted and the newly formed bicarbonate is conserved to maintain a proper buffering capacity.

10-3. Narrowing of the right renal artery was related to atherosclerosis. This narrowing restricts blood flow to the kidney and produces renal hypotension and ischemia. Decreased blood flow and ischemia would trigger the juxtaglomerular apparatus to produce more renin which would lead to elevated levels of angiotensin II and aldosterone. Both these substances can increase the blood pressure by vasoconstriction (angiotensin II) and by increasing the blood volume (aldosterone).

10-4. Her GFR would be reduced because (a) she has been hemorrhaging and her blood volume is reduced, and (b) she is probably in shock, and activation of the sympathetic nervous system will have reduced blood flow through the afferent arterioles.

10-5 Strenuous exercise causes sympathetic activation, which produces (1) powerful vasoconstriction of the afferent arterioles that deliver blood to the glomeruli and (2) a dilation of peripheral blood vessels, so that blood is shunted away from the renal vessels. As a result of these changes, the GFR is reduced. Potentially dangerous conditions develop as the circulating concentration of metabolic wastes increases and peripheral water losses mount.

10-6 Since the inulin is neither reabsorbed nor secreted, the rate at which the inulin is excreted is equal to the glomerular filtration rate (GFR). Thus if you find the rate at which the inulin is cleared from the plasma, you will have the GFR.

$$\text{GFR} \; \frac{0.10\text{mg/ml} \times 1.0\text{ml/min}}{0.05 \text{ mg/dl}} = 20$$

10-7 Lucy probably had strep throat, a bacterial infection. The bacteria produced a toxin that caused an inflammation of the filtering membranes of the kidneys (glomerulonephritis). The inflammed membranes leaked proteins into the filtrate which were not reabsorbed. This would account for the protein and traces of blood in Lucy's urine sample. The loss of proteins led to a decrease in blood osmotic pressure, leading to overall edema. This would account for the weight gain, swollen limbs, and general discomfort.

10-8 Since Mr. Smith is a diabetic, he is probably suffering from acidosis resulting from ketosis. Increased production of ketone bodies lowers the blood pH. The increased concentration of hydrogen ions causes more hydrogen ions to move into the cells and more potassium ions to leave the cells, so the level of potassium in the ECF is elevated. The urine pH should be quite acidic as the kidneys work to eliminate the excess hydrogen ions . The decreased pH would lead to increased rate and depth of breathing, so we would expect to see declining levels of CO_2, especially since the cause of the acidosis is not respiratory.

THE REPRODUCTIVE SYSTEM

11-1. (c).

11.2. Anabolic steroids are chemically related to the hormone testosterone. Elevated levels of the hormones suppress the release of GnRH from the hypothalamus and FSH and LH from the pituitary. This leads to low endogenous levels of testosterone, decreased spermatogenesis, and a low sperm count. Changes in male secondary sex characteristics may result from the conversion of anabolic steroids into estrogens.

11.3. In males, an infectious organism can move up the urethra to the urinary bladder or into the ejaculatory duct to the ductus deferens. There is no direct connection between the urinary or reproductive tract and abdominopelvic cavity in the male. In females, the urethral opening is in close proximity to the vaginal orifice; thus, the infectious organism could exit from the urethral opening, enter the vaginal orifice, proceed through the vagina to the uterus, into the uterine tubes, and into the abdominopelvic cavity via the infundibulum. This can lead to infection and inflammation of the peritoneal lining.

11.4. The blood-testis barrier is maintained by tight junctions between adjacent sustentacular cells. This barrier prevents sperm antigens from triggering an immune response. The breakdown in the blood-testis barrier caused by a traumatic blow to the scrotum would allow sperm antigens to escape into the interstitial spaces. This could cause an autoimmune response that would result in sterility.

Index

Notes

Notes

Surface Anatomy and Cadaver Atlas

Surface Anatomy and Cadaver Atlas

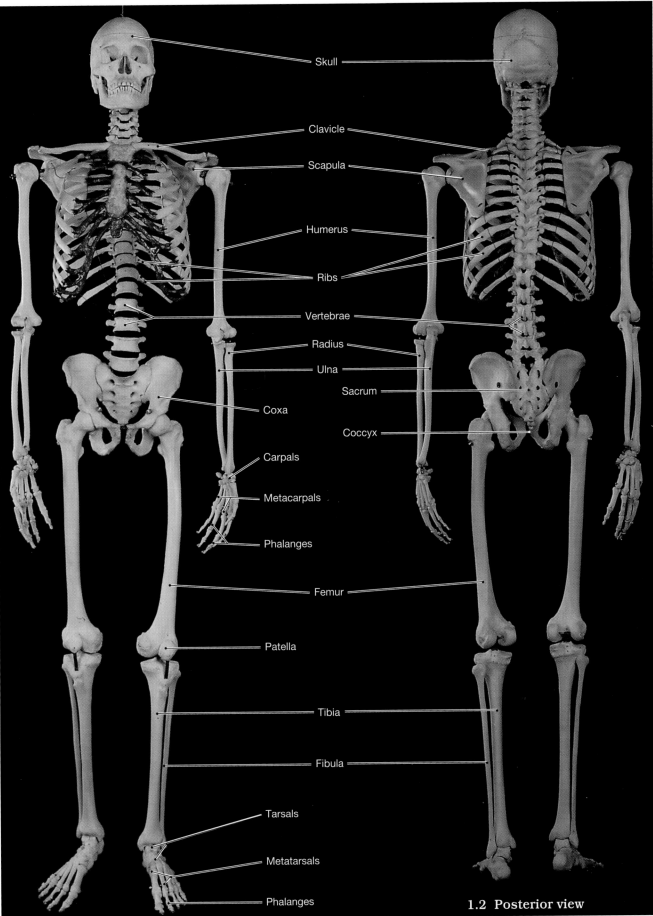

Skull

Clavicle

Scapula

Humerus

Ribs

Vertebrae

Radius

Ulna

Sacrum

Coxa

Coccyx

Carpals

Metacarpals

Phalanges

Femur

Patella

Tibia

Fibula

Tarsals

Metatarsals

Phalanges

1.2 Posterior view

PLATE 2 THE SKULL   The Adult Skull

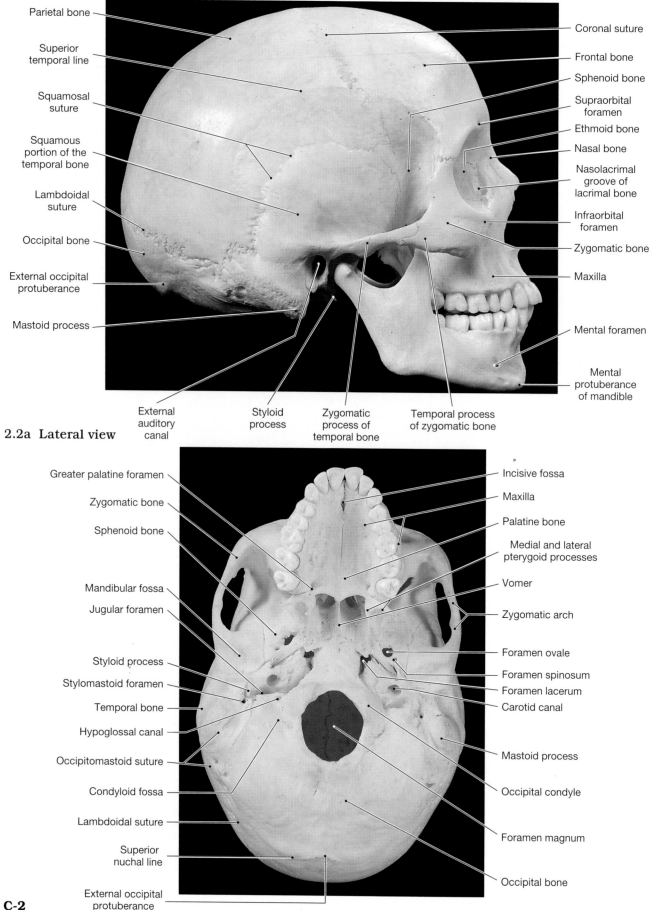

2.2a Lateral view

Parietal bone
Superior temporal line
Squamosal suture
Squamous portion of the temporal bone
Lambdoidal suture
Occipital bone
External occipital protuberance
Mastoid process

Coronal suture
Frontal bone
Sphenoid bone
Supraorbital foramen
Ethmoid bone
Nasal bone
Nasolacrimal groove of lacrimal bone
Infraorbital foramen
Zygomatic bone
Maxilla
Mental foramen
Mental protuberance of mandible

External auditory canal
Styloid process
Zygomatic process of temporal bone
Temporal process of zygomatic bone

Greater palatine foramen
Zygomatic bone
Sphenoid bone
Mandibular fossa
Jugular foramen
Styloid process
Stylomastoid foramen
Temporal bone
Hypoglossal canal
Occipitomastoid suture
Condyloid fossa
Lambdoidal suture
Superior nuchal line
External occipital protuberance

Incisive fossa
Maxilla
Palatine bone
Medial and lateral pterygoid processes
Vomer
Zygomatic arch
Foramen ovale
Foramen spinosum
Foramen lacerum
Carotid canal
Mastoid process
Occipital condyle
Foramen magnum
Occipital bone

2.2b Inferior view

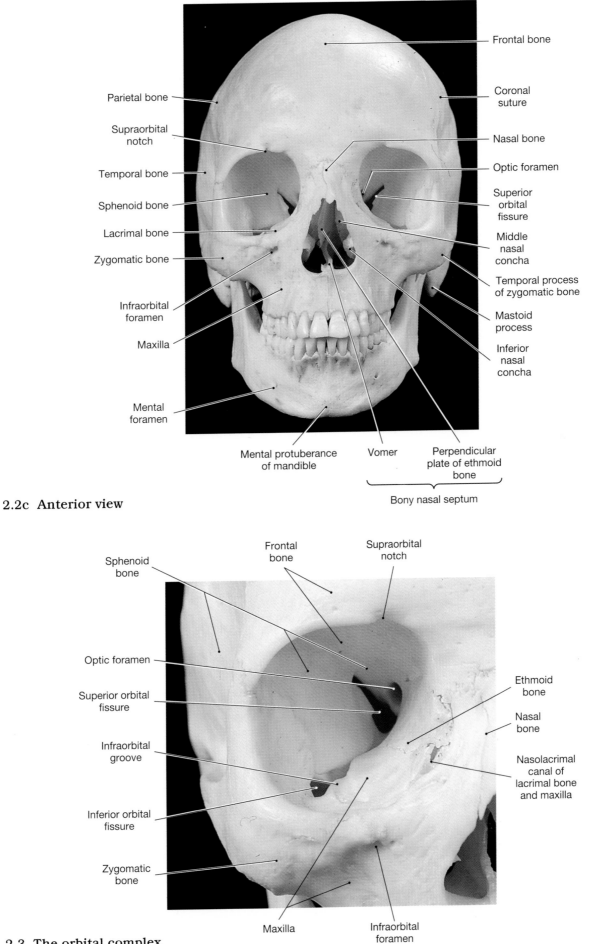

2.2c Anterior view

Frontal bone

Coronal suture

Nasal bone

Optic foramen

Superior orbital fissure

Middle nasal concha

Temporal process of zygomatic bone

Mastoid process

Inferior nasal concha

Parietal bone

Supraorbital notch

Temporal bone

Sphenoid bone

Lacrimal bone

Zygomatic bone

Infraorbital foramen

Maxilla

Mental foramen

Mental protuberance of mandible

Vomer

Perpendicular plate of ethmoid bone

Bony nasal septum

Sphenoid bone

Frontal bone

Supraorbital notch

Optic foramen

Superior orbital fissure

Infraorbital groove

Inferior orbital fissure

Zygomatic bone

Ethmoid bone

Nasal bone

Nasolacrimal canal of lacrimal bone and maxilla

Maxilla

Infraorbital foramen

2.3 The orbital complex

PLATE 3   THE HEAD AND NECK
Surface Anatomy of the Head and Neck

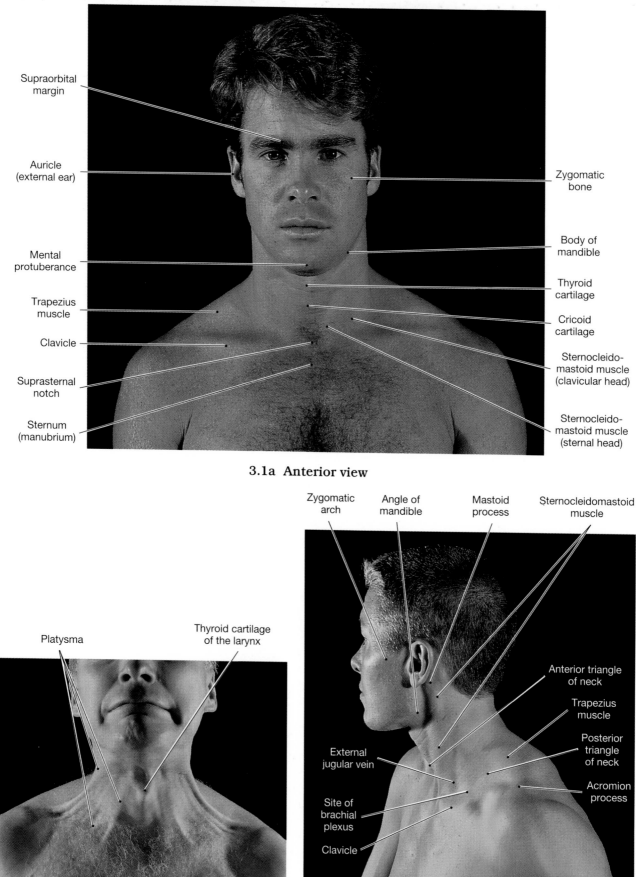

Supraorbital margin

Auricle (external ear)

Mental protuberance

Trapezius muscle

Clavicle

Suprasternal notch

Sternum (manubrium)

Zygomatic bone

Body of mandible

Thyroid cartilage

Cricoid cartilage

Sternocleido-mastoid muscle (clavicular head)

Sternocleido-mastoid muscle (sternal head)

3.1a Anterior view

Platysma

Thyroid cartilage of the larynx

3.1b The anterior neck

Zygomatic arch

Angle of mandible

Mastoid process

Sternocleidomastoid muscle

Anterior triangle of neck

Trapezius muscle

Posterior triangle of neck

Acromion process

External jugular vein

Site of brachial plexus

Clavicle

3.2a Posterior cervical triangle

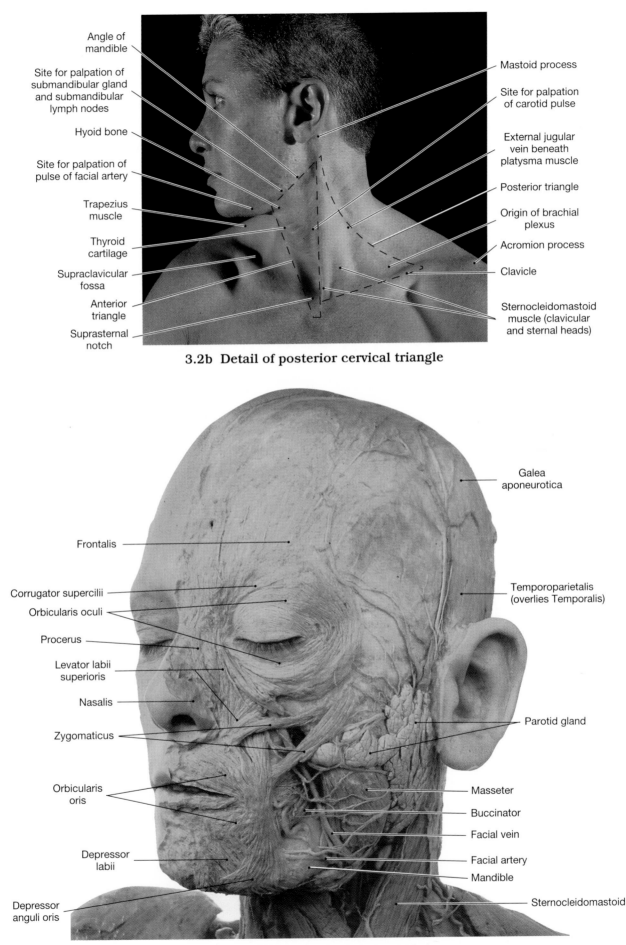

Angle of mandible

Site for palpation of submandibular gland and submandibular lymph nodes

Hyoid bone

Site for palpation of pulse of facial artery

Trapezius muscle

Thyroid cartilage

Supraclavicular fossa

Anterior triangle

Suprasternal notch

Mastoid process

Site for palpation of carotid pulse

External jugular vein beneath platysma muscle

Posterior triangle

Origin of brachial plexus

Acromion process

Clavicle

Sternocleidomastoid muscle (clavicular and sternal heads)

3.2b Detail of posterior cervical triangle

Frontalis

Corrugator supercilii

Orbicularis oculi

Procerus

Levator labii superioris

Nasalis

Zygomaticus

Orbicularis oris

Depressor labii

Depressor anguli oris

Galea aponeurotica

Temporoparietalis (overlies Temporalis)

Parotid gland

Masseter

Buccinator

Facial vein

Facial artery

Mandible

Sternocleidomastoid

3.3a Cadaver head and neck, lateral view

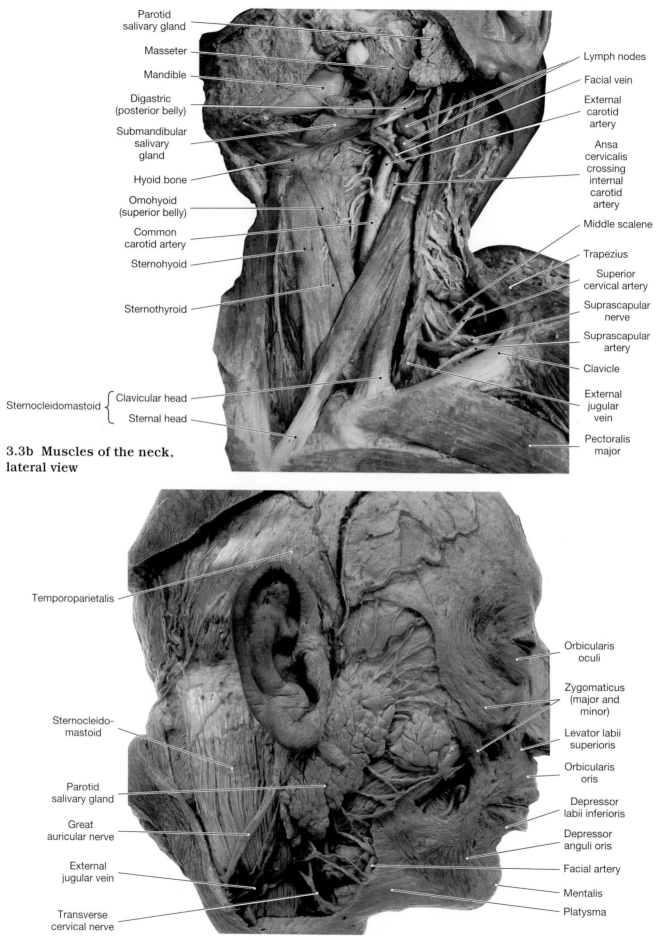

Parotid
salivary gland

Masseter

Mandible

Digastric
(posterior belly)

Submandibular
salivary
gland

Hyoid bone

Omohyoid
(superior belly)

Common
carotid artery

Sternohyoid

Sternothyroid

Sternocleidomastoid {
Clavicular head
Sternal head

Lymph nodes

Facial vein

External
carotid
artery

Ansa
cervicalis
crossing
internal
carotid
artery

Middle scalene

Trapezius

Superior
cervical artery

Suprascapular
nerve

Suprascapular
artery

Clavicle

External
jugular
vein

Pectoralis
major

**3.3b Muscles of the neck,
lateral view**

Temporoparietalis

Sternocleido-
mastoid

Parotid
salivary gland

Great
auricular nerve

External
jugular vein

Transverse
cervical nerve

Orbicularis
oculi

Zygomaticus
(major and
minor)

Levator labii
superioris

Orbicularis
oris

Depressor
labii inferioris

Depressor
anguli oris

Facial artery

Mentalis

Platysma

3.3c Muscles of the face

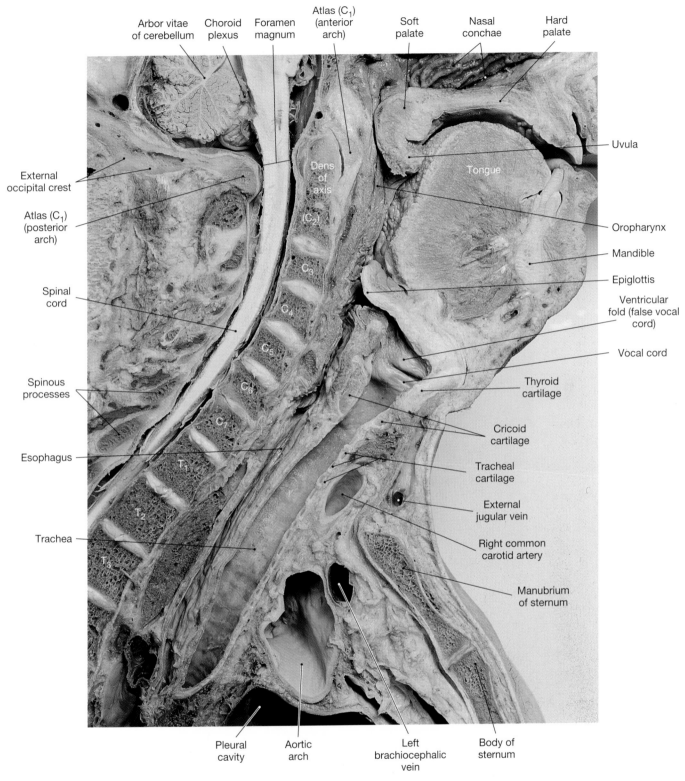

Arbor vitae of cerebellum — Choroid plexus — Foramen magnum — Atlas (C₁) (anterior arch) — Soft palate — Nasal conchae — Hard palate

External occipital crest

Atlas (C₁) (posterior arch)

Spinal cord

Spinous processes

Esophagus

Trachea

Dens of axis (C₂)

C₃

C₄

C₅

C₆

C₇

T₁

T₂

T₃

Uvula

Tongue

Oropharynx

Mandible

Epiglottis

Ventricular fold (false vocal cord)

Vocal cord

Thyroid cartilage

Cricoid cartilage

Tracheal cartilage

External jugular vein

Right common carotid artery

Manubrium of sternum

Pleural cavity — Aortic arch — Left brachiocephalic vein — Body of sternum

3.4 Sectional anatomy of the neck

PLATE 4   THE UPPER LIMBS
The Right Upper Limb

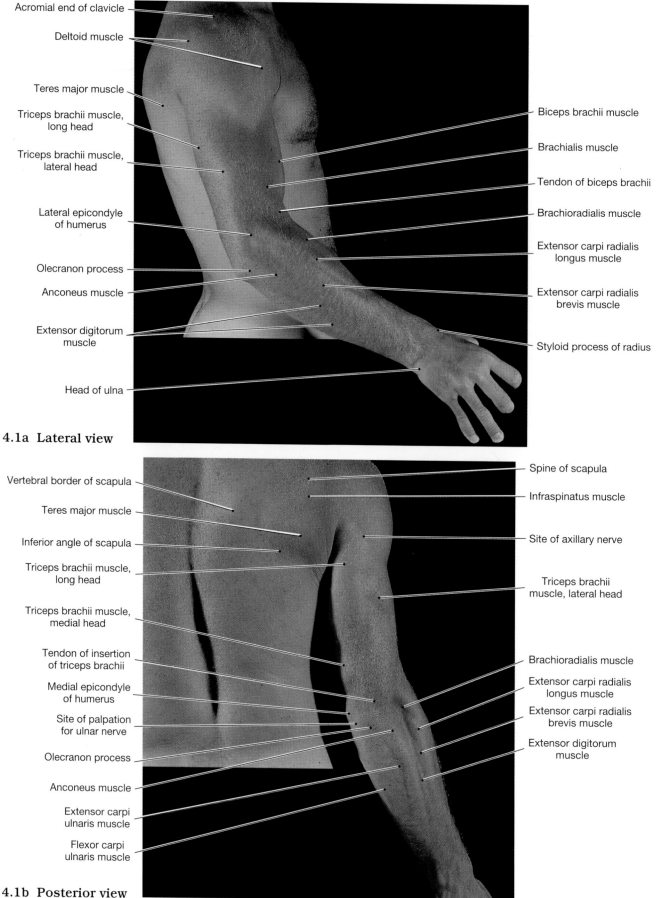

Acromial end of clavicle

Deltoid muscle

Teres major muscle

Triceps brachii muscle, long head

Triceps brachii muscle, lateral head

Lateral epicondyle of humerus

Olecranon process

Anconeus muscle

Extensor digitorum muscle

Head of ulna

Biceps brachii muscle

Brachialis muscle

Tendon of biceps brachii

Brachioradialis muscle

Extensor carpi radialis longus muscle

Extensor carpi radialis brevis muscle

Styloid process of radius

4.1a Lateral view

Vertebral border of scapula

Teres major muscle

Inferior angle of scapula

Triceps brachii muscle, long head

Triceps brachii muscle, medial head

Tendon of insertion of triceps brachii

Medial epicondyle of humerus

Site of palpation for ulnar nerve

Olecranon process

Anconeus muscle

Extensor carpi ulnaris muscle

Flexor carpi ulnaris muscle

Spine of scapula

Infraspinatus muscle

Site of axillary nerve

Triceps brachii muscle, lateral head

Brachioradialis muscle

Extensor carpi radialis longus muscle

Extensor carpi radialis brevis muscle

Extensor digitorum muscle

4.1b Posterior view
C-8

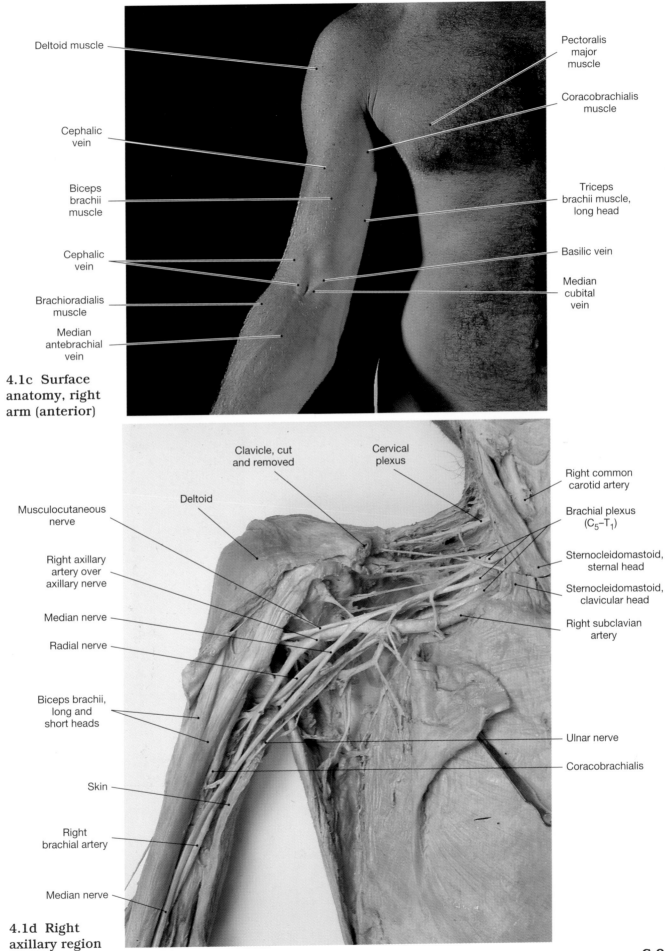

Deltoid muscle

Pectoralis
major
muscle

Coracobrachialis
muscle

Cephalic
vein

Biceps
brachii
muscle

Triceps
brachii muscle,
long head

Cephalic
vein

Basilic vein

Brachioradialis
muscle

Median
cubital
vein

Median
antebrachial
vein

**4.1c Surface
anatomy, right
arm (anterior)**

Clavicle, cut
and removed

Cervical
plexus

Right common
carotid artery

Musculocutaneous
nerve

Deltoid

Brachial plexus
(C_5-T_1)

Right axillary
artery over
axillary nerve

Sternocleidomastoid,
sternal head

Median nerve

Sternocleidomastoid,
clavicular head

Radial nerve

Right subclavian
artery

Biceps brachii,
long and
short heads

Ulnar nerve

Coracobrachialis

Skin

Right
brachial artery

Median nerve

**4.1d Right
axillary region**

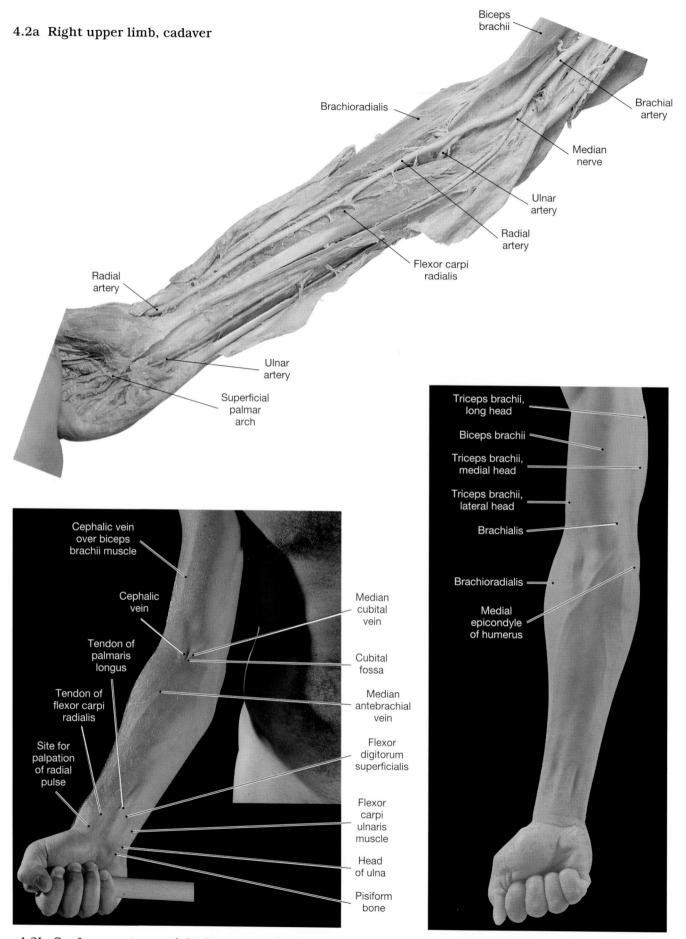

4.2a Right upper limb, cadaver

Biceps brachii

Brachioradialis

Brachial artery

Median nerve

Ulnar artery

Radial artery

Flexor carpi radialis

Radial artery

Ulnar artery

Superficial palmar arch

Cephalic vein over biceps brachii muscle

Cephalic vein

Tendon of palmaris longus

Tendon of flexor carpi radialis

Site for palpation of radial pulse

Median cubital vein

Cubital fossa

Median antebrachial vein

Flexor digitorum superficialis

Flexor carpi ulnaris muscle

Head of ulna

Pisiform bone

Triceps brachii, long head

Biceps brachii

Triceps brachii, medial head

Triceps brachii, lateral head

Brachialis

Brachioradialis

Medial epicondyle of humerus

4.2b Surface anatomy, right forearm and wrist, anterior view

C-10

4.3a Right arm, anterior view

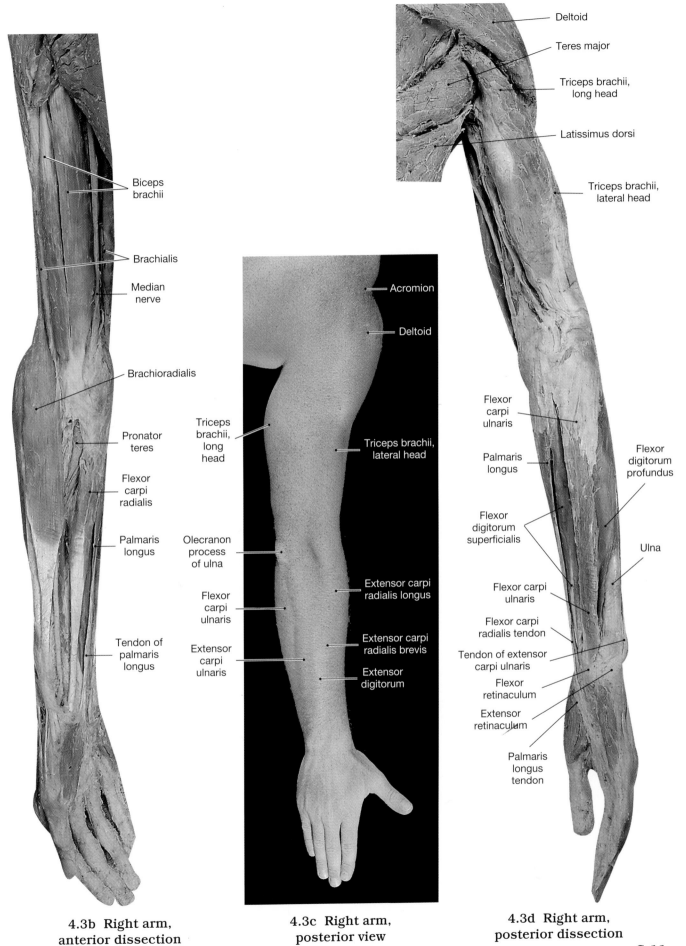

Deltoid

Teres major

Triceps brachii, long head

Latissimus dorsi

Triceps brachii, lateral head

Biceps brachii

Brachialis

Median nerve

Brachioradialis

Pronator teres

Flexor carpi radialis

Palmaris longus

Tendon of palmaris longus

Acromion

Deltoid

Triceps brachii, long head

Triceps brachii, lateral head

Olecranon process of ulna

Flexor carpi ulnaris

Extensor carpi ulnaris

Extensor carpi radialis longus

Extensor carpi radialis brevis

Extensor digitorum

Flexor carpi ulnaris

Palmaris longus

Flexor digitorum superficialis

Flexor carpi ulnaris

Flexor carpi radialis tendon

Tendon of extensor carpi ulnaris

Flexor retinaculum

Extensor retinaculum

Palmaris longus tendon

Flexor digitorum profundus

Ulna

4.3b Right arm, anterior dissection

4.3c Right arm, posterior view

4.3d Right arm, posterior dissection

C-11

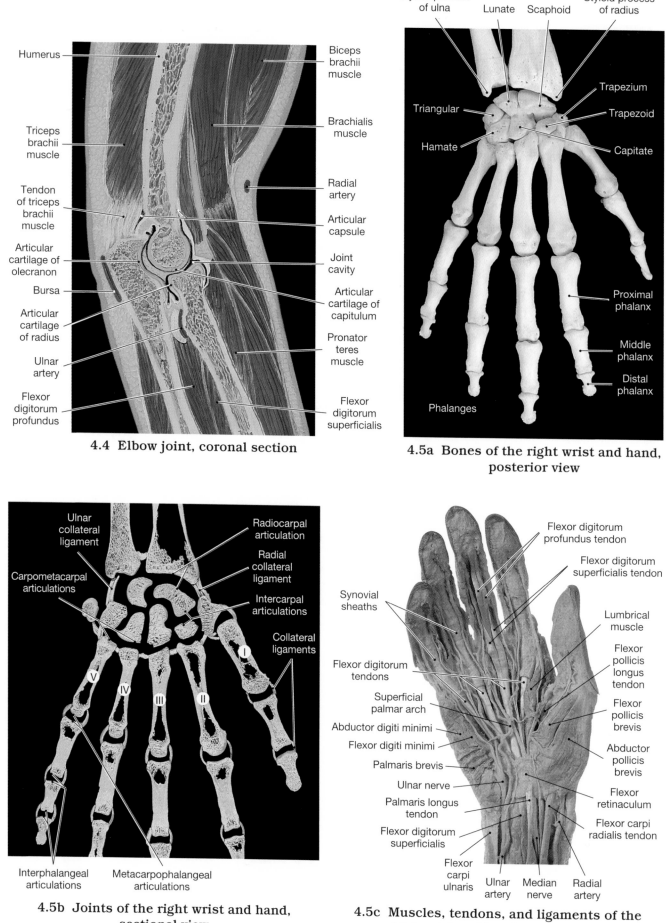

4.4 Elbow joint, coronal section

Humerus

Triceps brachii muscle

Tendon of triceps brachii muscle

Articular cartilage of olecranon

Bursa

Articular cartilage of radius

Ulnar artery

Flexor digitorum profundus

Biceps brachii muscle

Brachialis muscle

Radial artery

Articular capsule

Joint cavity

Articular cartilage of capitulum

Pronator teres muscle

Flexor digitorum superficialis

4.5a Bones of the right wrist and hand, posterior view

Styloid process of ulna

Lunate

Scaphoid

Styloid process of radius

Triangular

Trapezium

Trapezoid

Hamate

Capitate

Proximal phalanx

Middle phalanx

Distal phalanx

Phalanges

4.5b Joints of the right wrist and hand, sectional view

Ulnar collateral ligament

Carpometacarpal articulations

Radiocarpal articulation

Radial collateral ligament

Intercarpal articulations

Collateral ligaments

I

II

III

IV

V

Interphalangeal articulations

Metacarpophalangeal articulations

4.5c Muscles, tendons, and ligaments of the right wrist and hand, anterior view

Flexor digitorum profundus tendon

Flexor digitorum superficialis tendon

Lumbrical muscle

Flexor pollicis longus tendon

Flexor pollicis brevis

Abductor pollicis brevis

Flexor retinaculum

Flexor carpi radialis tendon

Synovial sheaths

Flexor digitorum tendons

Superficial palmar arch

Abductor digiti minimi

Flexor digiti minimi

Palmaris brevis

Ulnar nerve

Palmaris longus tendon

Flexor digitorum superficialis

Flexor carpi ulnaris

Ulnar artery

Median nerve

Radial artery

C-12

PLATE 5   THE TRUNK
The Thorax

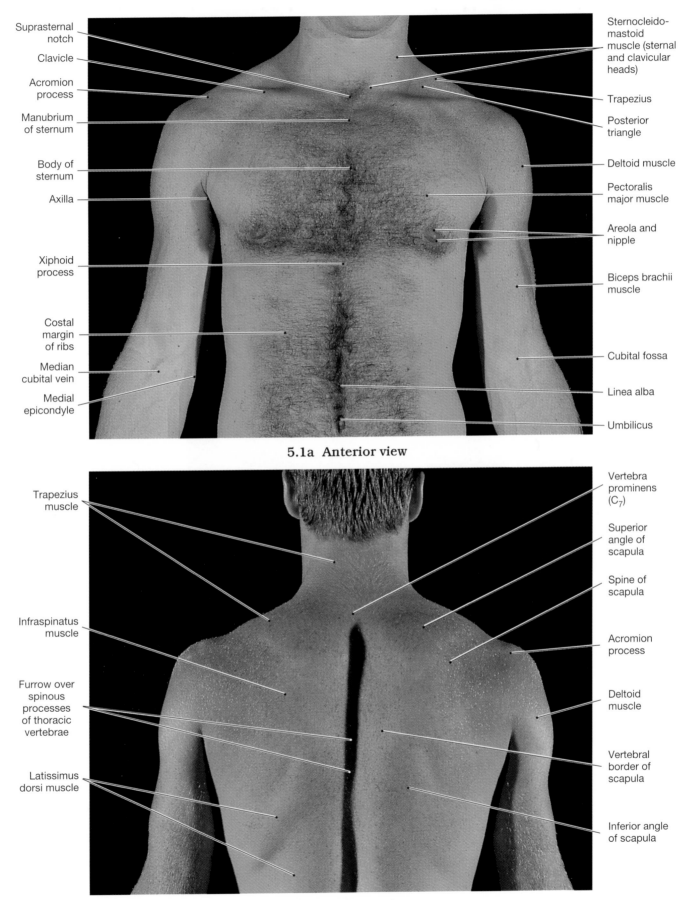

Suprasternal notch

Clavicle

Acromion process

Manubrium of sternum

Body of sternum

Axilla

Xiphoid process

Costal margin of ribs

Median cubital vein

Medial epicondyle

Sternocleido-mastoid muscle (sternal and clavicular heads)

Trapezius

Posterior triangle

Deltoid muscle

Pectoralis major muscle

Areola and nipple

Biceps brachii muscle

Cubital fossa

Linea alba

Umbilicus

5.1a Anterior view

Trapezius muscle

Infraspinatus muscle

Furrow over spinous processes of thoracic vertebrae

Latissimus dorsi muscle

Vertebra prominens (C$_7$)

Superior angle of scapula

Spine of scapula

Acromion process

Deltoid muscle

Vertebral border of scapula

Inferior angle of scapula

5.1b Posterior view

Muscles that Move the Arm

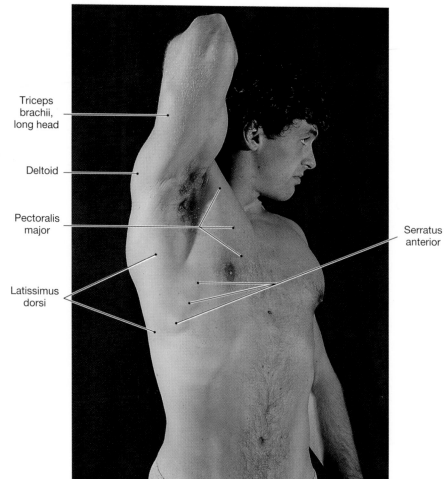

Triceps
brachii,
long head

Deltoid

Pectoralis
major

Latissimus
dorsi

Serratus
anterior

5.2a Anterolateral view

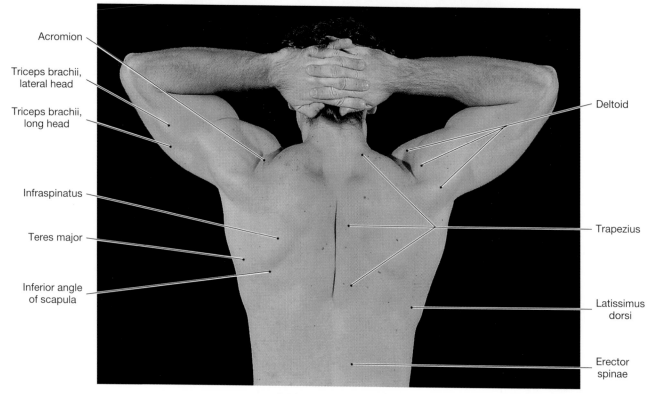

Acromion

Triceps brachii,
lateral head

Triceps brachii,
long head

Infraspinatus

Teres major

Inferior angle
of scapula

Deltoid

Trapezius

Latissimus
dorsi

Erector
spinae

5.2b Posterior view

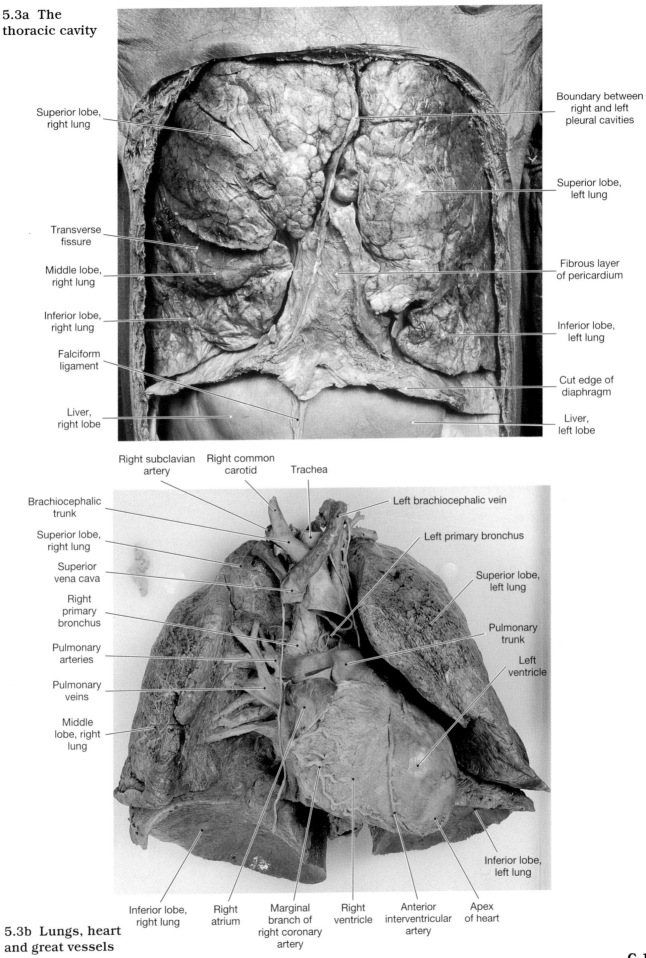

5.3a The thoracic cavity

Superior lobe, right lung

Transverse fissure

Middle lobe, right lung

Inferior lobe, right lung

Falciform ligament

Liver, right lobe

Boundary between right and left pleural cavities

Superior lobe, left lung

Fibrous layer of pericardium

Inferior lobe, left lung

Cut edge of diaphragm

Liver, left lobe

Right subclavian artery

Right common carotid

Trachea

Brachiocephalic trunk

Left brachiocephalic vein

Superior lobe, right lung

Left primary bronchus

Superior vena cava

Right primary bronchus

Superior lobe, left lung

Pulmonary arteries

Pulmonary trunk

Pulmonary veins

Left ventricle

Middle lobe, right lung

Inferior lobe, right lung

Right atrium

Marginal branch of right coronary artery

Right ventricle

Anterior interventricular artery

Apex of heart

Inferior lobe, left lung

5.3b Lungs, heart and great vessels

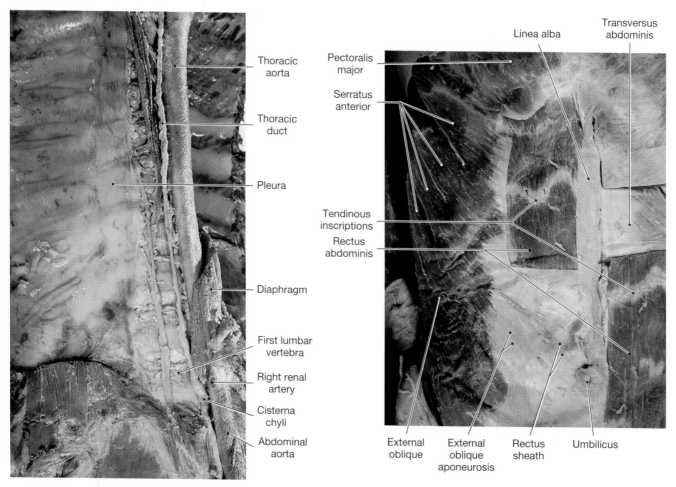

Thoracic aorta

Thoracic duct

Pleura

Diaphragm

First lumbar vertebra

Right renal artery

Cisterna chyli

Abdominal aorta

5.3j Major lymphatic vessels of the trunk

Pectoralis major

Serratus anterior

Linea alba

Transversus abdominis

Tendinous inscriptions

Rectus abdominis

External oblique

External oblique aponeurosis

Rectus sheath

Umbilicus

5.4a Cadaver, superficial anterior view of the abdominal wall

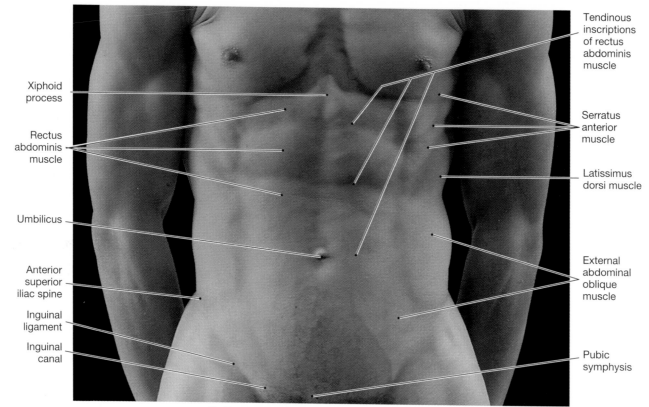

Tendinous inscriptions of rectus abdominis muscle

Xiphoid process

Rectus abdominis muscle

Umbilicus

Anterior superior iliac spine

Inguinal ligament

Inguinal canal

Serratus anterior muscle

Latissimus dorsi muscle

External abdominal oblique muscle

Pubic symphysis

5.4b Abdominal wall, anterior view

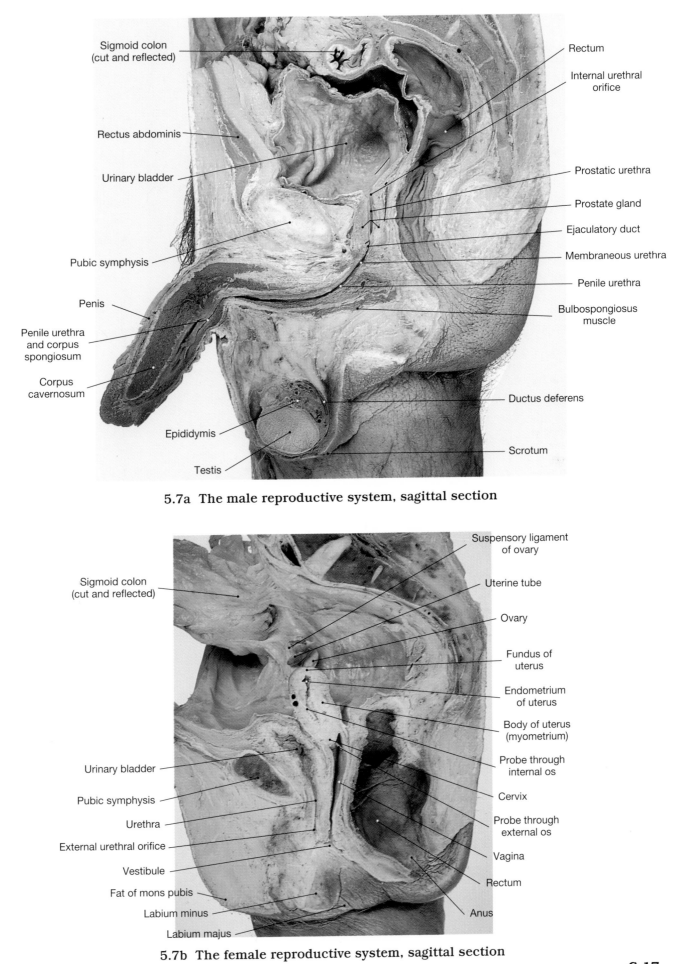

Sigmoid colon (cut and reflected)

Rectus abdominis

Urinary bladder

Pubic symphysis

Penis

Penile urethra and corpus spongiosum

Corpus cavernosum

Epididymis

Testis

Rectum

Internal urethral orifice

Prostatic urethra

Prostate gland

Ejaculatory duct

Membraneous urethra

Penile urethra

Bulbospongiosus muscle

Ductus deferens

Scrotum

5.7a The male reproductive system, sagittal section

Sigmoid colon (cut and reflected)

Urinary bladder

Pubic symphysis

Urethra

External urethral orifice

Vestibule

Fat of mons pubis

Labium minus

Labium majus

Suspensory ligament of ovary

Uterine tube

Ovary

Fundus of uterus

Endometrium of uterus

Body of uterus (myometrium)

Probe through internal os

Cervix

Probe through external os

Vagina

Rectum

Anus

5.7b The female reproductive system, sagittal section

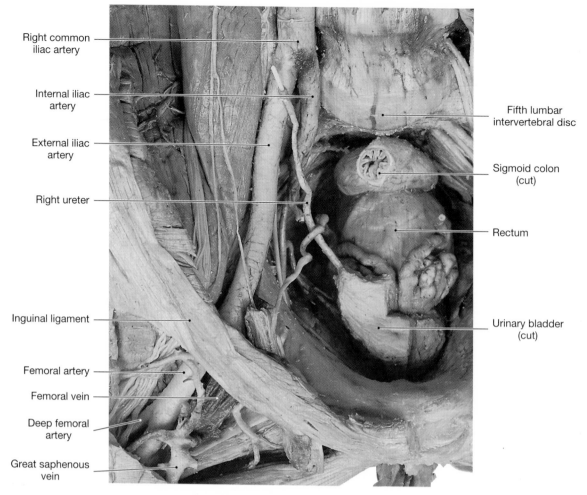

Right common
iliac artery

Internal iliac
artery

External iliac
artery

Right ureter

Inguinal ligament

Femoral artery

Femoral vein

Deep femoral
artery

Great saphenous
vein

Fifth lumbar
intervertebral disc

Sigmoid colon
(cut)

Rectum

Urinary bladder
(cut)

5.8 Major vessels of the pelvis

PLATE 6   THE LOWER LIMBS

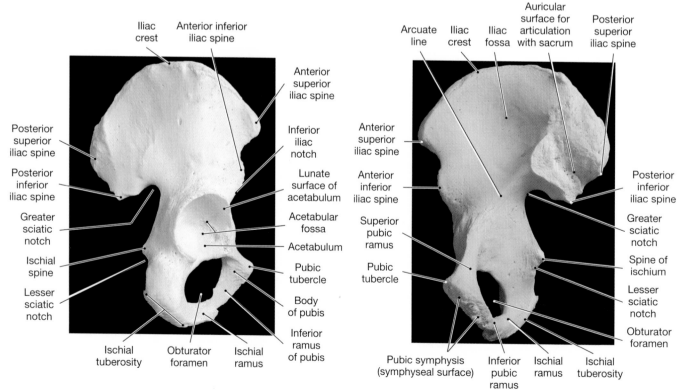

Iliac
crest

Anterior inferior
iliac spine

Posterior
superior
iliac spine

Posterior
inferior
iliac spine

Greater
sciatic
notch

Ischial
spine

Lesser
sciatic
notch

Anterior
superior
iliac spine

Inferior
iliac
notch

Lunate
surface of
acetabulum

Acetabular
fossa

Acetabulum

Pubic
tubercle

Body
of pubis

Inferior
ramus
of pubis

Ischial
tuberosity

Obturator
foramen

Ischial
ramus

6.1a The pelvic girdle, lateral view

Arcuate
line

Iliac
crest

Iliac
fossa

Auricular
surface for
articulation
with sacrum

Posterior
superior
iliac spine

Anterior
superior
iliac spine

Anterior
inferior
iliac spine

Superior
pubic
ramus

Pubic
tubercle

Posterior
inferior
iliac spine

Greater
sciatic
notch

Spine of
ischium

Lesser
sciatic
notch

Obturator
foramen

Pubic symphysis
(symphyseal surface)

Inferior
pubic
ramus

Ischial
ramus

Ischial
tuberosity

6.1b The pelvic girdle, medial view

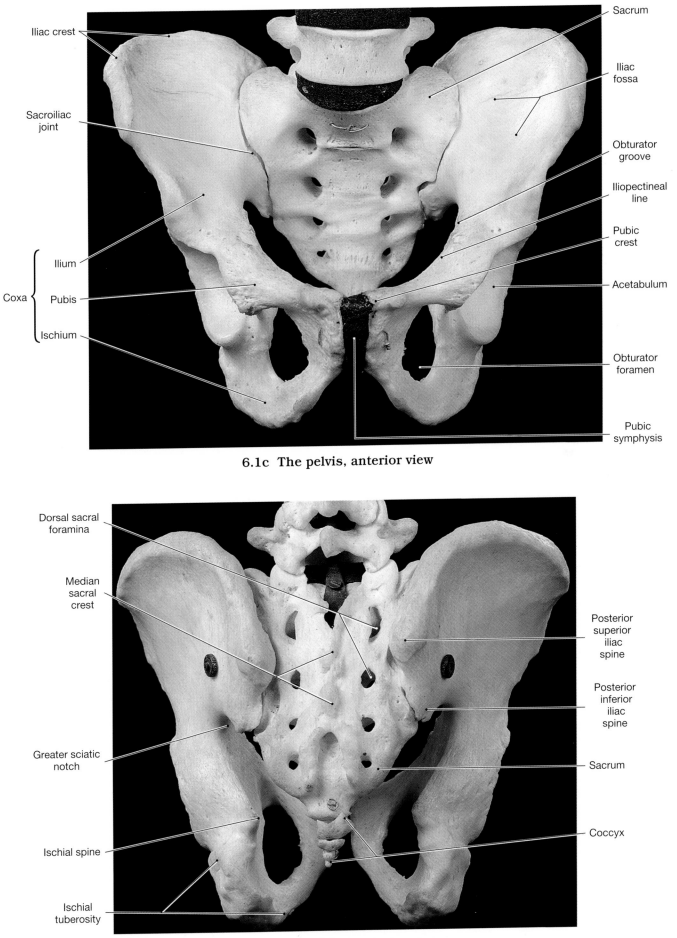

Sacrum

Iliac crest

Iliac fossa

Sacroiliac joint

Obturator groove

Iliopectineal line

Pubic crest

Coxa { Ilium

Pubis

Acetabulum

Ischium

Obturator foramen

Pubic symphysis

6.1c The pelvis, anterior view

Dorsal sacral foramina

Median sacral crest

Posterior superior iliac spine

Posterior inferior iliac spine

Greater sciatic notch

Sacrum

Coccyx

Ischial spine

Ischial tuberosity

6.1d The pelvis, posterior view

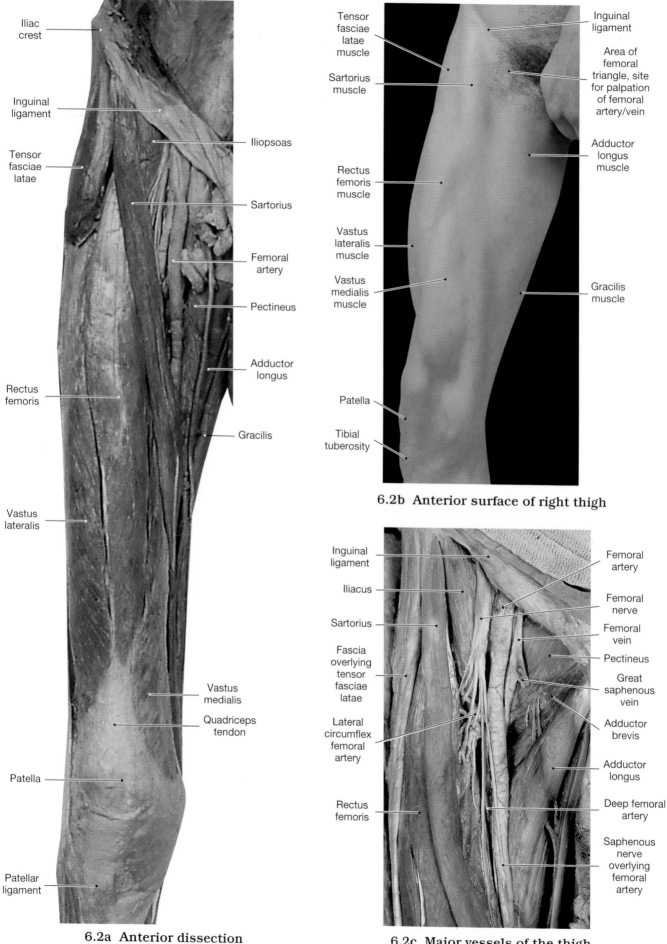

Iliac crest

Inguinal ligament

Tensor fasciae latae

Rectus femoris

Vastus lateralis

Vastus medialis

Quadriceps tendon

Patella

Patellar ligament

Iliopsoas

Sartorius

Femoral artery

Pectineus

Adductor longus

Gracilis

6.2a Anterior dissection

C-20

Tensor fasciae latae muscle

Sartorius muscle

Rectus femoris muscle

Vastus lateralis muscle

Vastus medialis muscle

Patella

Tibial tuberosity

Inguinal ligament

Area of femoral triangle, site for palpation of femoral artery/vein

Adductor longus muscle

Gracilis muscle

6.2b Anterior surface of right thigh

Inguinal ligament

Iliacus

Sartorius

Fascia overlying tensor fasciae latae

Lateral circumflex femoral artery

Rectus femoris

Femoral artery

Femoral nerve

Femoral vein

Pectineus

Great saphenous vein

Adductor brevis

Adductor longus

Deep femoral artery

Saphenous nerve overlying femoral artery

6.2c Major vessels of the thigh

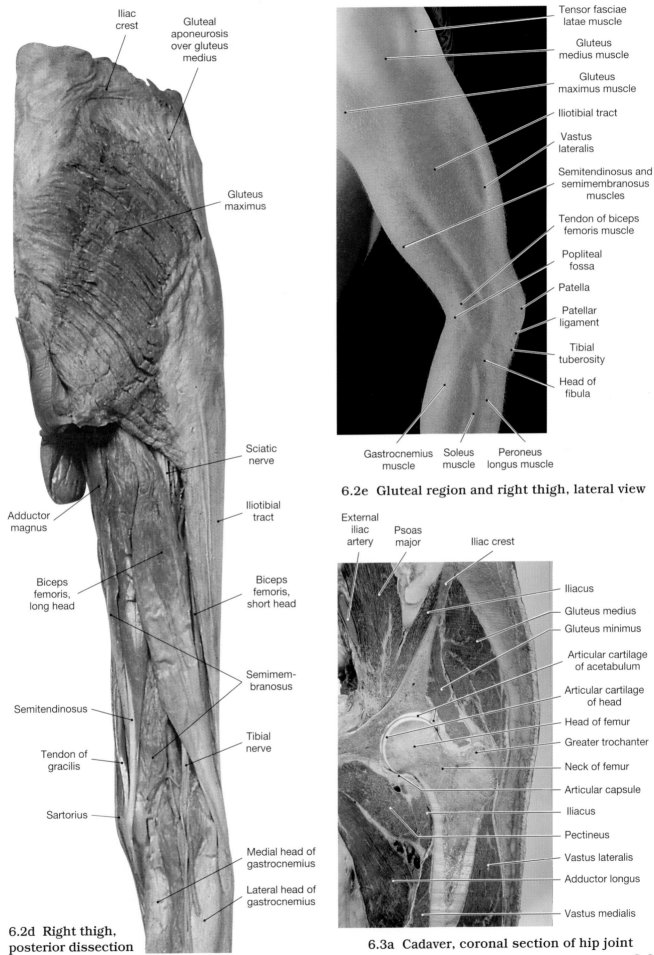

Iliac crest

Gluteal aponeurosis over gluteus medius

Gluteus maximus

Sciatic nerve

Iliotibial tract

Adductor magnus

Biceps femoris, long head

Biceps femoris, short head

Semimembranosus

Semitendinosus

Tibial nerve

Tendon of gracilis

Sartorius

Medial head of gastrocnemius

Lateral head of gastrocnemius

6.2d Right thigh, posterior dissection

Tensor fasciae latae muscle

Gluteus medius muscle

Gluteus maximus muscle

Iliotibial tract

Vastus lateralis

Semitendinosus and semimembranosus muscles

Tendon of biceps femoris muscle

Popliteal fossa

Patella

Patellar ligament

Tibial tuberosity

Head of fibula

Gastrocnemius muscle

Soleus muscle

Peroneus longus muscle

6.2e Gluteal region and right thigh, lateral view

External iliac artery

Psoas major

Iliac crest

Iliacus

Gluteus medius

Gluteus minimus

Articular cartilage of acetabulum

Articular cartilage of head

Head of femur

Greater trochanter

Neck of femur

Articular capsule

Iliacus

Pectineus

Vastus lateralis

Adductor longus

Vastus medialis

6.3a Cadaver, coronal section of hip joint

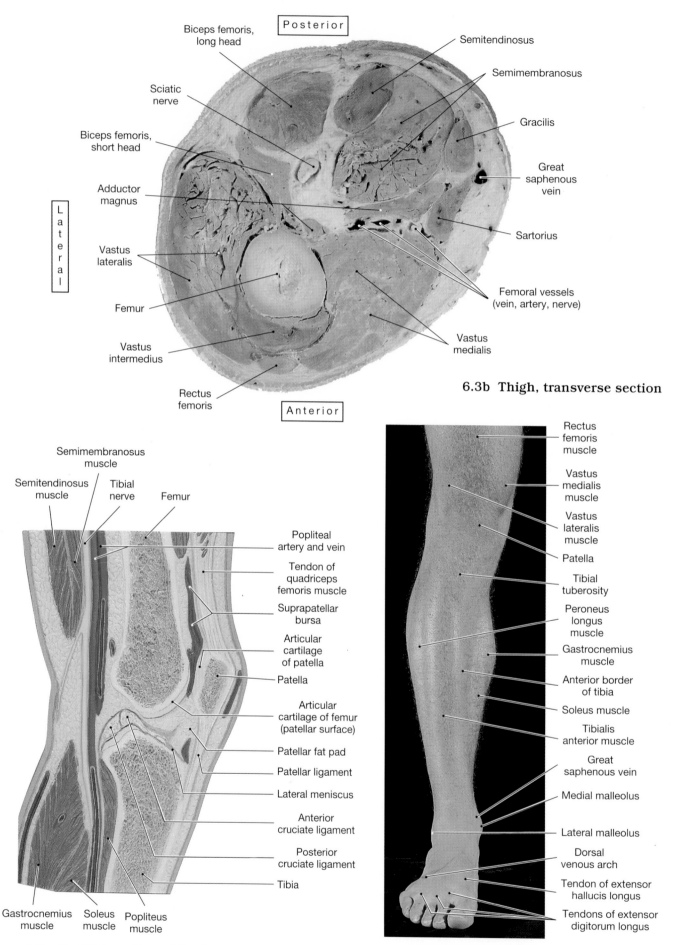

Posterior

Biceps femoris, long head

Sciatic nerve

Biceps femoris, short head

Adductor magnus

Vastus lateralis

Femur

Vastus intermedius

Rectus femoris

Anterior

Lateral

Semitendinosus

Semimembranosus

Gracilis

Great saphenous vein

Sartorius

Femoral vessels (vein, artery, nerve)

Vastus medialis

6.3b Thigh, transverse section

Semimembranosus muscle

Semitendinosus muscle

Tibial nerve

Femur

Popliteal artery and vein

Tendon of quadriceps femoris muscle

Suprapatellar bursa

Articular cartilage of patella

Patella

Articular cartilage of femur (patellar surface)

Patellar fat pad

Patellar ligament

Lateral meniscus

Anterior cruciate ligament

Posterior cruciate ligament

Tibia

Gastrocnemius muscle

Soleus muscle

Popliteus muscle

6.3c Sagittal section of the knee

Rectus femoris muscle

Vastus medialis muscle

Vastus lateralis muscle

Patella

Tibial tuberosity

Peroneus longus muscle

Gastrocnemius muscle

Anterior border of tibia

Soleus muscle

Tibialis anterior muscle

Great saphenous vein

Medial malleolus

Lateral malleolus

Dorsal venous arch

Tendon of extensor hallucis longus

Tendons of extensor digitorum longus

6.4a Knee and leg, anterior view

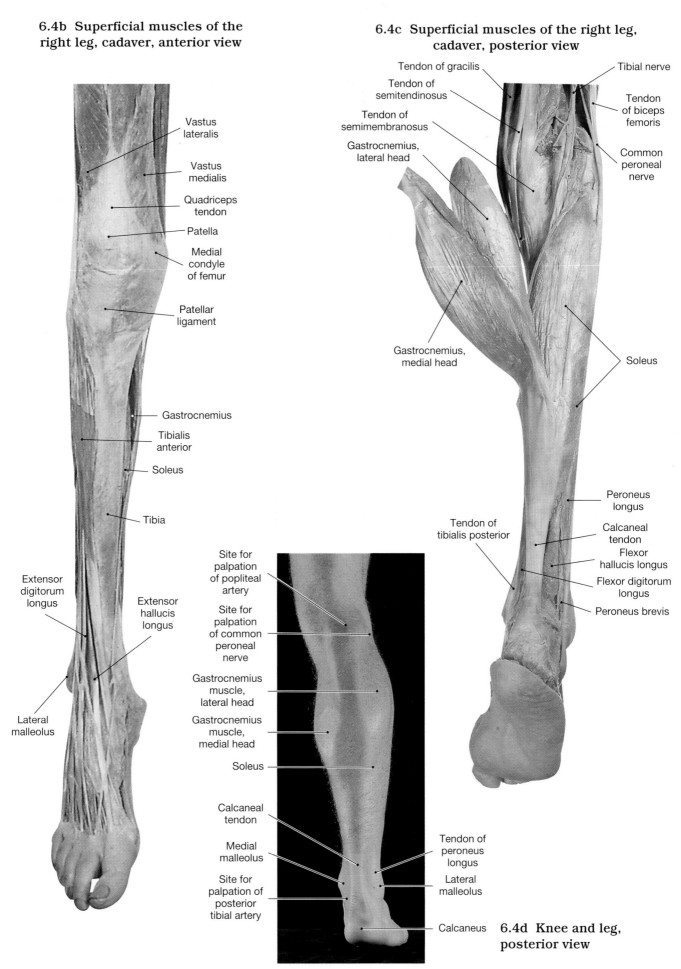

6.4b Superficial muscles of the right leg, cadaver, anterior view

Vastus lateralis

Vastus medialis

Quadriceps tendon

Patella

Medial condyle of femur

Patellar ligament

Gastrocnemius

Tibialis anterior

Soleus

Tibia

Extensor digitorum longus

Extensor hallucis longus

Lateral malleolus

6.4c Superficial muscles of the right leg, cadaver, posterior view

Tendon of gracilis

Tendon of semitendinosus

Tendon of semimembranosus

Gastrocnemius, lateral head

Tibial nerve

Tendon of biceps femoris

Common peroneal nerve

Gastrocnemius, medial head

Soleus

Tendon of tibialis posterior

Peroneus longus

Calcaneal tendon

Flexor hallucis longus

Flexor digitorum longus

Peroneus brevis

Site for palpation of popliteal artery

Site for palpation of common peroneal nerve

Gastrocnemius muscle, lateral head

Gastrocnemius muscle, medial head

Soleus

Calcaneal tendon

Medial malleolus

Site for palpation of posterior tibial artery

Tendon of peroneus longus

Lateral malleolus

Calcaneus

6.4d Knee and leg, posterior view

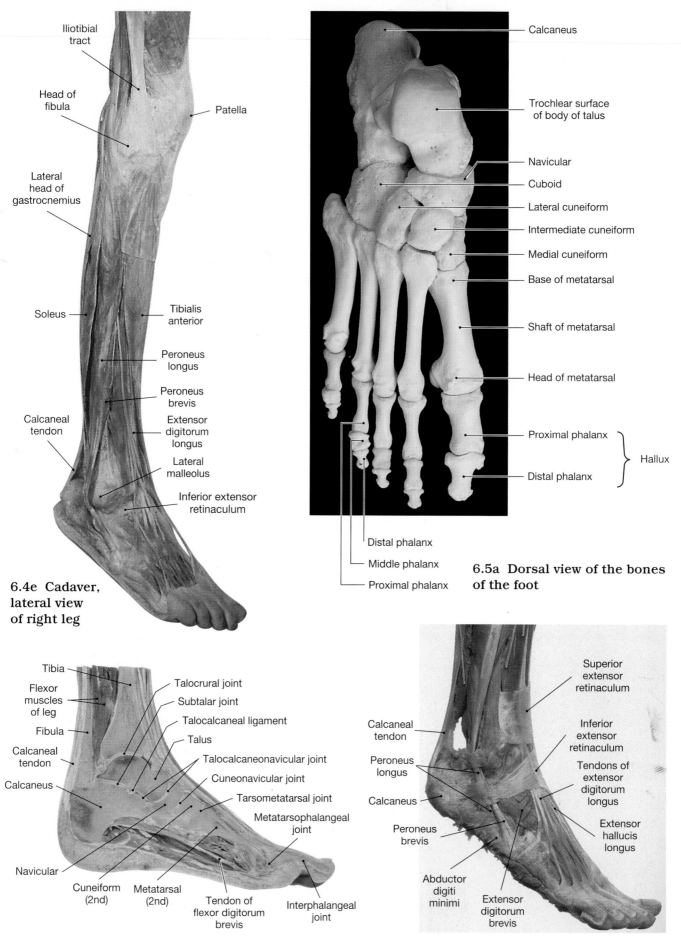

6.4e Cadaver, lateral view of right leg

Iliotibial tract
Head of fibula
Patella
Lateral head of gastrocnemius
Soleus
Tibialis anterior
Peroneus longus
Peroneus brevis
Extensor digitorum longus
Calcaneal tendon
Lateral malleolus
Inferior extensor retinaculum

6.5a Dorsal view of the bones of the foot

Calcaneus
Trochlear surface of body of talus
Navicular
Cuboid
Lateral cuneiform
Intermediate cuneiform
Medial cuneiform
Base of metatarsal
Shaft of metatarsal
Head of metatarsal
Proximal phalanx
Distal phalanx
Hallux
Distal phalanx
Middle phalanx
Proximal phalanx

6.5b The foot, dissection

Tibia
Flexor muscles of leg
Fibula
Calcaneal tendon
Calcaneus
Navicular
Cuneiform (2nd)
Metatarsal (2nd)
Talocrural joint
Subtalar joint
Talocalcaneal ligament
Talus
Talocalcaneonavicular joint
Cuneonavicular joint
Tarsometatarsal joint
Metatarsophalangeal joint
Tendon of flexor digitorum brevis
Interphalangeal joint

6.5c Muscles of the foot, dissection

Calcaneal tendon
Peroneus longus
Calcaneus
Peroneus brevis
Abductor digiti minimi
Extensor digitorum brevis
Superior extensor retinaculum
Inferior extensor retinaculum
Tendons of extensor digitorum longus
Extensor hallucis longus